AF812623

Diarylethene Molecular Photoswitches

Diarylethene Molecular Photoswitches

Concepts and Functionalities

Masahiro Irie

WILEY-VCH

Author

Masahiro Irie
Professor *emeritus*
Kyushu University
Japan

All books published by **Wiley-VCH** are carefully produced. Nevertheless, authors, editors, and publisher do not warrant the information contained in these books, including this book, to be free of errors. Readers are advised to keep in mind that statements, data, illustrations, procedural details or other items may inadvertently be inaccurate.

Library of Congress Card No.: applied for

British Library Cataloguing-in-Publication Data
A catalogue record for this book is available from the British Library.

Bibliographic information published by the Deutsche Nationalbibliothek
The Deutsche Nationalbibliothek lists this publication in the Deutsche Nationalbibliografie; detailed bibliographic data are available on the Internet at <http://dnb.d-nb.de>.

© 2021 WILEY-VCH GmbH, Boschstr. 12, 69469 Weinheim, Germany

All rights reserved (including those of translation into other languages). No part of this book may be reproduced in any form – by photoprinting, microfilm, or any other means – nor transmitted or translated into a machine language without written permission from the publishers. Registered names, trademarks, etc. used in this book, even when not specifically marked as such, are not to be considered unprotected by law.

Print ISBN: 978-3-527-34640-0
ePDF ISBN: 978-3-527-34642-4
ePub ISBN: 978-3-527-82286-7
oBook ISBN: 978-3-527-82285-0

Typesetting SPi Global, Chennai, India
Printing and Binding CPI Group (UK) Ltd, Croydon, CR0 4YY

Printed on acid-free paper

C094969_210321

Contents

Preface

Molecules capable of reversible photoswitching between two isomers having different absorption spectra are called photochromic molecules or molecular photoswitches. The two isomers differ from each other not only in their absorption and fluorescence spectra, but also in their refractive indices, dielectric constants, oxidation/reduction potentials, and geometrical structures. These photoswitchable bistable molecules are applied to construct photonic devices, such as erasable optical memory media and optical switch elements. Although the first finding of photoswitchable molecules can be traced back to the middle of nineteenth century, they are still awaiting their time to go on the stage of photonics devices in wide use.

Digital cameras take photos by using physical phenomena of inorganic materials. Photodiodes, such as CCD (charge-coupled device) and CMOS (complementary metal oxide semiconductor), detect photons based on photovoltaic effects and construct photo-images. Animals and plants have no such inorganic semiconductors. In biological systems, molecular photoswitches are extensively employed in photoreceptors. Vision, for example, uses the *cis*-to-*trans* photoisomerization of retinal to control the conformation of rhodopsin and initiate the transduction cascade to generate neural signals, while phototaxis of *Chlamydomonas* is activated by the *trans*-to-*cis* photoisomerization of retinal in the channel rhodopsin. In plants, the photoisomerization of phytochromes plays a key role in controlling their biological activity. These ingenious uses of organic molecules for the detection of photons in biological systems indicate that molecular photoswitches have the potential to be applied in the construction of various types of photon-working reagents and devices.

A characteristic feature of molecules is their small size (~ 1 nm). If a single molecule would work as one-bit memory, ultimate high-density (1 P bit/inch^2) optical memory could be realized. Conductance switches are key components of almost all electronic devices. Fabrication of single-molecule photoswitches is the first step toward molecular electronics. Photoswitching of single-molecule fluorescence has revolutionized fluorescence microscopy imaging. The super-resolution technique realizes a resolution of a few tens of nanometers. For such applications, molecular photoswitches are required to possess superior properties, such as thermal stability of both isomers, fatigue resistance, high sensitivity, rapid response, and reactivity in the solid state. Among them, thermal stability and fatigue resistance are indispensable properties. Although tremendous efforts were made in the 1970–1980s to provide the thermal irreversibility to molecular photoswitches in order to apply them to optical memory media, all attempts to modify existing photoswitchable molecules failed, because there was no guiding principle on how to prepare such thermally stable molecular photoswitches.

In 1988, the serendipitous discovery of diarylmaleic anhydrides, which undergo thermally irreversible photoswitching reactions, paved the way to solve the problem. Inferring from experimental as well as theoretical analysis, the molecular design principle of thermally irreversible molecular photoswitches was established. This new class of molecular photoswitches is named "Diarylethene." The well-designed diarylethenes provide outstanding photoswitching performance: both isomers are thermally stable for more than 470 000 years at 30 °C, photocyclization(coloration)/photocycloreversion(decoloration) can be repeated for more than 10^4 cycles, the quantum yield of cyclization reaction is close to 1 (100%), and the response times of both photocyclization and photocycloreversion reactions are less than 20 ps. Many of the diarylethene derivatives undergo photoswitching reactions even in the crystalline phase. In this book, the discovery and development of diarylethenes are described comprehensively from the basic concepts to their applications.

I wish to express my deep appreciation to my colleagues, Prof. M. Morimoto, Prof. S. Kobatake, Prof. K. Matsuda, Prof. T. Fukaminato, Prof. K. Uchida, Prof. T. Kawai, Prof. T. Tsujioka, and Dr. K. Uno for their kind help and support to prepare the manuscript. Finally, I also express my thanks to my family, Setsuko, Fumi, and Hisafumi for their continual encouragement to complete this book.

December 2020 *Masahiro Irie*

1

Introduction

1.1 General Introduction

Biological systems have developed various kinds of photoactive organs to adapt themselves to environmental electromagnetic radiation, sunlight. In biological systems, light is used in two ways as shown in Figure 1.1. In plants, for example, photosynthetic systems have evolved to utilize light as an energy source. Carbohydrates are produced from water and carbon dioxide in plants by using light as an energy source. Another category is the use of light for information access and transmission. Light is used to seek out optimum conditions for their life. Photoresponsive biological systems for vision, phototaxis, and phototropism have evolved to recognize surrounding conditions and to access external information. In the former, photoinduced electron transfer is a key chemical reaction. On the other hand, photoisomerization plays a key role in the latter.

Photoisomerization is one of the fundamental reactions in photochemistry [1–3]. *Trans–cis* isomerization, sigmatropic rearrangements, and electrocyclic rearrangements are typical examples. Molecules capable of these reversible photoisomerization reactions are called photochromic molecules or molecular photoswitches [4–10]. The two isomers differ from each other not only in their absorption and fluorescence spectra but also in their geometrical structures, oxidation/reduction potentials, refractive indices, and dielectric constants.

Scheme 1.1 shows typical examples of photoswitchable molecules and when they were discovered. The three upper molecules, azobenzene [11], spirobenzopyran [12], and bridged imidazole dimer [13], belong to T-type (thermally reversible) molecular photswitches, in which photogenerated right-side colored isomers are thermally unstable and spontaneously revert back to left-side colorless isomers in the dark. Azobenzene undergoes a large change in geometrical structure during the photoisomerization reaction from the *trans*- to the *cis*-form. The distance between 4 and 4′ carbon atoms (the long axis of the molecule) decreases from 0.90 to 0.55 nm and the dipole moment increases from 0.5 to 3.1 D [14, 15]. A spirobenzopyran derivative, 6-nitro-1′,3′,3′-trimethylspiro[2*H*-1-benzopyran-2,2′-indoline], converts from a less polar spiro-form to a polar merocyanine-form upon irradiation with UV light. It is reported that the dipole moment of the spiro-form is 6.2 D, while it

Diarylethene Molecular Photoswitches: Concepts and Functionalities, First Edition. Masahiro Irie.
© 2021 WILEY-VCH GmbH. Published 2021 by WILEY-VCH GmbH.

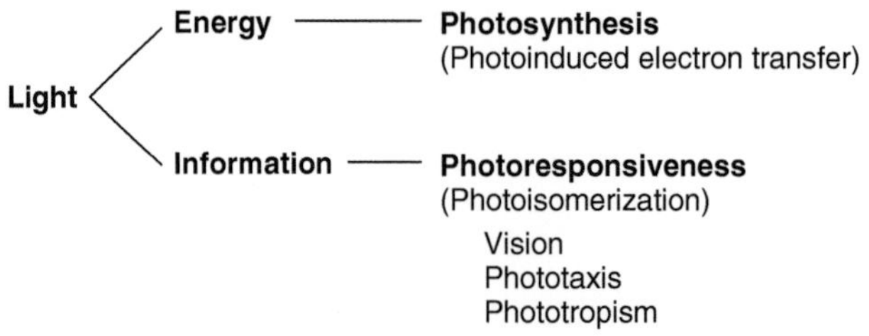

Figure 1.1 The use of light in biological systems.

increases to 13.9 D in the merocyanine form [16]. The blue color of the merocyanine form disappears in less than an hour even in high-*Tg* polymer matrices in the dark at room temperature [17]. A colorless bridged imidazole dimer, 1,8-TPID-naphthalene (TPID: dimer of triphenylimidazolyl radicals), turns green in toluene upon irradiation with UV light, and the green color disappears in less than a few seconds after switching off the UV light [18]. The imidazole dimer exhibits extremely quick response.

The two lower molecules, furylfulgide [19] and diarylethene [20–22], undergo P-type (thermally irreversible but photochemically reversible) photoswitching reactions. In the P-type molecular photoswitches, photogenerated right-side colored isomers are thermally stable and practically never return to the right-side colorless isomers in the dark at room temperature. Although many molecular photoswitches have been so far reported, P-type chromophores are very rare. The families, furylfulgides and diarylethenes, are two such rare examples exhibiting P-type reactivity. The primary difference between furylfulgides and diarylethenes is fatigue resistance. Photoinduced coloration/decoloration cycles of well-designed diarylethene derivatives can be repeated more than 10^4 times maintaining adequate photoswitching ability (see Section 3.3), whereas in most cases the corresponding cycles of furylfulgides are limited to less than 10^2 times.

The instant property changes of photoswitchable molecules by photoirradiation without any additional process lead to their use in various photoresponsive materials and photonic devices. When the bistable molecules are incorporated into materials, the electronic structure changes can be applied to optical memory media and conductance photoswitches, while the geometrical structure changes can be applied to light-driven actuators and others. Molecular photoswitches used in such applications are required to fulfill following properties.

1. Thermal stability of both isomers
2. Fatigue-resistance
3. High sensitivity
4. Rapid response
5. Reactivity in the solid state.

Molecular photoswitches belonging to the diarylethene family fulfill the above requirements simultaneously. Diarylethenes are derivatives of stilbene. When the phenyl rings of stilbene are replaced with five-membered heterocyclic rings, such

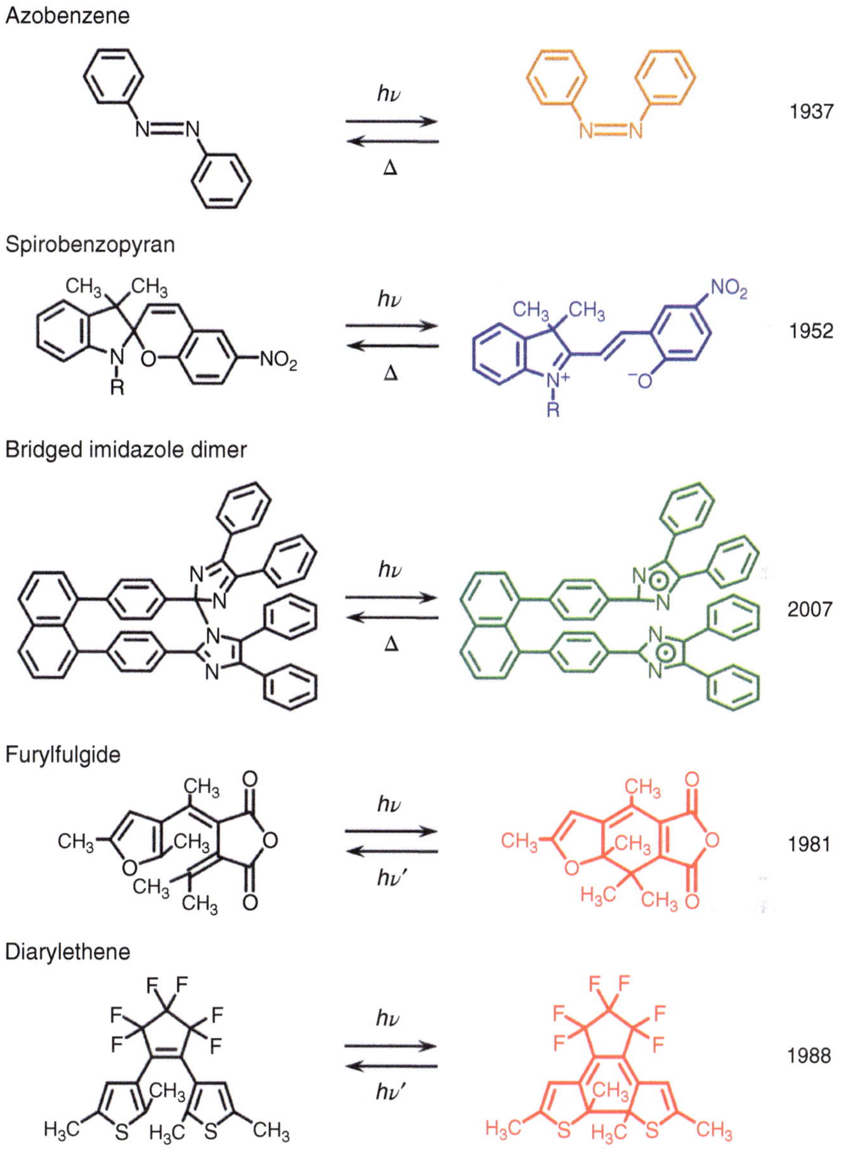

Scheme 1.1 Molecular photoswitches and years when they were discovered.

as thiophene or furan rings, both open- and closed-ring isomers become thermally stable and photoinduced coloration/decoloration cycles can be repeated many times. The best photoswitching performance of well-designed diarylethenes is summarized as follows.

1. Both isomers are thermally stable: half-life times at room temperature are as long as 470 000 years at 30 °C.
2. Photoinduced coloration/decoloration can be repeated for more than 10^4 cycles.

3. The quantum yield of cyclization (coloration) reaction is close to 1 (100%).
4. Response times of both coloration and decoloration reactions are less than 20 ps.
5. Many of diarylethene derivatives undergo photoswitching even in the single crystalline phase.

1.2 Discovery of Diarylethene Molecular Photoswitches

The diarylethene molecular photoswitches were serendipitously discovered during the course of a study on photoresponsive polymers [23]. Various polymers having molecular photoswitches, such as spirobenzopyran, azobenzene, or stilbene, in the side groups or main chains have been prepared in an attempt to modulate their conformations by photoirradiation. When azobenzene chromophores are incorporated into a polymer backbone, the solution viscosity was found to reversibly change upon alternate irradiation with UV and visible light [24]. Before UV light irradiation, the polymer has a rod-like extended conformation. Upon UV light irradiation, the azobenzene units convert from the *trans*- to the *cis*-form and the isomerization kinks the polymer chain, resulting in a compact conformation and a decrease in the viscosity, as shown in Figure 1.2. Not only viscosity but also other properties, such as pH, solubility, and sol–gel phase transition temperature, were successfully modulated upon photoirradiation [23–28].

Just like azobenzene, stilbene also undergoes the *trans–cis* photoisomerization reaction. The photoresponsive polymer research was extended to polymers having stilbene units. A polymer having stilbene units in the backbone can be prepared by 1,4-addition radical polymerization of 2,3-diphenylbutadiene, which is prepared from acetophenone, as shown in Figure 1.3 [29]. Upon irradiation with 313-nm light, the poly(2,3-diphenylbutadiene) efficiently underwent photocyclization reactions to produce a polymer having yellow colored dihydrophenanthrene units in a deaerated dichloromethane solution, instead of the *trans–cis* photoisomerization. The *trans–cis* photoisomerization of stilbene units in the backbone was strongly suppressed due to rigidity of the polymer chain. The dihydrophenanthrene units readily returned to the initial 2,3-diphenyl-2-butene units and the yellow color disappeared in less than 10 minutes at room temperature.

On the other hand, in the presence of air the dihydrophenanthrene units converted to phenanthrene units by hydrogen elimination and the reversibility was lost. To prevent hydrogen elimination and provide reversibility even under aerated conditions, 2,3-dimesitylbutadiene was designed (Figure 1.4A(a)). The synthesis of 2,3-dimesitylbutadiene was attempted by photoreduction of 2,4,6-trimethylacetophenone, as shown in Figure 1.4B(a). But, the synthesis of pinacol failed because of the bulky size of the mesityl group. To reduce steric hindrance, the mesitylene was replaced with 2,5-dimethylthiophene (Figure 1.4A(b)). According to the synthetic route shown in Figure 1.4B(b), 2,3-bis(2,5-dimethyl-3-thienyl)butadiene was successfully synthesized from 2,5-dimethyl-3-acetylthiophene. The butadiene was polymerized to poly(2,3-bis(2,5-dimethyl-3-thienyl)butadiene) by 1,4-addition radical polymerization.

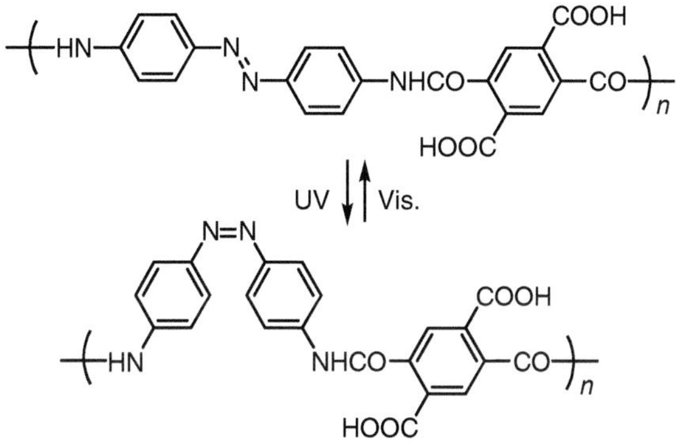

Figure 1.2 Schematic illustration of the photoinduced conformational change of a polymer having azobenzene units in the backbone.

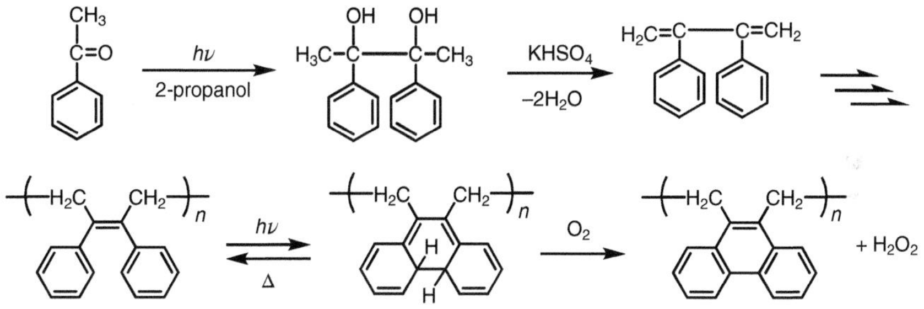

Figure 1.3 A synthesis route of poly(2,3-diphenylbutadiene) and its photochemical and thermal reactions.

The polymer having 2,3-dithienyl-2-butene units was dissolved in benzene and the solution was irradiated with 313-nm light. The colorless solution turned yellow ($\lambda_{max} \sim 430$ nm) along with the formation of cyclized closed-ring isomers. The yellow color disappeared upon irradiation with visible light. In contrast to poly(2,3-diphenylbutadiene), the yellow color of the closed-ring isomer units was found to remain stable overnight in the dark. The yellow closed-ring units were stable even at 100 °C and returned to the initial colorless open-ring isomer units with visible light. The dithienylethene unit in the polymer was unprecedentedly found to undergo a thermally irreversible photoswitching reaction. Poly(2,3-bis(2,5-dimethyl-3-furyl)butadiene) also underwent the thermally irreversible photoswitching reaction. The amazing result led us to study the photochemistry of the monomer unit, 2,3-di(2,5-dimethyl-3-thienyl)-2-butene and its derivatives in detail. This is the course of serendipitous discovery of diarylethene molecular photoswitches.

Since the discovery of thermally irreversible diarylethene molecular photoswitches in the middle of 1980s, various types of diarylethene derivatives have been synthesized to improve their photoswitching performance. Figure 1.5 shows a list of main diarylethene derivatives developed in Kyushu University and Rikkyo University until 2017. Upon irradiation with UV light, 2,3-di(2,5-dimethyl-3-thienyl)-2-butene underwent a *cis–trans* isomerization in addition to the cyclization reaction. To prevent the unfavorable

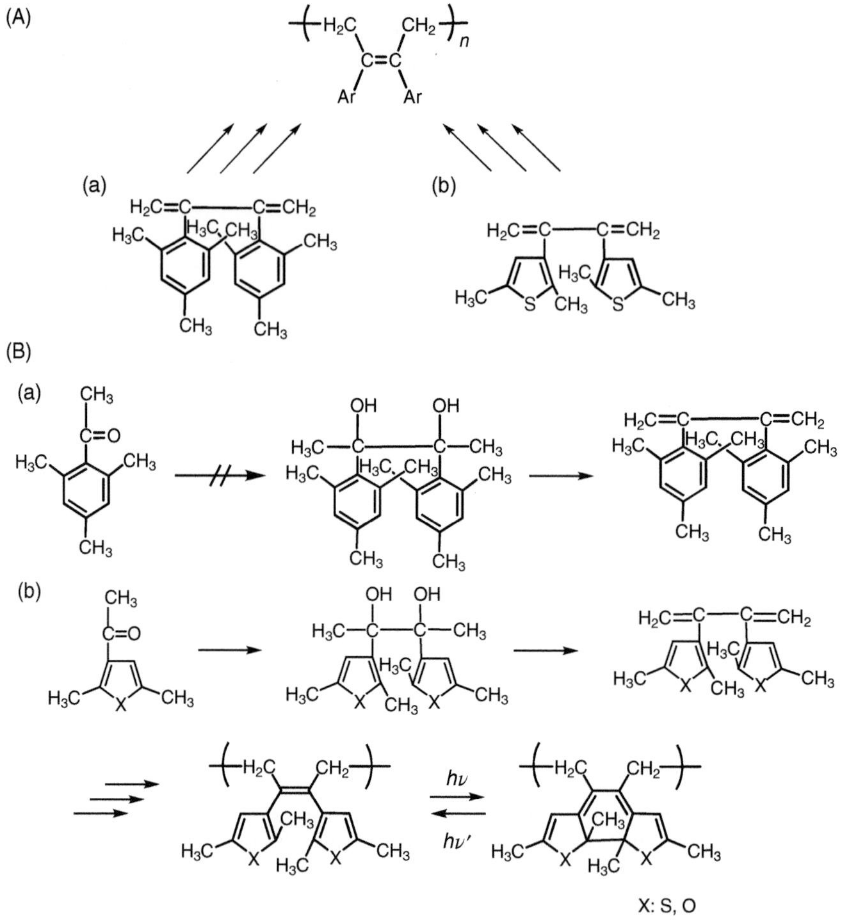

Figure 1.4 (A) Synthesis of polymers having (a) 2,3-dimesitylbutene units and (b) 2,3-di(2,5-dimethyl-3-thienyl)butene units in the backbone. (B) (a) A synthetic route to prepare 2,3-dimesitylbutadiene. (b) Synthetic routes and photochemical reactions of poly(2,3-di(2,5-dimethyl-3-thienyl)butadiene) and poly(2,3-di(2,5-dimethyl-3-furyl)butadiene).

cis–trans photoisomerization, a cyclic bridge, such as maleic anhydride or maleimide, was introduced. Although diarylethene derivatives with the maleic anhydride or maleimide bridge showed photocyclization reactivity in less polar solvents, the reactivity was strongly suppressed in polar solvents, such as methanol or acetonitrile. To provide photoswitching reactivity even in polar solvents, the ethene bridges were replaced with perfluorocycloalkenes with four-, five-, and six-membered rings [30]. The 1,2-bis(2-methyl-1-benzothiophen-3-yl)perfluorocycloalkenes underwent reversible photoinduced cyclization/cycloreversion reactions in polar methanol and acetonitrile. Among the three derivatives having four-, five-, and six-membered rings, five-membered 1,2-bis(2-methyl-1-benzothiophen-3-yl)perfluorocyclopentene was found to offer

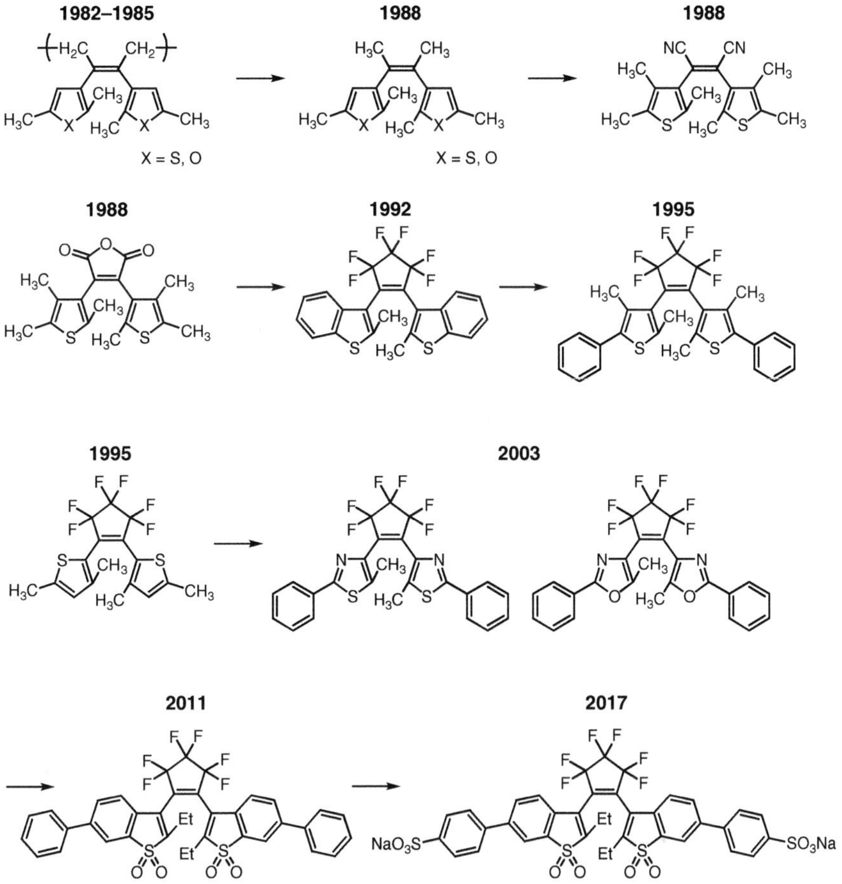

Figure 1.5 Development of diarylethene molecular photoswitches.

the highest resistance to photofatigue. Since then, perfluorocyclopentene derivatives have been mainly studied.

Although diarylethene photoswitches exhibit brilliant color changes upon photoirradiation, most of them are nonfluorescent or very weakly fluorescent in both isomer forms. It was a long-standing ambition to prepare photoswitchable fluorescent diarylethenes without attaching fluorescent chromophores to the diarylethenes. In 2011, sulfone derivatives of 1,2-bis(2-ethyl-6-aryl-1-benzothiophen-3-yl)perfluorocyclopentene were found to exhibit very strong fluorescence (fluorescence quantum yield ~ 0.9) in the closed-ring isomers [31]. The turn-on mode fluorescent diarylethenes are now extensively applied to super-resolution fluorescence microscopy in materials science and biological systems. Diarylethenes are able to switch both absorption (color) and fluorescence emission upon photoirradiation.

At first sight, the most striking phenomenon observed in molecular photoswitches is a photoinduced instantaneous color change. Figure 1.6 shows photos of the color

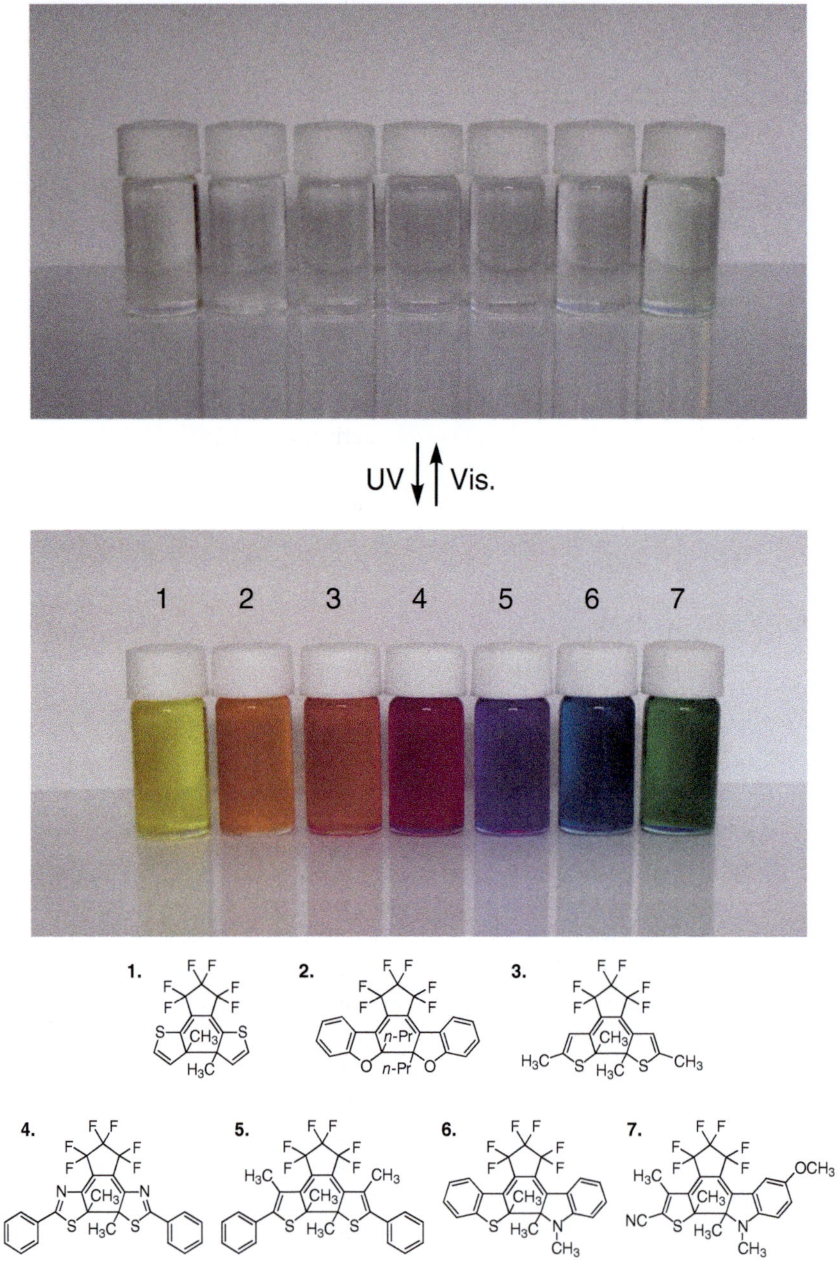

Figure 1.6 Color changes of diarylethene derivatives **1–7** in toluene upon irradiation with UV and visible light.

changes of diarylethene derivatives in solution. When the toluene solutions of the derivatives are irradiated with UV light, the colorless solutions turn yellow, orange, red, violet, blue, cyan, and green. The chemical structures of the derivatives are shown below in the photos. These colors disappear upon irradiation with visible light. The photoinduced coloration/decoloration cycles upon alternate irradiation with UV and visible light can be repeated many times.

The color changes are ascribed to the electronic structure changes of the derivatives from the open- to the closed-ring isomers. Two typical examples of the electronic structure changes are shown in Figure 1.7. 1,2-Bis(2,5-dimethyl-3-thienyl) perfluorocyclopentene (**3**) and 1,2-bis(3,5-dimethyl-2-thienyl)perfluorocyclopentene (**1**) undergo reversible electrocyclic rearrangements. The electrocyclic reactions involve rearrangements of positions of single and double bonds in a molecule. During the reactions, a new single bond is made between the central reactive carbon atoms by the cyclization reaction and the bond is broken as the ring is opened.

Figures 1.7 shows the chemical structures of the open- and the closed-ring isomers and their absorption spectra. In both derivatives **3** and **1**, upon irradiation with appropriate wavelength of light (λ_1 or λ_3) a single bond is formed between the central reactive carbon atoms and the double bonds change the position. Upon irradiation with another wavelength of light (λ_2 or λ_4) the single bond is broken and the molecule returns to the initial structure. The color is controlled by the length of π-conjugation. In the open-ring isomers, two thiophene rings have no particular interaction and the spectra are comparable to substituted thiophenes. In the closed-ring isomers, the π-conjugation length depends on the attached position of thiophene rings to the ethene bridge. When the thiophene rings are attached to the ethene bridge at 3-position, such as derivative **3**, π-conjugation is delocalized throughout the molecule in the closed-ring isomer and the delocalization results in red color. The π-conjugation is further extended when phenyl groups are substituted at 5- and 5′-positions of the thiophene rings, such as **5**. The long π-conjugation shifts the absorption band to longer wavelengths, resulting in blue color. On the other hand, when the thiophene rings are attached to the ethene bridge at 2-position, such as derivative **1**, π-conjugation is localized in the central part. The short π-conjugation results in yellow color in the closed-ring isomer.

The photoswitching between two discrete states and thermal irreversibility of the two states are indispensable for applications to memory media and switching devices. The bistability is a basic characteristic of diarylethenes. Although the chemical structures of the two isomers suggest photoswitching between two discrete states, in general, the absorption and fluorescence spectra gradually change upon photoirradiation in ensemble systems. Figure 1.8b shows the photoswitching performance of 1,2-bis(2-ethyl-6-phenyl-1-benzothiophene-1,1-dioxide-3-yl)perfluorocyclopentene (**8**, Figure 1.8a) in the ensemble system in 1,4-dioxane. The gradual analog increase in the fluorescence intensity upon irradiation with UV light indicates a change in the concentrations of the two isomers. Upon UV irradiation

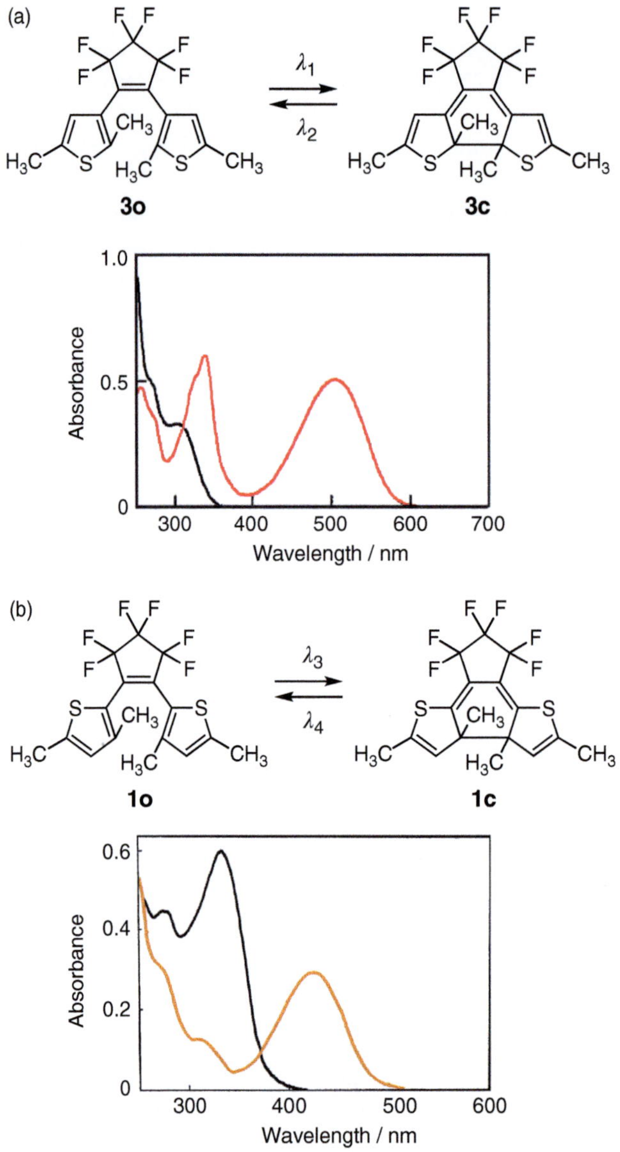

Figure 1.7 Chemical structures and absorption spectra of open- and closed-ring isomers of (a) **3** and (b) **1** in *n*-hexane.

the open-ring isomers convert to the fluorescent closed-ring isomers and the concentration ratio of the closed-ring isomers increases, causing the gradual increase in the fluorescence intensity. Subsequently, upon irradiation with visible light the ratio of the closed-ring isomers decreases, resulting in disappearance of the fluorescence. The switching between the two discrete states cannot be discerned from the photoswitching performance in the ensemble system.

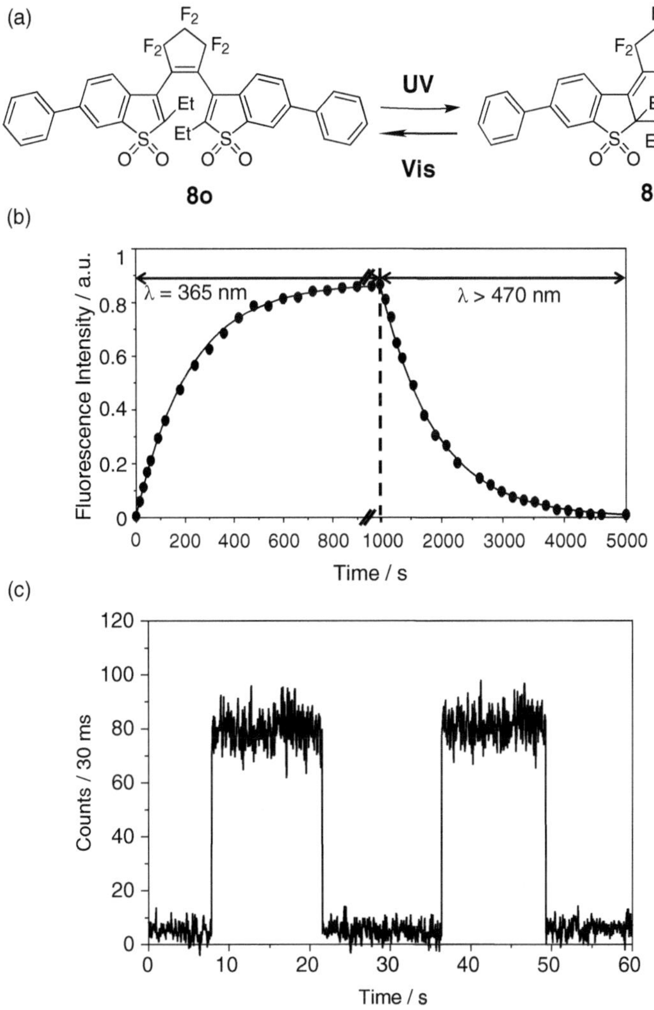

Figure 1.8 (a) Chemical structures of open- and closed-ring isomers of **8**. (b) Fluorescence photoswitching of **8** upon irradiation with UV and visible light in the ensemble system in 1,4-dioxane. (c) Fluorescence photoswitching of **8** at the single-molecule level in a Zeonex polyolefin film.

Digital *on/off* photoswitching between two discrete states was confirmed by measuring the switching response at a single-molecule level. Figure 1.8c shows the fluorescence photoswitching of a single molecule of derivative **8** upon alternate irradiation with UV and visible light. Upon irradiation with UV light, the fluorescence abruptly switches from the *off*-state to the *on*-state, while upon irradiation with visible light the *on*-state abruptly returns to the *off*-state. The digital photoswitching response definitely indicates that diarylethene photoswitch **8** has bistable states. The photoisomerization between two discrete isomer states expressed by the two chemical structures is experimentally evidenced.

References

1 Barltrop, J.A. and Coyle, J.D. (1975). *Excited States in Organic Chemistry*. London: Wiley.
2 Turro, N.J. (1978). *Modern Molecular Photochemistry*. Menlo Park, CA: Benjamin/Cummings Pub.
3 Balzani, V., Ceroni, P., and Juris, A. (2014). *Photochemistry and Photophysics: Concept, Research, Applications*. Weinheim: Wiley-VCH.
4 Brown, G.H. (1971). *Photochromism*. New York: Wiley-Interscience.
5 Crano, J.C. and Guglielmetti, R.J. (1998). *Organic Photochromism and Thermochromic Compounds*. New York: Plenum Press.
6 Bouas-Laurent, H. and Dürr, H. (2001). Organic photochromism. *Pure Appl. Chem.* 73: 639–665.
7 Dürr, H. and Bouas-Laurent, H. (eds.) (2003). *Photochromism: Molecules and Systems*. Amsterdam: Elsevier.
8 Irie, M., Yokoyama, Y., and Seki, T. (eds.) (2013). *New Frontiers in Photochromism*. Tokyo: Springer.
9 Tian, H. and Zhang, J. (eds.) (2016). *Photochromic Materials: Preparation, Properties and Applications*. Weinheim: Wiley-VCH.
10 Yokoyama, Y. and Nakatani, K. (eds.) (2017). *Photon-Working Switches*. Tokyo: Springer.
11 Hartley, G.S. (1937). The cis-form of azobenzene. *Nature* 140: 281.
12 Fischer, E. and Hirshberg, Y. (1952). Formation of coloured forms of spiropyrans by low-temperature irradiation. *J. Chem. Soc.*: 4522–4524.
13 Iwahori, F., Hatano, S., and Abe, J. (2007). Rational design of a new class of diffusion-inhibited HABI with fast back-reaction. *J. Phys. Org. Chem.* 20: 857–863.
14 Hampson, G.C. and Robertson, J.M. (1941). Bond lengths and resonance in the cis-azobenzene molecule. *J. Chem. Soc.*: 409–413.
15 Bullock, D.J.W., Cumper, C.W.N., and Vogel, A.I. (1965). Physical properties and chemical constitution. Part XLLIII. The electric dipole moments of azobenzene, azopyridines, and azoquinolines. *J. Chem. Soc.*: 5316–5323.
16 Malic, E., Weber, C., Richter, M. et al. (2011). Microscopic model of the optical absorption of carbon nanotubes functionalized with molecular spiropyran photoswitches. *Phys. Rev. Lett.* 106: 097401.
17 Kryszewski, M., Nadolski, B., North, A.M., and Pethrick, R.A. (1980). Kinetic matrix effects (response and density distribution function): ring closure reaction of indolinobenzospiropyrans in glassy poly(alkyl methacrylates). *J. Chem. Soc., Faraday Trans II* 76: 351–368.
18 Fujita, K., Hatano, S., Kato, D., and Abe, J. (2008). Photochromism of a radical diffusion-inhibited hexaarylbiimidazole derivative with intense coloration and fast decolouration performance. *Org. Lett.* 10 (14): 3105–3108.
19 Heller, H.G. and Oliver, S. (1981). Photochromic heterocyclic fulgides. Part 1. Rearrangement reactions of (*E*)-α-3-furylethylidene(isopropylidene)succinic anhydride. *J. Chem. Soc. Perkin Trans. 1*: 197–201.

20 Irie, M. and Mohri, M. (1988). Thermally irreversible photochromic systems. Reversible photocyclization of diarylethene derivatives. *J. Org. Chem.* 53 (4): 803–808.

21 Irie, M. (2000). Diarylethenes for memories and switches. *Chem. Rev.* 100 (5): 1685–1716.

22 Irie, M., Fukaminato, T., Matsuda, K., and Kobatake, S. (2014). Photochromism of diarylethene molecules and crystals: memories, switches and actuators. *Chem. Rev.* 114: 12174–12277.

23 Irie, M. (1990). Photoresponsive polymers. *Adv. Polym. Sci.* 94: 27–67.

24 Irie, M., Hirano, Y., Hashimoto, S., and Hayashi, K. (1981). Photoresponsive polymers. 2. Reversible solution viscosity change of polyamides having azobenzene residues in the main chain. *Macromolecules* 14 (2): 262–267.

25 Irie, M. and Schnabel, W. (1981). Photoresponsive polymers. On the dynamics of conformational changes of polyamides with backbone azobenzene groups. *Macromolecules* 14 (5): 1246–1249.

26 Irie, M. and Tanaka, H. (1983). Photoresponsive polymers. 5. Reversible solubility change of polystyrene having azobenzene groups. *Macromolecules* 16 (2): 210–214.

27 Irie, M. and Schnabel, W. (1985). On the dynamics of photostimulated conformational change of polystyrene with pendant azobenzene groups in solution. *Macromolecules* 18 (3): 394–398.

28 Irie, M. and Iga, R. (1986). Photoresponsive polymers. 9. Photostimulated reversible sol-gel transition of polystyrene with pendant azobenzene groups in carbon disulfide. *Macromolecules* 19 (10): 2480–2484.

29 Irie, M. and Schnabel, W. (1982). Photochemical conversion of poly-2,3-diphenylbutadiene to poly-9,10-dimethylenephenanthrene. *Eur. Polym. J.* 18: 15–18.

30 Hanazawa, M., Sumiya, R., Horikawa, Y., and Irie, M. (1992). Thermally irreversible photochromic systems. Reversible photocyclization of 1,2-bis(2-methylbenzo[*b*]thiophen-3-yl)perfluorocycloalkene derivatives. *J. Chem. Soc., Chem. Commun.* 3: 206–207.

31 Uno, K., Niikura, H., Morimoto, M. et al. (2011). In situ preparation of highly fluorescent dyes upon photoirradiation. *J. Am. Chem. Soc.* 133: 13558–13564.

2
Reaction Mechanism

2.1 Basic Concepts

Thermal irreversibility is an essential and indispensable property for the applications of molecular photoswitches to memory media and switches. Although tremendous efforts were made in the 1970–1980s to provide the thermal irreversibility to molecular photoswitches, all attempts to modify existing photoswitchable molecules failed. We had to wait until the thermally stable molecular photoswitches were serendipitously discovered. In the beginning and the middle of the 1980s, it was found that furylfulgides and diarylethenes undergo thermally irreversible photoswitching. The photogenerated colored isomers practically never revert back to the colorless isomers at room temperature. Although they undergo thermally irreversible photoswitching, the reason why the molecules show the thermal stability was not understood. It was a crucial task to reveal the reason. The basic principle behind the thermally irreversible photoswitching reactivity was elucidated using both theoretical and experimental approaches, as follows.

The 2,3-diphenyl-2-butene unit, shown in Figure 1.3, underwent a thermally reversible photoswitching reaction in a deaerated solution, while the 2,3-di(2,5-dimethyl-3-thienyl)-2-butene unit, shown in Figure 1.4B(b) exhibited a thermally irreversible reactivity. The photogenerated closed-ring form of 2,3-di(2,5-dimethyl-3-thienyl)-2-butene was found to remain stable and practically never returned to the open-ring form at room temperature. In addition, the open-ring isomer was stable even at elevated temperatures and no thermochromic reaction was observed. To reveal the reason why the diarylethene having phenyl rings and that having thiophene rings exhibit such a different reactivity, semiempirical modified neglect of diatomic overlap (MNDO) calculation was carried out for diarylethene derivatives **9–12** (Scheme 2.1) [1].

For electrocyclic reactions, two modes of geometrical structure changes, conrotatory and disrotatory, are distinguished, as shown in Scheme 2.2. According to the Woodward-Hoffmann rule [2] based on π-orbital symmetries for 1,3,5-hexatriene (HT), which is the simplest molecular framework of the above molecules, the conrotatory cyclization reaction to cyclohexadiene (CHD) is brought about by light and the disrotatory cyclization by heat.

Diarylethene Molecular Photoswitches: Concepts and Functionalities, First Edition. Masahiro Irie.
© 2021 WILEY-VCH GmbH. Published 2021 by WILEY-VCH GmbH.

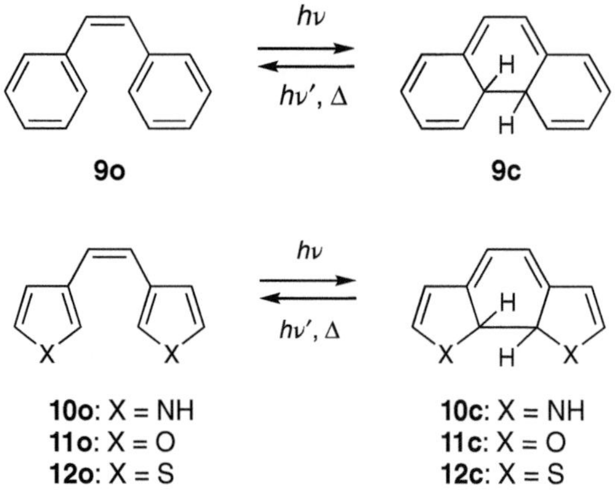

Scheme 2.1 Electrocyclic reactions of diarylethenes **9–12**.

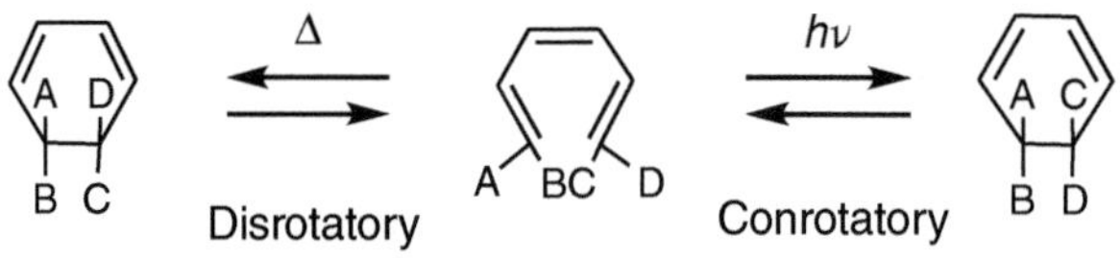

Scheme 2.2 Disrotatory and conrotatory cyclization reactions of hexatriene.

The cycloreversion reaction is allowed photochemically in the conrotatory mode and thermally in the disrotatory mode. From this simple symmetry consideration of the HT molecular framework, the thermal stability of the open-ring isomer of 2,3-di(2,5-dimethyl-3-thienyl)-2-butene and thermal irreversibility in the cycloreversion reaction cannot be explained. A state energy calculation is necessary to discuss the thermal stability.

Figures 2.1 and 2.2 show the state correlation diagrams for the electrocyclic reactions of **9** and **11** in disrotatory and conrotatory modes, respectively. Full lines in the figures show that interconnecting states belong to the same symmetry groups. The relative ground state energy differences between the open- and closed-ring isomers are shown in Table 2.1. The two heterocyclic rings were assumed to be in the parallel orientation for the disrotatory reaction and in the antiparallel orientation for the conrotatory reaction. As can be seen from Figure 2.1, orbital symmetry allows the disrotatory cyclizations in the ground states from **9o** to **9c** and from **11o** to **11c**. The relative ground state energies of the closed-ring isomers of **9** and **11** are, however, 175 and 113 kJ/mol higher than the respective energies of the open-ring isomers. This indicates that the open-ring isomers are stable and the thermal cyclization reaction does not take place practically in both cases.

On the contrary, orbital symmetry forbids the conrotatory cyclizations in the ground states from **9o** to **9c** and from **11o** to **11c**, because each S_0 open-ring isomer state correlates with a highly excited state of the closed-ring isomer, as shown in Figure 2.2. On the other hand, no such large barrier exists in the S_1 state for **9o** and the S_2 state for **11o**. This indicates that electrocyclic reactions of both

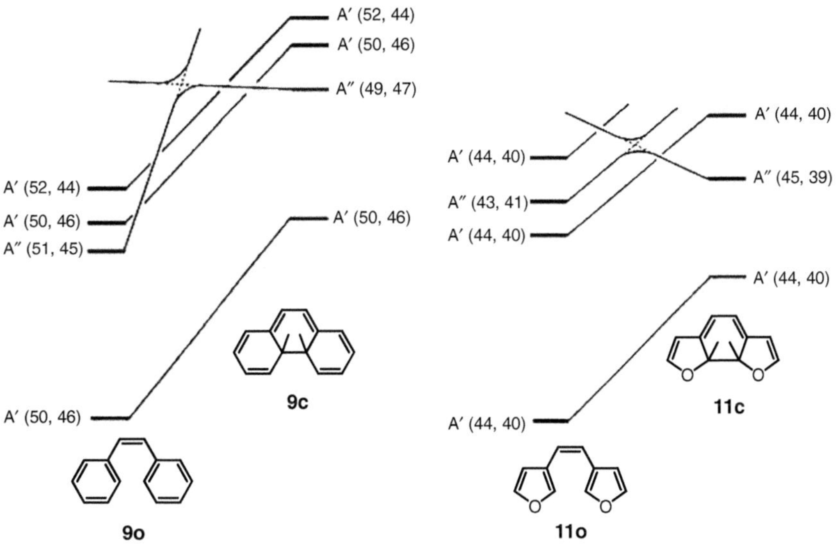

Figure 2.1 State correlation diagrams for the electrocyclic reactions in disrotatory mode. **9c** and **11c** are the closed-ring isomers, in which two hydrogens attached to the reactive central carbons are in a cis configuration.

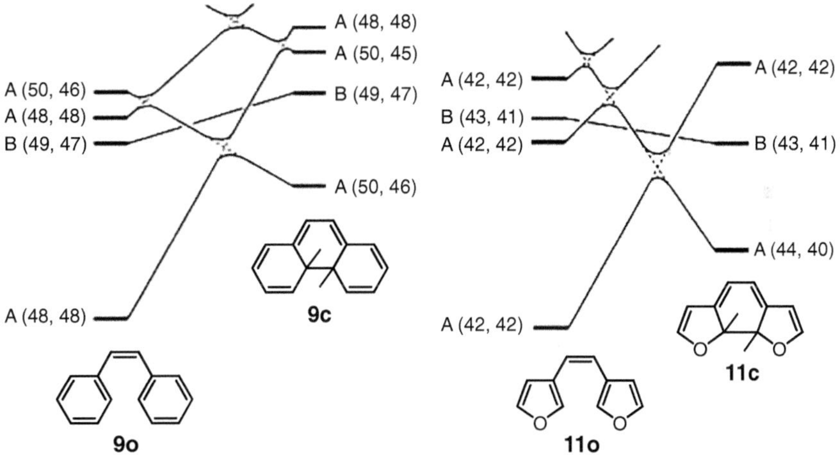

Figure 2.2 State correlation diagrams for the electrocyclic reactions in conrotatory mode. **9c** and **11c** are closed-ring isomers, in which two hydrogens attached to the reactive central carbons are in a trans configuration.

1,2-diphenylethene and 1,2-bis(3-furul)ethene are allowed in the photochemically excited states.

What should be discussed here is the stability of the closed-ring isomers. Figure 2.2 shows that in both **9c** and **11c,** the cycloreversion reactions in the ground state have to overcome energy barriers, and the barriers correlate with ground state energy differences between the open- and closed-ring isomers. The calculated

Table 2.1 Relative ground state energy differences between the open- and closed-ring isomers.

Compound	Disrotatory (kJ/mol)	Conrotatory (kJ/mol)
1,2-Diphenylethene (**9**)	175	114
1,2-Di(3-pyrrolyl)ethene (**10**)	135	65
1,2-Di(3-furyl)ethene (**11**)	113	38
1,2-Di(3-thienyl)ethene (**12**)	51	−14

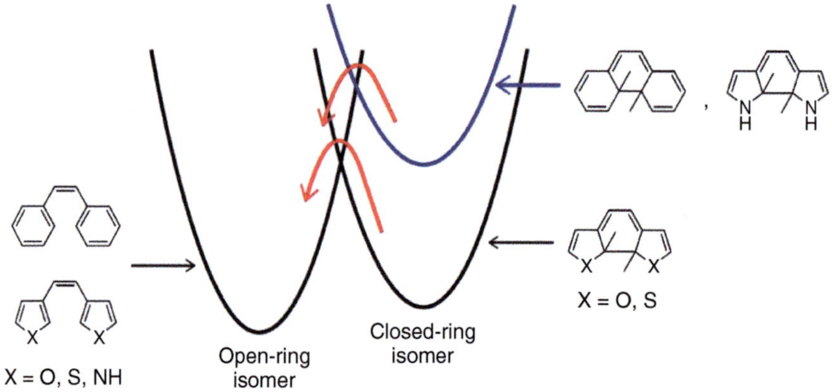

Figure 2.3 Correlation between the ground state energy difference between open- and closed-ring isomers and the energy barrier.

energy differences are shown in Table 2.1. When the energy difference is large, as in the case of **9**, the energy barrier becomes small and the cycloreversion reaction takes place readily. On the other hand, the energy barrier becomes large when the energy difference is small. In this case, the cycloreversion reaction hardly takes place. The correlation between the ground state energy difference and the energy barrier is well explained by the Horiuti–Polanyi rule as shown in Figure 2.3. The energy difference in the ground states between the open- and closed-ring isomers controls the stability of the closed-ring isomers.

The next question is what causes the difference in the ground state energy levels of the two isomers. First, strain energies of the six-membered rings of the closed-ring isomers were compared. The optimized geometries of the closed-ring isomers, **9c** and **11c**, however, showed almost identical six-membered ring structures and the ring-strain could not explain the energy difference. Next, the aromaticity change from the open- to the closed-ring isomers was examined. During the cyclization reaction, phenyl and heterocyclic rings change the structures as shown in Scheme 2.3. The aromaticity of the rings is lost during the cyclization reactions. The energy differences between the right- and left-side groups were calculated and are shown in Table 2.2. The aromatic stabilization energy of the aryl groups correlates well with

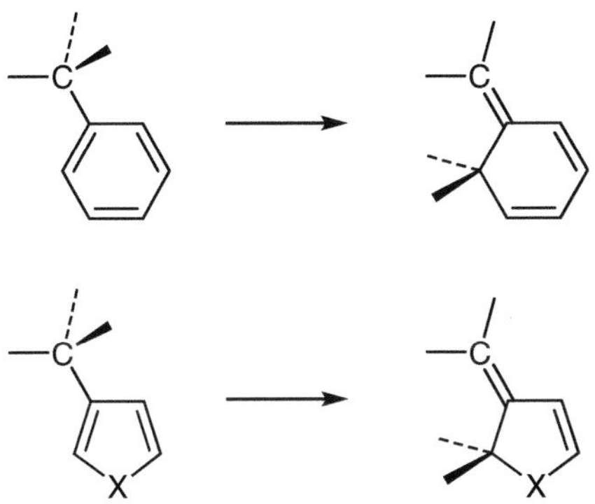

Scheme 2.3 The structure changes of phenyl and five-membered heterocyclic rings along with the cyclization reactions.

Table 2.2 Aromatic stabilization energy differences.

Group	Energy (kJ/mol)
Phenyl	116
Pyrrol	58
Furyl	38
Thienyl	20

the ground state energy difference. The highest energy difference was calculated for the phenyl group and the lowest one for the thienyl group. Destabilization due to destruction of the aromatic ring during the cyclization reaction increases the energy of the closed-ring form. The aromaticity is the key molecular property that controls the thermal stability of the closed-ring isomers.

The theoretical prediction was confirmed by the synthesis of diarylethene derivatives with various types of aryl groups as shown in Figure 2.4. When the aryl groups are thiophene, benzothiophene, thiazole, or oxazole rings, which have low aromatic stabilization energy, the closed-ring isomers are stable (more than 12 hours at 80 °C). On the other hand, photogenerated closed-ring isomers of diarylethenes with indole rings, which have intermediate aromatic stabilization energy, undergo thermally reversible photoswitching (half-life time at 80 °C of **16c**: 2.5 hours). The closed-ring isomers of diarylethene derivatives with phenyl rings readily returned back to open-ring isomers (half-life time at 20 °C of **18c**: 1.5 minutes).

From the above theoretical and experimental results, the guiding principle for the synthesis of thermally irreversible diarylethenes is defined as follows.

> The thermally irreversible photoswitching diarylethenes can be prepared by employing aryl groups with low aromatic stabilization energy.

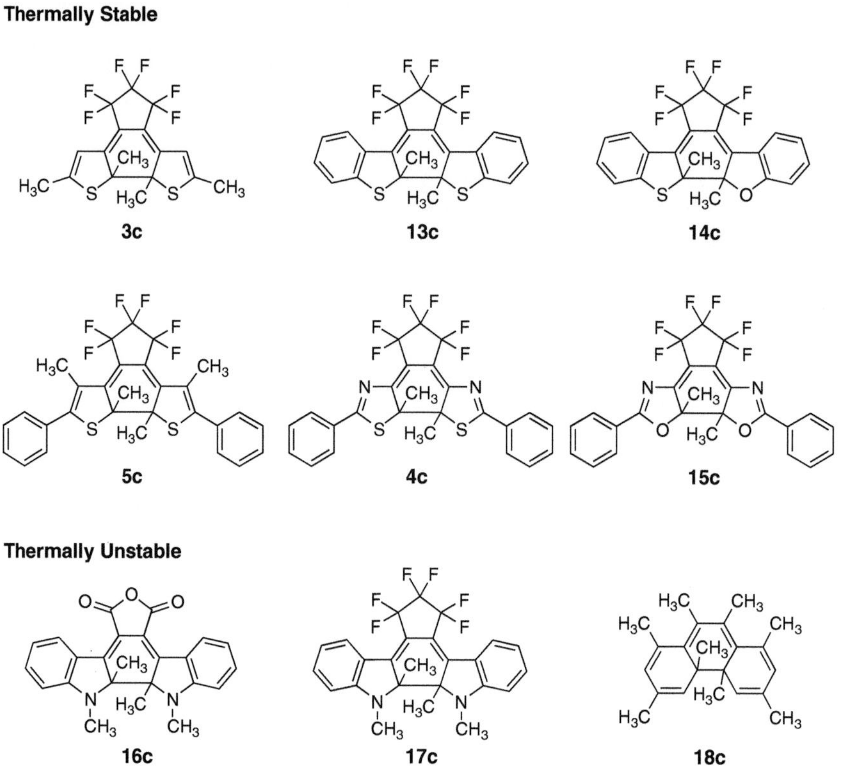

Figure 2.4 Thermal stability of diarylethene derivatives. Any appreciable change of the absorption intensity of the closed-ring isomer was not observed in the thermally stable derivatives for more than 12 hours at 80 °C.

2.2 Theoretical Study

The well-studied photoinduced cyclization and cycloreversion reactions between 1,3,5-hexatriene (HT) and cyclohexadiene (CHD) provide a useful framework for understanding the basic reaction mechanism of diarylethenes. As the first step, potential energy surfaces of a model diarylethene, 1,2-bis(cyclopenta-1,3-dien-2-yl)-ethene, were calculated using a complete active space self-consistent field (CASSCF) method [3]. Figure 2.5 shows S_0 and S_1 potential energy surfaces along the reaction coordinate. On both S_0 and S_1 surfaces there exist two minima, a closed-ring isomer (CHD and CHD*) and an open-ring isomer (HT and HT*). Transition structures (TS_0 and TS_1) were also characterized on each potential energy surface. Several conical intersection minima (indicated by crosses) exist on both closed- and open-ring sides of the potential surface in addition to CI_3. CI_3 is the most important conical intersection, because it is the only one that provides a pathway toward both open and closed minima on the ground state. At the conical intersection geometry, one has a triangular arrangement between three unpaired electrons belonging to carbon atoms in the two 5-membered rings. These three electrons are weakly

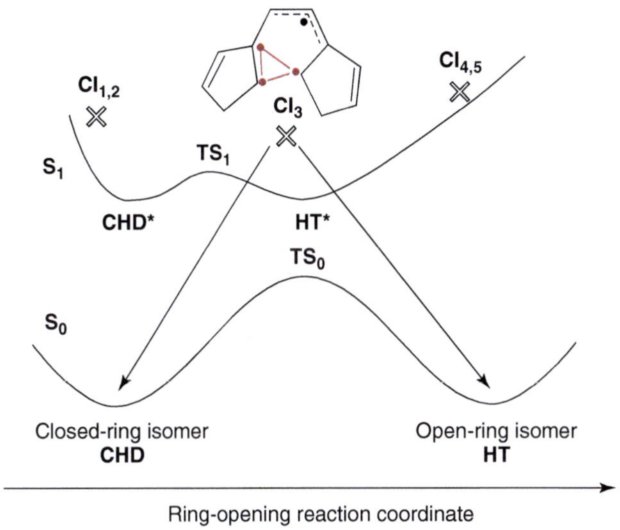

Figure 2.5 Potential energy surfaces of a model diarylethene. *Source*: Reprinted with permission from Ref. [3]. Copyright 2003 American Chemical Society.

coupled π-electrons, and a fourth (belonging to a three-electron allyl fragment) is an uncoupled spectator. It is worth noting that the TS_1 barrier does not exist in the CHD/HT system. The barrier is probably due to steric constraint from the sigma-bond framework.

Figure 2.5 shows the potential energy surfaces along the reaction coordinate of the hydrocarbon model system. A more realistic model is dithienylethene. The potential energy surfaces of 1,2-di(3-thienyl)ethene (**12**) were calculated as a function of the distance between two reactive carbon atoms, R(C—C), using the CASSCF method, as shown in Figure 2.6 [4]. To understand the reaction mechanism, a molecular mechanics-valence bond (MMVB) computation of the dynamics was also carried out. Three-dimensional representation of the structures of S_0 and S_1 potential energy surfaces and two-dimensional simplified one of the structures are shown in Figures 2.7 and 2.8, respectively [5]. These figures also outline minimum energy paths corresponding to ring-closing and ring-opening reactions of the dithienylethene.

Upon irradiation with UV light, the open-ring isomer is excited to the allowed Frank–Condon state (1BFC(o)), which is close in energy to a 2A surface, and fast internal conversion to thc 2A statc takes place. The excited wavepacket moves down along the 2A surface and the conical intersection (2A/1ACI(o)) can then be accessed via vibrations orthogonal to the reaction pass, as shown in Figure 2.7. At the conical intersection, efficient decay to the ground state potential energy surface takes place, leading to the formation of the closed-ring isomer. This mechanism accounts for the highly efficient and ultrafast ring-closing reaction.

The ring-opening reaction takes place as follows. Upon irradiation with visible light, the closed-ring isomer is excited to the allowed 1B Frank–Condon state (1BFC(c)). The excited wavepacket moves away from the Frank–Condon region

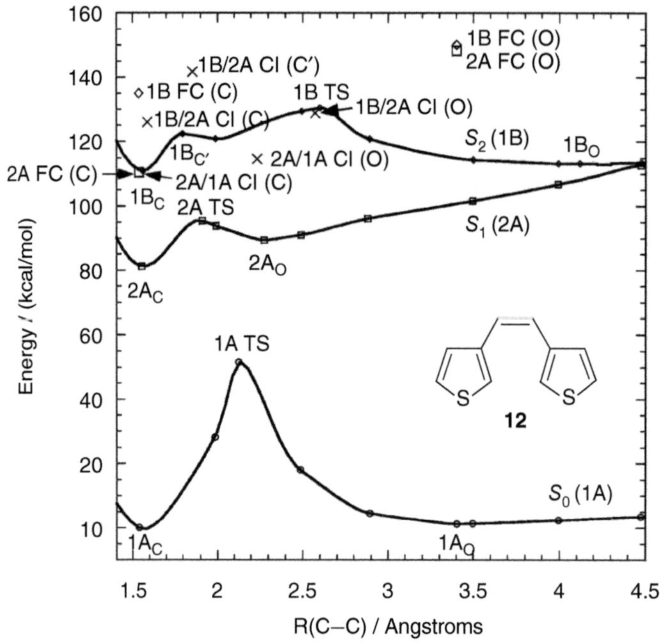

Figure 2.6 Potential energy surfaces of ground (1A) and two excited (2A and 1B) states of **12** as a function of the distance between two reactive carbon atoms. *Source*: Reprinted with permission from Ref. [4]. Copyright 2004 American Chemical Society.

along the 1B surface and rapidly falls from the 1B to 2A surface via a conical intersection (1B/2ACI(c)). The molecule excited with light of shorter wavelengths having vibrational excess energy on the 2A surface goes over the energy barrier $2ATS_1$ more readily and reaches 2A/1A conical intersection (2A/1ACI(o)). The conical intersection is reached via vibrations orthogonal to the initial motion in the valley. At the conical intersection, fast deactivation to the ground state takes place and thus the open-ring isomer is formed. The route suggests that thermal activation or excess kinetic energy on the 2A surface is required for the ring-opening reaction to take place. This agrees with experimental observation that the ring-opening reaction has an activation energy and that the quantum yield increases upon irradiation with light of shorter wavelengths (see Section 3.1.3) [7, 8].

2.3 Reaction Dynamics

2.3.1 Cyclization Reaction

As discussed in the theoretical study in Section 2.2, diarylethenes are anticipated to undergo very fast ring-closing (cyclization) reactions. The rapid cyclization dynamics was confirmed using femtosecond absorption spectroscopies in solution as well as in the single crystalline phase [9].

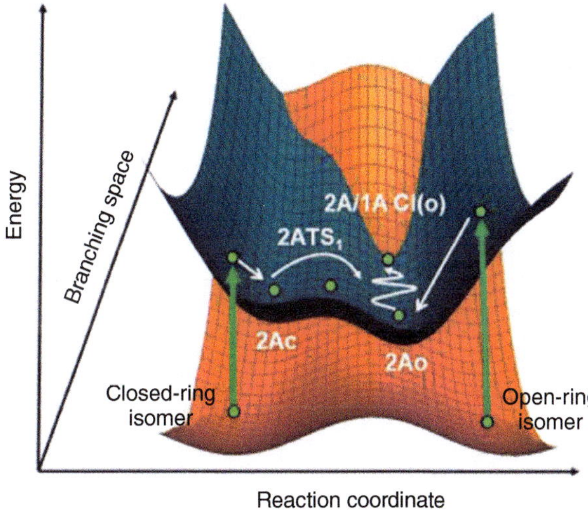

Figure 2.7 Schematic representation of the structures of S_0 (orange) and S_1 (green) potential energy surfaces corresponding to the photoisomerization between open- and closed-ring isomers of a diarylethene. Reaction coordinate is the distance between the two reactive carbon atoms. *Source*: Adapted from Ref. [5] with permission from the PCCP Owner Societies.

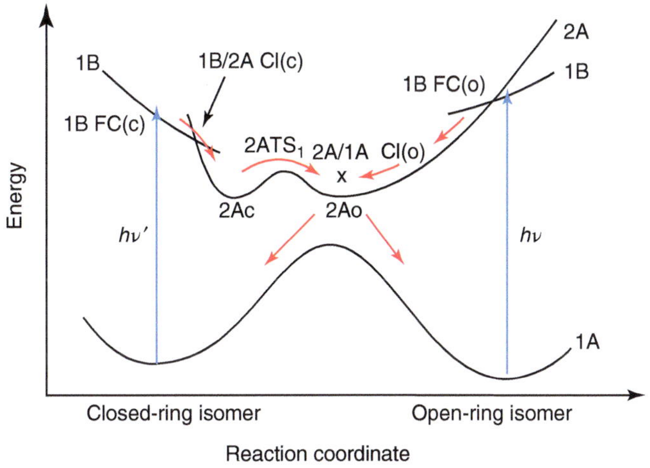

Figure 2.8 Outline of two reaction paths corresponding to ring-opening and ring-closing reactions of a diarylethene. The letters in parentheses, c and o, represent closed- and open-ring isomer structures, respectively. *Source*: Reprinted with permission from Ref. [6]. Copyright 2014 American Chemical Society.

The cyclization reaction of 1,2-bis(2,4-dimethyl-5-phenyl-3-thienyl)perfluorocyclopentene (**5**) was investigated in the single crystalline phase. In solution, the open-ring isomer has two conformations, photoactive antiparallel and photoinactive parallel ones, as shown in Scheme 2.4. Coexistence of the two conformations in solution prevents clear resolution of spectral features associated with the

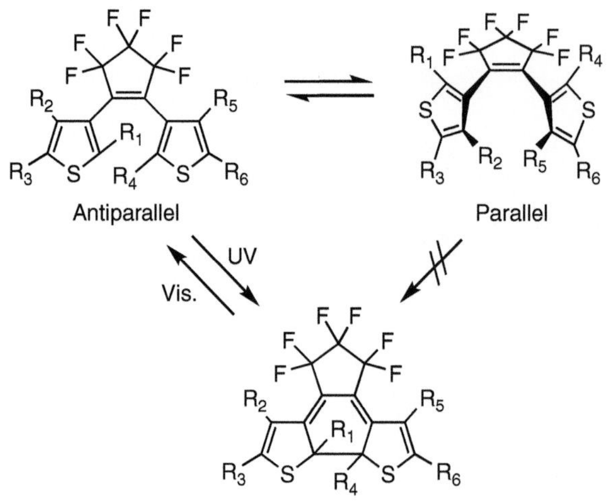

Scheme 2.4 A diarylethene open-ring isomer has two conformations with two aryl groups in mirror symmetry and C_2 symmetry, which are called parallel and antiparallel conformations, respectively.

photocyclization reaction. On the other hand, in the single crystalline phase of **5o,** all molecules are fixed in the photoactive antiparallel conformation and this simplifies the analysis of reaction dynamics.

Figure 2.9 shows the transient absorption spectra associated with the cyclization reaction in the single crystalline phase upon irradiation with a 343 nm femtosecond laser pulse. The formation of the spectrum ascribed to the closed-ring isomer is clearly observed. After approximately 20 ps, the transient spectrum is essentially fully converged to that of the final product. The subpicosecond dynamics following excitation with a 343 nm pulse indicates that the initially broad photoinduced absorption seen at 75 fs progressively shifts from the red (635 nm) to the blue (480 nm) end of the visible spectrum in the first 500 fs following the UV excitation. The progressive spectral shift occurs prior to the ring-closure itself and is attributed to the evolution on the excited state potential of the open-ring isomer due to the difference in the equilibrium geometry between ground and excited states. The absorption band at 635 nm is assigned to the excited state absorption of the open-ring isomer initially in the equilibrium geometry associated with ground state charge distribution. The absorption in the blue end is associated with the relaxed structure of the open-ring isomer in the excited state. The formation of this intermediate occurs with a time constant of 200 fs.

Figure 2.10a shows the convergence on the picosecond time scale of the transient absorption toward that of the closed-ring photoproduct. Initially, at 0.5 ps after the pulse the absorption associated with the relaxed open-ring isomer in the excited state dominates the signal. In the time delay between 0.5 and 20 ps, the spectral feature associated with the intermediate decays in concert with the growth of the ground state absorption associated to the closed-ring isomer centered at 635 nm. Figure 2.10b shows the transient time traces at 490 and 635 nm. The latter is the

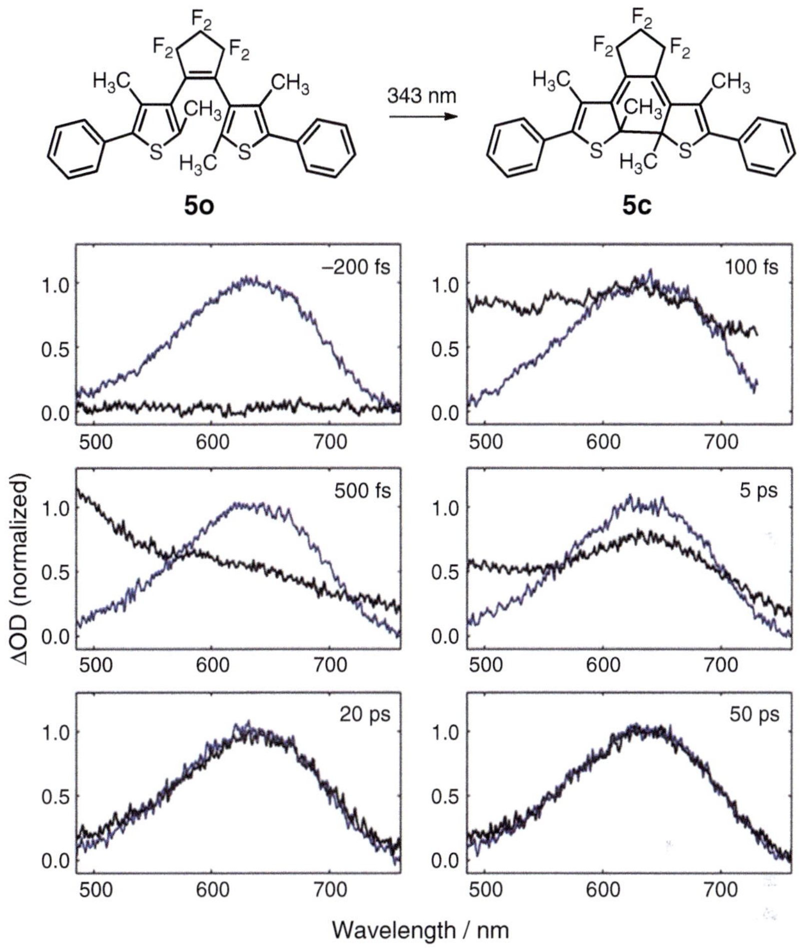

Figure 2.9 Transient absorption spectra of **5** during the ring-closing reaction in the single crystal excited with a 343 nm femtosecond laser pulse. Selected spectra for various time delays between −200 fs and +50 ps (black curves), as compared to their associated τ_{∞} measurements (blue curves). The 100-fs spectrum was extracted from a separated data set taken with higher resolution time steps. *Source*: Adapted with permission from Ref. [9]. Copyright 2011 American Chemical Society.

central wavelength of the absorption associated with ground state absorption of the closed-ring isomer and follows the growth of the photoproduct. The former follows the decay of the excited state absorption associated with the relaxed open-ring isomer. The time constants for the rise at 635 nm and the decay at 490 nm are 7.3 ± 0.8 and 5.3 ± 0.3 ps, respectively. It was anticipated that the rise and decay of the signals would be equally matched, as the closed-ring photoproduct is thought to be directly formed from the relaxed excited state of the open-ring isomer through the conical intersection. The discrepancy in the time constants is attributed to the vibrational relaxation of the nascent closed-ring isomer. The growth of the signal at 635 nm is the convolution of the cyclization reaction, vibrational cooling and lattice relaxation.

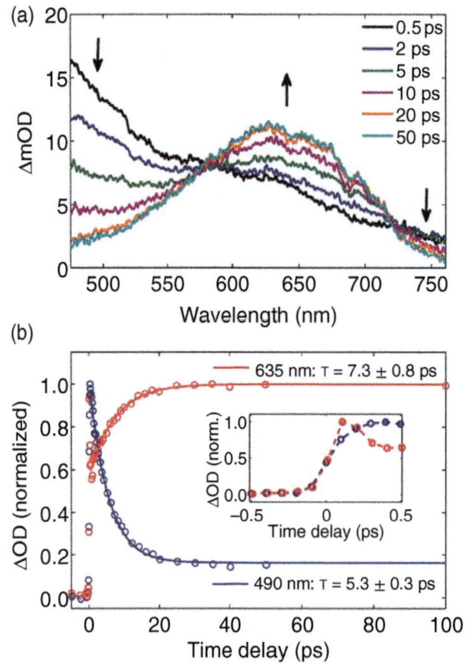

Figure 2.10 Transient absorption spectra of **5** during the ring-closing reaction in the single crystal excited with a 343 nm femtosecond laser pulse. (a) Transient absorption spectra for time delay between 0.5 and 50 ps. (b) Transient traces for the probe wavelengths at 635 and 490 nm. Monoexponential fits are shown, with time constants of 7.3 and 5.3 ps for 635 and 490 nm, respectively. The inset demonstrates the sub-picosecond behavior. *Source*: Reprinted with permission from Ref. [9]. Copyright 2011 American Chemical Society.

Therefore, the decay of the excited absorption of the open-ring isomer is assigned to the time constant of the ring-closing (cyclization) reaction, that is 5.3 ± 0.3 ps. The cyclization time constant in the single crystalline phase is slightly slower than the values in solution, ~ subpicoseconds. This is presumably attributed due to the impeding influence of the close-packed environment in crystal.

Although in solution both types of conformers, antiparallel and parallel ones, are equally excited by the pump pulse and the coexistence of the two conformers prevents clear resolution of the transient absorption spectra, it is still possible to appropriately analyze the spectra, because of the distinct dynamics of the two conformers. Femtosecond laser spectroscopy study was carried out in cyclohexane solutions containing 1,2-bis(2-methyl-5-phenyl-3-thienyl)perfluorocyclopentene (**20**) and nonfluorinated (perhydro) analogue 1,2-bis(2-methyl-5-phenyl-3-theinyl)cyclopentene (**21**), and the effect of fluorination in the photcyclization reaction was examined [10]. The dynamics after photoexcitation of **20o** and **21o** can be expressed in three stages: (i) pre-switching due to the excited-state mixing and relaxation, (ii) the ring-closure, and (iii) post-switching related to the vibrational cooling. In all stages, the fluorinated diarylethene **20o** was found to switch faster than its nonfluorinated analogue **21o**. The mixing and relaxation time constants of **21**, 70 ± 15 and 150 ± 30 fs, respectively, were accelerated to the time constants of 50 ± 10 and 120 ± 30 fs in **20**. The ring-closure reaction rate also increased from 4.2 to 0.9 ps. The results indicate that fluorinated switch is faster and more efficient than the nonfluorinated switch.

The cyclization dynamics of 1,2-bis(2-methyl-1-benzothiophen-3-yl)perfluorocyclopentene (**13**) was also studied in detail in a *n*-hexane solution [11]. Figure 2.11

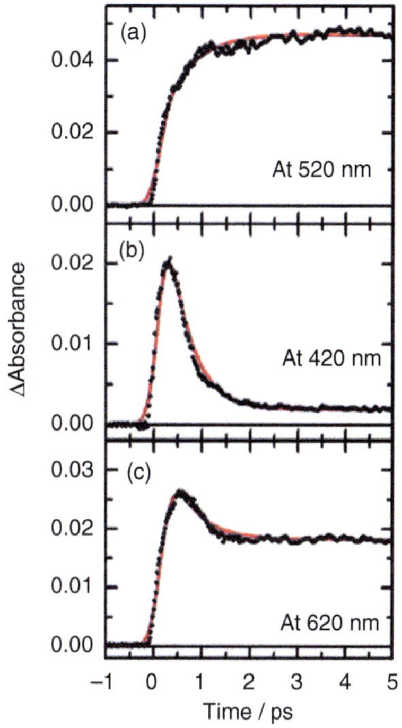

Figure 2.11 Time profiles of transient absorbance of **13** in *n*-hexane excited with a 310 nm femtosecond laser pulse. The detection wavelength is 520 nm for (a), 420 nm for (b), and 620 nm for (c), respectively. Solid lines in each of the frames are calculated curves by taking into account the pulse durations and the time constants. *Source*: Adapted with permission from Ref. [11]. Copyright 2011 American Chemical Society.

shows the time profile of the transient absorbance in *n*-hexane excited with a 310 nm femtosecond laser pulse. The time profile monitored at 520 nm in the initial 5 ps after the excitation shows that the positive signal appears within the response of the apparatus and gradually rises in a few picosecond time regions. The spectral shape and absorption maximum around 520 nm are the same as those of the ground state absorption of **13c**. The solid line is the result calculated with a time constant of 450 fs. On the other hand, the time profiles monitored at 420 and 620 nm decay with the time constants of 480 and 420 fs, respectively. These time constants agree with the rise constant at 520 nm. The decreasing signals observed at 420 and 620 nm are assigned to the decay of the excited state of **13o** undergoing the cyclization. The rise and decay profiles indicate that the open-ring isomer having a broad absorption from 420 to 620 nm converts to the closed-ring isomer having the absorption maximum at 520 nm with the time constant of 450 fs.

These transient spectroscopic studies in solution indicate that the cyclization reactions take place in less than 1 ps. Although in the crystalline phase the cyclization time constant is slowed to several picosecond region, it is safe to say that the central carbon–carbon bond is made within 20 ps. The very fast cyclization dynamics in solution and in the single crystalline phase is consistent with the prediction of theoretical study.

2.3.2 Cycloreversion Reaction

In contrast to the cyclization reaction described in Section 2.3.1, it is difficult to directly measure the time constant of the ring-opening (cycloreversion) reaction,

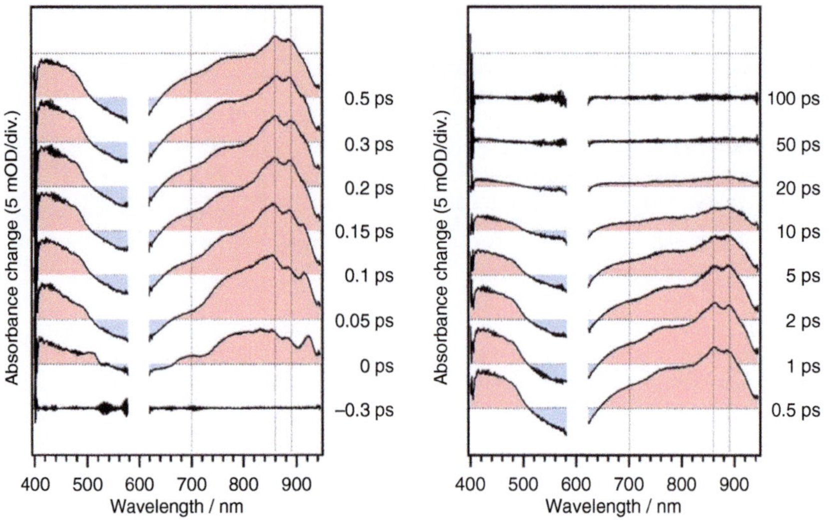

Figure 2.12 Time-resolved transient absorption spectra of **20c** in *n*-hexane excited with a 600 nm femtosecond laser pulse. *Source*: Reprinted with permission from Ref. [12]. Copyright 2017 American Chemical Society.

because the absorption spectrum of the photogenerated open-ring isomer and that of the closed-ring isomer overlap in the UV region. The time constant was estimated from the decay of the transient absorption spectra as well as the fluorescence signal of the closed-ring isomer.

Figure 2.12 shows time-resolved transient absorption spectra of the closed-ring isomer of 1,2-bis(2-methyl-5-phenyl-3-thienyl)perfluorocyclopentene (**20c**) in a *n*-hexane solution excited with a femtosecond laser pulse at 600 nm [12]. Time evolution of the spectra following the excitation can be divided into three stages. In the time region immediately after the excitation (0–0.5 ps), the absorption at 860 nm appears together with the negative signal due to the bleaching of the steady-state absorption at around 600 nm. In the initial 0.5-ps time range, the intensity of the absorption at 860 nm slightly decreases and that at 890 nm relatively increases, together with the appearance of the absorption around 700 nm. The absorption at 860 nm is due to the initially formed 1B state, while the subsequently growing bands at 890 and 700 nm are attributable to the 2A state. In the following second stage at the time region of 0.5–5 ps, the relative ratio of the transient absorption at 890 and 700 nm to that at 860 nm gradually increases. Finally, after 5 ps following the excitation, the decay of the positive and negative absorption intensities to the baseline was observed with no remarkable evolution of the spectral band shape.

The time profiles of the transient absorbance monitored at 840 and 700 nm are reproduced by a triple exponential function with time constants of 200 fs, 3.0 ps, and 12 ps. As shown in Figures 2.7 and 2.8, the 1BFC(c) state produced by photoexcitation of the closed-ring isomer **20c** undergoes the internal conversion to the local minimum of 2A state (2Ac) via 1B/2A CI(c). The 2Ac state goes over the energy barrier, $2ATS_1$, in competition with the direct non-radiative as well as radiative decay processes to the ground state. According to this mechanism, the shortest time constant of 200 fs is attributed to the internal conversion from the 1B state to the 2Ac state. The bands at 890 and 700 nm are ascribable to the 2Ac state, because the rise of these bands is accompanied by the decay of the band at 860 nm assigned to 1B state. To elucidate the origin of the second 3-ps component, the time profile of the transient absorbance at 700 nm divided by that at 840 nm was plotted. The relative intensity rapidly increased within 1 ps and was followed by the gradual increase in several picosecond time region. This time profile indicates that the rapid internal conversion yields the 2Ac state with the time constant of 200 fs and the slow 3-ps process leads to the spectral evolution in the 2Ac state. The time constant of 3.0 ps is attributed to the geometrical rearrangement in the excited state, because 3 ps is in the typical temporal scale of vibrational cooling in solution. Finally, the longest time constant of 12 ps is ascribable to the lifetime of the 2Ac state, because the recovery of the transient bleaching of the ground state and the decay of the positive transient absorption have the same time constant of 12 ps. Temperature dependence of the transient absorption decay and the fluorescence decay at 700 nm was measured and it was found that both transient absorption and fluorescence decay constants become longer with decrease in the temperature. The time constant of the transient absorption decay was almost identical to the fluorescence lifetime. The agreement of these two time constants indicates that the 2Ac state gives the fluorescence. The 2Ac state goes over the energy barrier ($2ATS_1$) with a time constant of 12 ps at room temperature and finally leads to the photoproduct via 2A/1ACI(o). Both cyclization and cycloreversion reactions take place via the common CI(o). The cyclization reaction in solution completes within 1 ps, as described before. Therefore, the staying time in the 2Ao state is considered to be less than 1 ps. The time constant of 12 ps reflects the time taken to overcome the barrier $2ATS_1$.

Inferring from the experimental results, the cycloreversion dynamics of **20c** is summarized as follows. The 1B state produced by the excitation of the closed-ring isomer undergoes internal conversion into the 2Ac state with the time constant of 200 fs, which is followed by geometrical rearrangement with vibrational cooling in 3.0 ps. The resultant 2Ac state goes over the activation barrier ($2ATS_1$) into the 2Ao state with the time constant of 12 ps and finally leads to the open-ring isomer via the 2A/1ACI(o).

References

1 Nakamura, S. and Irie, M. (1988). Thermally irreversible photochromic systems. A theoretical study. *J. Org. Chem.* 53 (26): 6136–6138.

2 Woodward, R.B. and Hoffman, R. (1970). *The Conservation of Orbital Symmetry*. Weinheim: Verlag Chemie.

3 Boggio-Pasqua, M., Ravaglia, M., Bearpark, M.J. et al. (2003). Can diarylethene photochromism be explained by a reaction path alone? A CASSCF study with model MMVB dynamics. *J. Phys. Chem. A* 107 (50): 11139–11152.

4 Asano, Y., Murakami, A., Kobayashi, T. et al. (2004). Theoretical study on the photochromic cycloreversion reactions of dithienylethenes; on the role of the conical intersections. *J. Am. Chem. Soc.* 126 (38): 12112–12120.

5 Mendive-Tapia, D., Perrier, A., Bearpark, M.J. et al. (2014). New insights into the by-product fatigue mechanism of the photo-induced ring-opening in diarylethenes. *Phys. Chem. Chem. Phys.* 16: 18463–18471.

6 Irie, M., Fukaminato, T., Matsuda, K., and Kobatake, S. (2014). Photochromism of diarylethene molecules and crystals: memories, switches and actuators. *Chem. Rev.* 114: 12174–12277.

7 Irie, M., Lifka, T., Kobatake, S., and Kato, N. (2000). Photochromism of 1,2-bis(2-methyl-5-phenyl-3-thienyl)perfluorocyclopentene in a single-crystalline phase. *J. Am. Chem. Soc.* 122 (20): 4871–4876.

8 Sumi, T., Takagi, Y., Yagi, A. et al. (2014). Photoirradiation wavelength dependence of cycloreversion quantum yields of diarylethenes. *Chem. Commun.* 50: 3928–3930.

9 Jean-Ruel, H., Cooney, R.R., Gao, M. et al. (2011). Femtosecond dynamics of ring closing process diarylethene: a case study of electrocyclic reactions in photochromic single crystals. *J. Phys. Chem. A* 115: 13158–13168.

10 Hania, P.R., Pugzlys, A., Lucas, L.N. et al. (2005). Ring closure dynamics of BTE-based photochromic switch: perfluoro- versus perhydrocyclopentene derivatives. *J. Phys. Chem. A* 109 (42): 9437–9442.

11 Ishibashi, Y., Fujiwara, M., Umesato, T. et al. (2011). Cyclization reaction dynamics of a photochromic diarylethene derivative as revealed by femtosecond to microsecond time-resolved spectroscopy. *J. Phys. Chem. C* 115: 4265–4272.

12 Ishibashi, Y., Fujiwara, M., Umesato, T. et al. (2017). Cycloreversion reaction of a diarylethene derivative at higher excited states attained by two-color, two-photon femtosecond pulsed excitation. *J. Am. Chem. Soc.* 139: 17159–17167.

3

Photoswitching Performance

3.1 Quantum Yield

The rate of the following photochemical reaction is expressed as follows.

$$\text{Isomer-}\mathbf{A} \rightarrow \text{Isomer-}\mathbf{B}$$

$$-\frac{dC_A}{dt} = \Phi_{AB} I_0 (1 - 10^{-\varepsilon_A C_A d}) \frac{10^3}{d}$$

where Φ_{AB}, ε_A, C_A, d, and I_0 are the photoreaction quantum yield, the absorption coefficient of isomer-**A** at the illumination wavelength, the concentration of isomer-**A** (M), the cell length (cm), and the light intensity (einstein/cm^2 s), respectively. The quantum yield is the most important parameter in photochemical reactions and defined as follows:

$$\Phi_{AB} = \frac{\text{Number of molecules, which undergo the reaction from } \mathbf{A} \text{ to } \mathbf{B}}{\text{Number of photons of radiation absorbed}}$$

The rate increases linearly with the increase in the quantum yield and the light intensity. Table 3.1 summarizes photocyclization and photocycloreversion quantum yields in n-hexane together with absorption maximum wavelengths and coefficients of typical diarylethenes.

3.1.1 Photocyclization Quantum Yield

As shown in Scheme 2.4, a diarylethene open-ring isomer has two conformations with the two aryl groups in mirror symmetry and C_2 symmetry, which are called parallel and antiparallel conformations, respectively. Only the antiparallel conformer is allowed to undergo the conrotatory cyclization reaction. In solution, the two conformers are interconvertible with each other by the rotation of single bonds between the ethene bridge and aryl groups. The population ratio is, in most cases, 1 : 1. For example, ^{1}H NMR of diarylethene **5o** shows two doublet signals in almost equal intensity at 2.09 and 2.36 ppm, as shown in Figure 3.1a [3]. The two signals are assigned to methyl protons at 4- and 2-positions of the thiophene rings in antiparallel and parallel conformations, respectively. The cyclization quantum yield was measured to be 0.45. The quantum yield indicates that almost all antiparallel conformers

Diarylethene Molecular Photoswitches: Concepts and Functionalities, First Edition. Masahiro Irie.
© 2021 WILEY-VCH GmbH. Published 2021 by WILEY-VCH GmbH.

Table 3.1 Photocyclization and photocycloreversion quantum yields of typical diarylethenes in *n*-hexane.

	Open-ring isomer		Closed-ring isomer				
	λ_{max} (nm)	ε (M^{-1} cm^{-1})	λ_{max} (nm)	ε (M^{-1} cm^{-1})	Φ_{OC}	Φ_{CO}	References
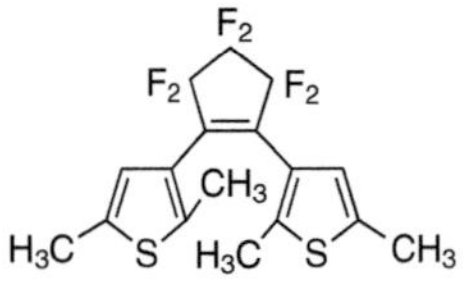 **3**	303	5 300	505	8000	0.40	0.12	[1, 2]
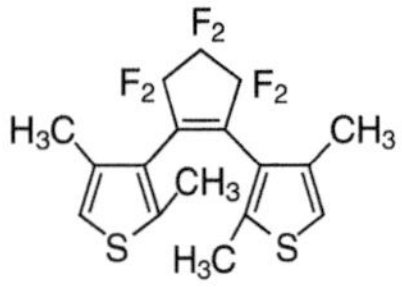 **22**	234	13 000	534	5000	0.21	0.13	[3]
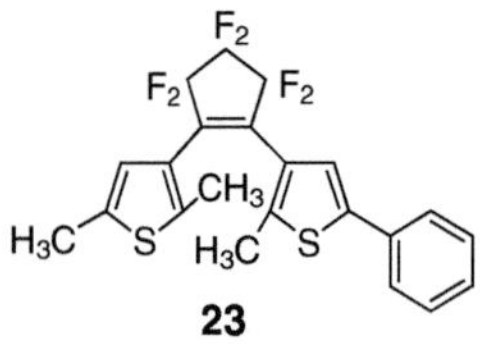 **23**	289	24 700	541	3680	0.53	0.119	[4]
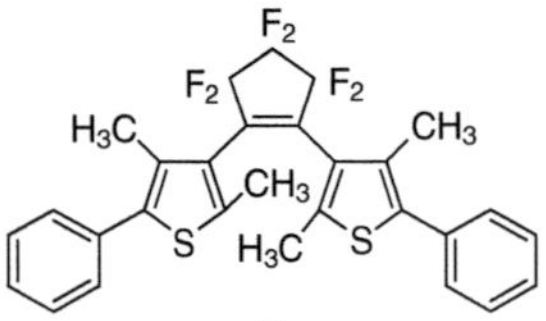 **5**	268	28 400	562	10 900	0.45	0.018	[3, 5]
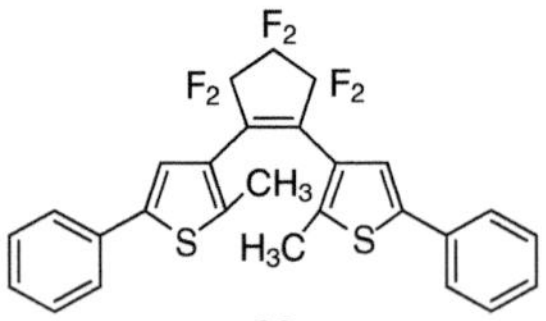 **20**	280	35 600	575	15 600	0.59	0.013	[6]

(Continued)

Table 3.1 (Continued)

	Open-ring isomer		Closed-ring isomer				
	λ_{max} (nm)	ε (M^{-1} cm^{-1})	λ_{max} (nm)	ε (M^{-1} cm^{-1})	Φ_{OC}	Φ_{CO}	References
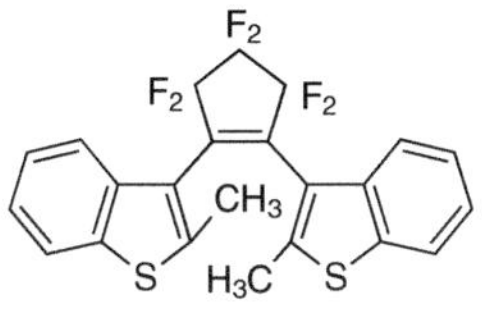 **13**	258	15 200	516	10 000	0.31	0.29	[5, 7]
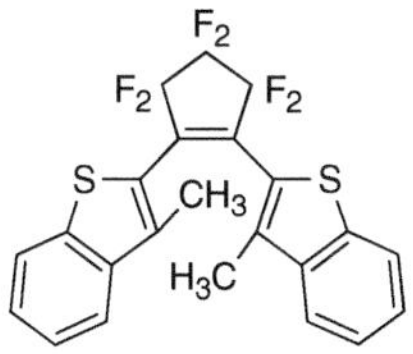 **24**	325	13 700	438	6500	0.46	0.47	[8]
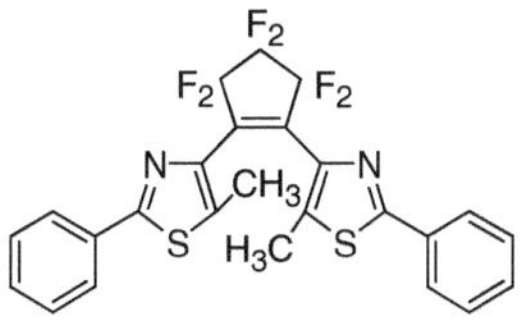 **4**	305	37 000	534	14 500	0.42	0.017	[9]
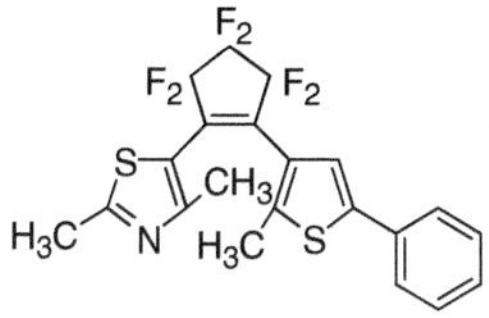 **25**	291	19 000	483	8100	0.39	0.11	[10]
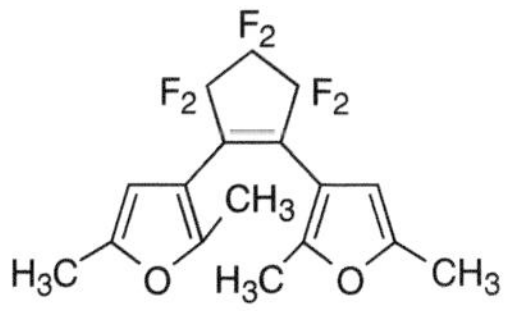 **26**	302	6 100	449	8700	0.35	0.32	[11]

(Continued)

Table 3.1 (Continued)

	Open-ring isomer		Closed-ring isomer				
	λ_{max} (nm)	ε (M^{-1} cm^{-1})	λ_{max} (nm)	ε (M^{-1} cm^{-1})	Φ_{OC}	Φ_{CO}	References
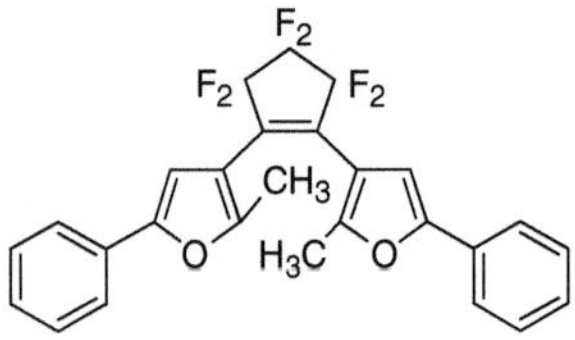 **27**	285	38 700	525	16 500	0.53	0.077	[11]
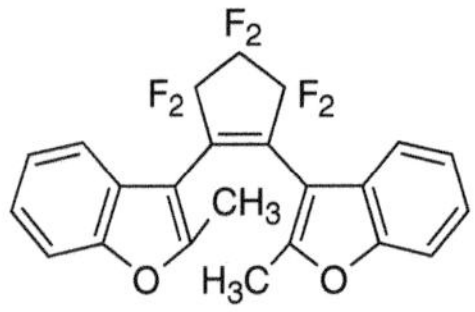 **28**	274	10 000	469	14 400	0.38	0.35	[12]
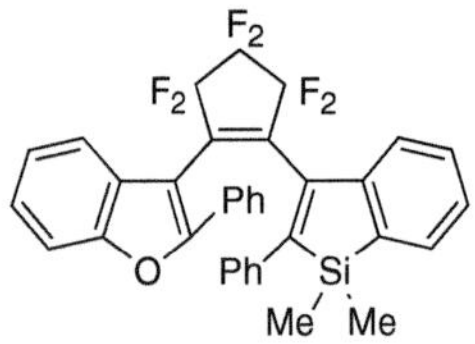 **29**	280	23 500	435	16 800	0.46	0.38	[13]
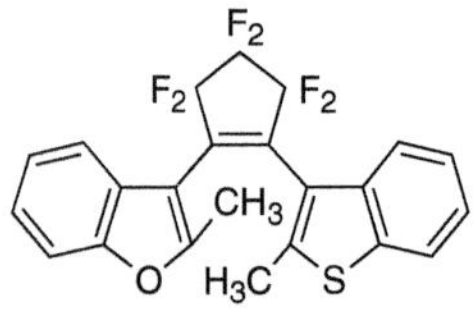 **30**	299	6 900	403	15 300	0.32	0.25	[14, 15]
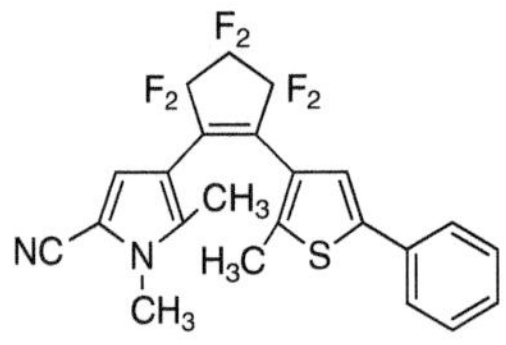 **31**	290	24 900	584	10 000	0.20	0.054	[16, 17]

(Continued)

Table 3.1 (Continued)

	Open-ring isomer		Closed-ring isomer				
	λ_{max} (nm)	ε (M^{-1} cm^{-1})	λ_{max} (nm)	ε (M^{-1} cm^{-1})	Φ_{OC}	Φ_{CO}	References
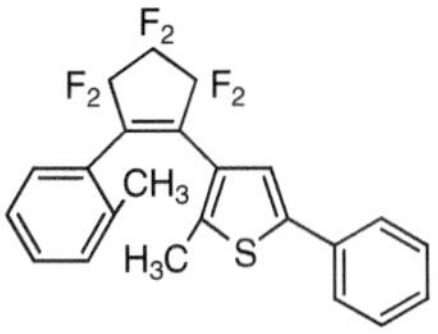 **32**	264	23 000	552	4800	0.25	0.31	[18]
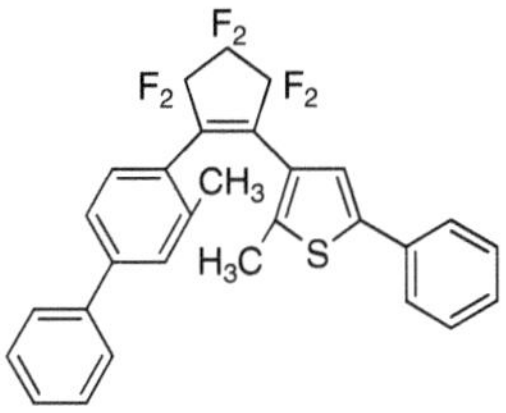 **33**	285	34 200	556	6110	0.37	0.11	[19]
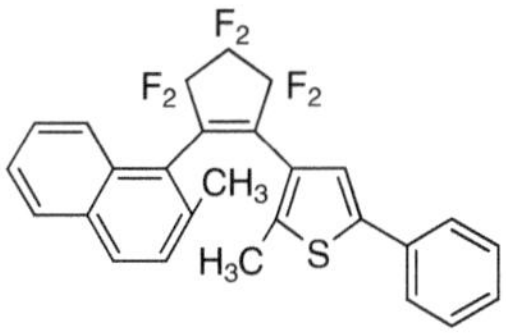 **34**	261	27 600	508	15 000	0.31	0.21	[20]
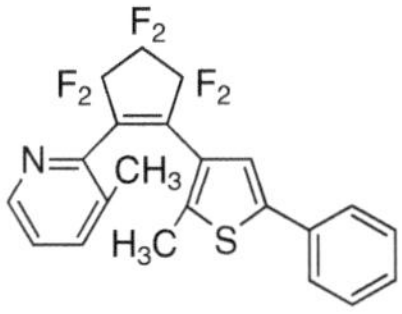 **35**	284	17 000	555	6590	0.33	0.096	[21]
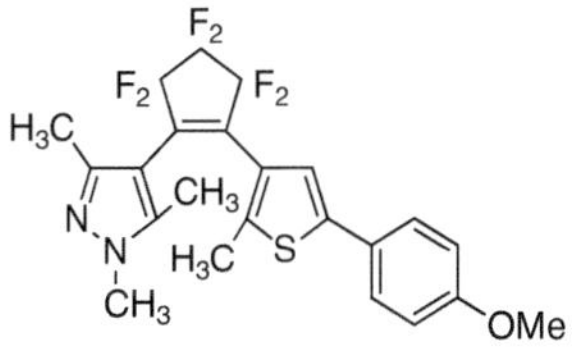 **36**	290	29 000	573	10 000	0.47	0.57	[22]

(Continued)

Table 3.1 (Continued)

	Open-ring isomer λ_{max} (nm)	Open-ring isomer ε (M^{-1} cm^{-1})	Closed-ring isomer λ_{max} (nm)	Closed-ring isomer ε (M^{-1} cm^{-1})	Φ_{OC}	Φ_{CO}	References
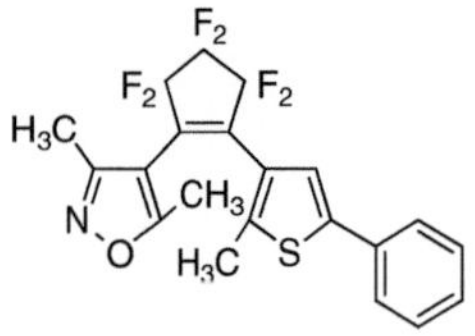 **37**	276	17 700	526	8510	0.15	0.20	[23]
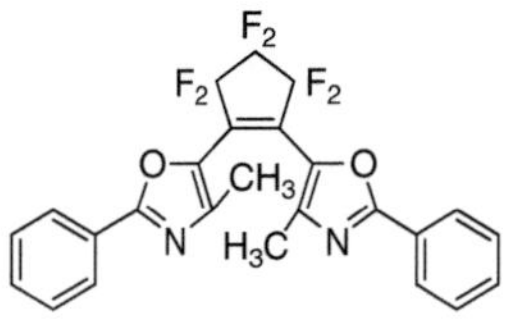 **38**	373	17 000	353	2400	0.10	0.33	[24]
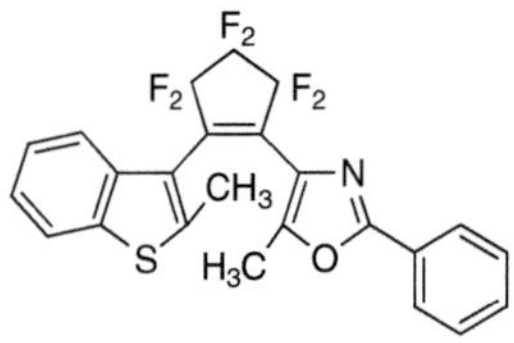 **38**	267	21 500	482	10 500	0.19	0.08	[25]
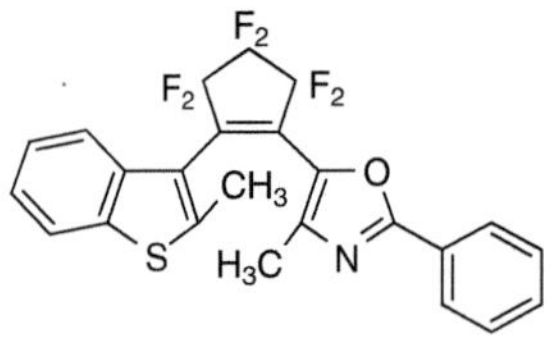 **40**	325	19 700	440	5700	0.04	0.49	[25]
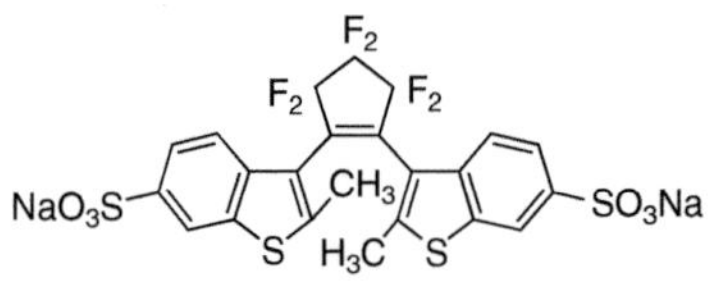 **41**	233	49 000	529	9800	0.32 (in water)	—	[26]
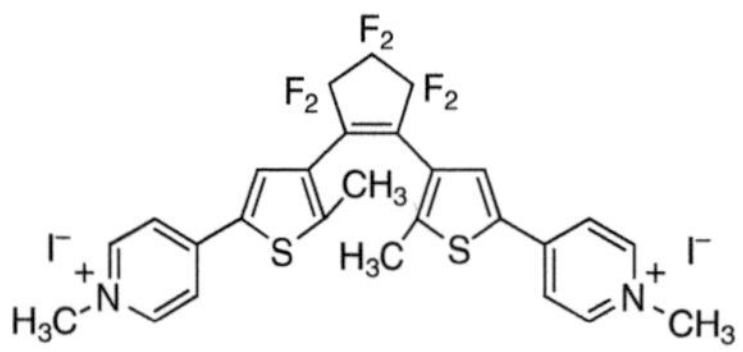 **43**	353	50 500	655	16 200	0.71 (in water)	—	[27]

Figure 3.1 ^{1}H NMR methyl signals of (a) **5o** and (b) **13o** in $CDCl_3$.

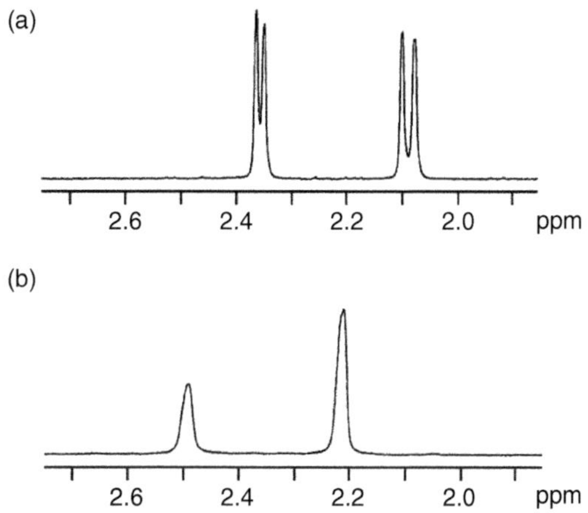

participate in the photocyclization reaction and the cyclization yield of the antiparallel conformers is close to 1 (100%). Therefore, the increase in the population of the antiparallel conformers is expected to lead to the increase in the cyclization quantum yield. Various molecular design approaches have been proposed and examined to increase the population of the antiparallel conformers.

A simple approach is to introduce long alkyl or bulky substituents at the central reactive carbon atoms [28]. Figure 3.1b shows the ^{1}H NMR spectrum of 1,2-bis(2-methyl-1-benzothiophen-3-yl)perfluorocyclopentene (**13o**) in $CDCl_3$. The signal at 2.21 ppm is assigned to the antiparallel conformer, while the signal at 2.49 ppm to the parallel conformer. The ratio (antiparallel/parallel) is 65/35. The ratio of the two conformers was found to depend on the long alkyl or bulky substituents at the reactive carbon atoms, as shown in Table 3.2. Table 3.2 also shows the cyclization quantum yields of the derivatives. When bulky isopropyl substituents are introduced as **52**, the ratio of the antiparallel conformers increases to 94% and the cyclization quantum yield increases to 0.52, which is 1.7 times larger than that of **13** [7].

The ratio of the antiparallel conformers can also be increased by including the chromophore into a cyclodextrin cavity [26] as well as introducing pyridinium cations at both sides [27]. Methyl protons of water-soluble derivative **41o** at 2.27 and 2.55 ppm in D_2O are assigned to antiparallel and parallel conformers, respectively. The ratio of the signal ascribed to the antiparallel conformers was found to increase from 64 to 94% by the addition of 10 equiv of β-cyclodextrin. At the same time, the cyclization quantum yield increased as large as 1.5 times. **41o** is considered to be included in the β-cyclodextrin cavity in the antiparallel conformation and undergo the efficient photocyclization reaction.

Coulombic repulsion between side group cations suppresses formation of parallel conformers and enriches the population of the antiparallel conformers [27]. Pyridinium cations were introduced at both sides of 1,2-bis(2-methyl-5-pyridyl-3-thienyl)-perfluorocyclopentene (**42**) by treating with methyl iodide. The pyridinium cations derivative **43o** underwent the photocyclization reaction in methanol upon irradiation with 365 nm light and the cyclization quantum yield was found to be 0.71, which

Table 3.2 Ratios of antiparallel and parallel conformers and photocyclization quantum yields in *n*-hexane.

	R	Antiparallel[a)] (%)	Parallel[a)] (%)	Φ_{OC}	References
13	CH_3	65	35	0.31	[5, 7]
44	C_2H_5	69	31	0.39	[28]
45	C_3H_7	74	26	0.41	[28]
46	C_4H_9	77	23	0.44	[28]
47	C_5H_{11}	78	21	0.49	[28]
48	C_6H_{13}	82	18	0.49	[28]
49	C_7H_{15}	82	18	0.49	[28]
50	C_8H_{17}	82	18	0.48	[28]
51	$C_{12}H_{25}$	82	18	0.47	[28]
52	$CH(CH_3)_2$	94	6	0.52	[7]

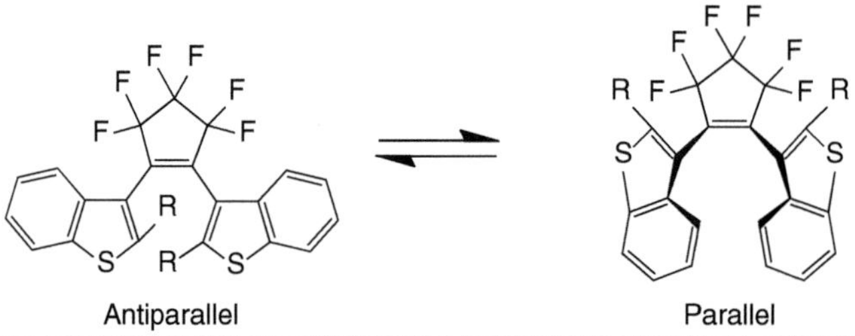

a) The ratios of two conformers were determined by ^{1}H NMR analysis in $CDCl_3$.

is larger than the yield (0.57) of the corresponding pyridyl derivative **42**. The high quantum yield is ascribed to the preference of the photoactive antiparallel conformation due to Coulombic repulsion between the pyridinium cations.

Another approach to increase the cyclization quantum yield is to replace the perfluorocyclopentene bridge with heterocyclic aryl groups. Table 3.3 shows the photocyclization and photocycloreversion quantum yields of bisarylbenzothiophenes, which were prepared by Kawai and co-workers [29, 30]. Although the cyclization quantum yield of the derivative **56o** having 2-methyl-5-phenyl-3-thienyl substituents was slightly less than that of perfluorocyclopentene derivative **20o**, the yield was found to dramatically increase by replacing the thiophene groups with thiazolyl ones. When 5-methyl-2-phenyl-4-thiazolyl group was employed as Ar_1 and Ar_2, such as **53o**, the yield increased to 0.98. The increase is ascribed to intramolecular noncovalent interactions, such as S/N and CH/N interactions. The interactions between the benzothiophene bridge and thiazolyl groups stabilize the photoreactive antiparallel conformation, as shown in Table 3.3, in less polar solvents. The high quantum yield, however, decreased to an ordinary value of 0.54 in polar methanol, wherein the hydrogen bonding interactions are less effective. The yield was improved by replacing one of the thiazolyl groups with a thiophene

Table 3.3 Photocyclization and photocycloreversion quantum yields of bisarylbenzothiophenes.

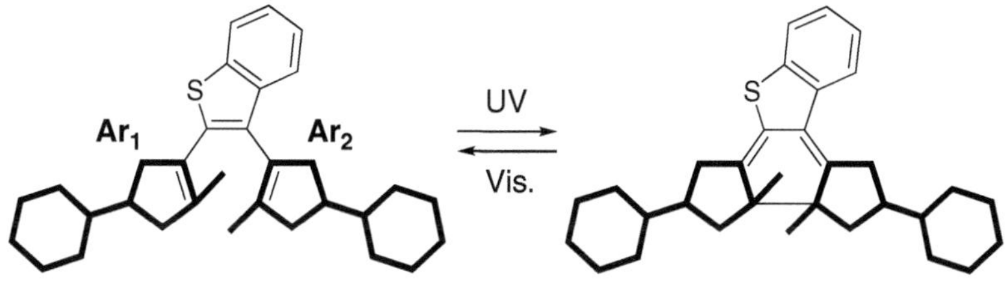

	Ar_1	Ar_2	Φ_{OC}		Φ_{CO}
			In *n*-hexane	In methanol	In *n*-hexane
53	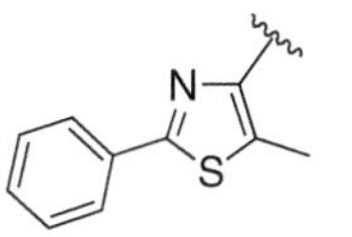	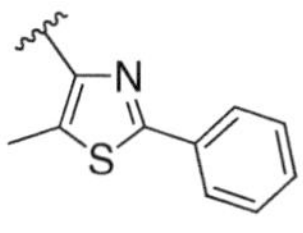	0.98	0.54	0.008
54	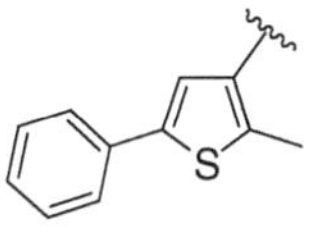	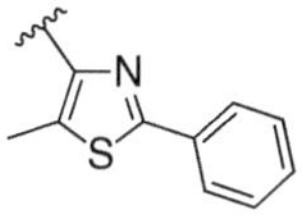	0.50	0.91	0.007
55	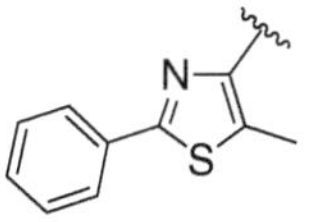	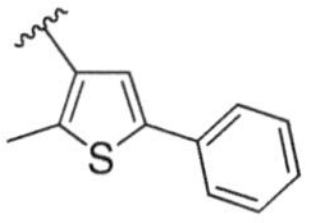	0.71	0.99	0.036
56	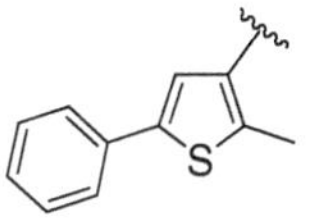	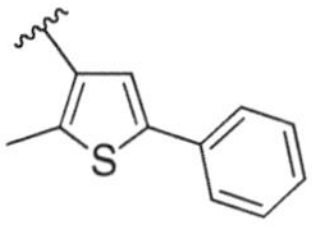	0.49	0.38	0.029

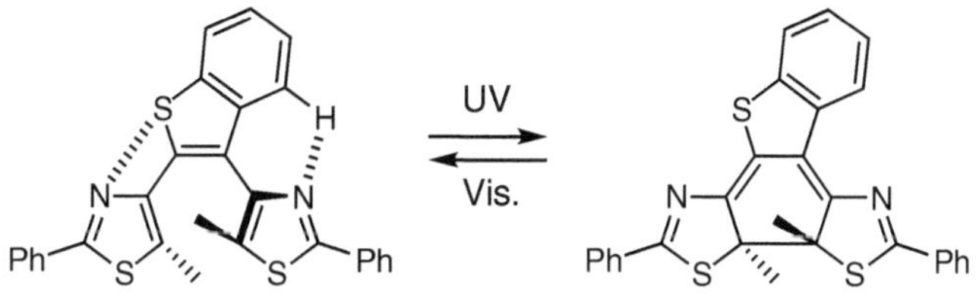

group. The derivatives **54** and **55** having nonsymmetrical side-aryl groups increased the photocyclization quantum yield in polar methanol, while ordinary values were observed in nonpolar *n*-hexane. The increase in the yield in polar methanol is ascribed to CH/π interactions between methyl groups at reactive carbon atoms and aryl rings, which contribute to maintain the antiparallel conformation. The nature of the CH/π interaction characterized by the dispersion force is likely to work more efficiently in polar solvents than in less polar ones with the aid the solvophobic effect.

Although the above approach to use intramolecular noncovalent interactions between the bridged moiety and the aryl group is effective to increase the photocyclization quantum yield, a fundamental drawback is their intrinsic solvent dependency, which is reflected in the quantum yield. Hecht and co-workers [31] employed steric repulsion, which is largely independent on environments, to influence the equilibrium between antiparallel and parallel conformers. Sterically demanding substituents can be incorporated in the ethene bridge. Introduction of bulky substituents in the bridge enables flexible functionalization of the aryl groups.

Table 3.4 shows the cyclization and cycloreversion quantum yields of the derivatives having bulky ethene bridges. It is clear that compound **58** containing α,α′-disubstituted cyclopentene moieties has a higher cyclization quantum yield around 0.6 than the parent unsubstituted derivative **57**. By introducing four methyl groups, α,α,α′,α′-tetrasubstituted derivative **59** shows a remarkably high cyclization quantum yield of 0.83. Most importantly, the large quantum yield was found to be independent on the solvent polarity, the solvents tested being nonpolar cyclohexane to polar acetonitrile and the protic methanol. The observed increase can be rationalized by considering a repulsion between the substituents in the cyclohexane bridge and 2-methyl and 4-hydrogen substituents of the thiophen groups that would shift the conformational equilibrium in favor of the antiparallel conformers. The fraction of the antiparallel conformer in the open-ring isomer of **59** was estimated to be 81 ± 10% using ^{1}HNMR spectroscopy. This value is in good agreement with the cyclization quantum yield of 0.83, which supports the hypothesis that the ratio of antiparallel conformers is the key factor that determines the photocyclization quantum yields.

Table 3.4 Photocyclization and photocycloreversion quantum yields of diarylethene derivatives having bulky ethene bridges in acetonitrile.

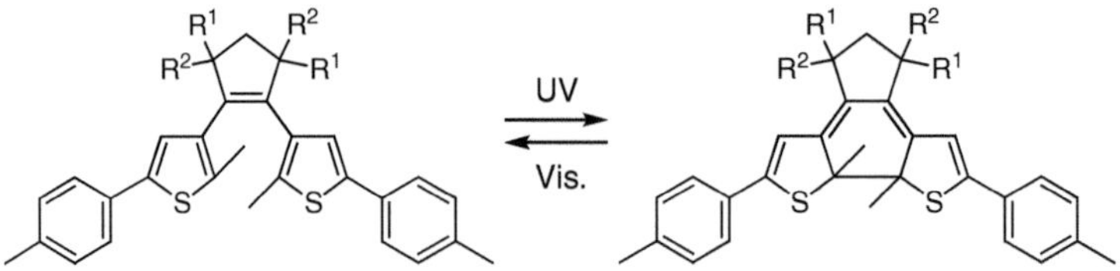

	R^1	R^2	Φ_{OC}	Φ_{CO}
57	H	H	0.50	0.01
58	Me	H	0.57	0.01
59	Me	Me	0.83	0.01

Zhu, Tian, Nakatani and co-workers [32] introduced new electron-withdrawing bulky bridges to drive molecules to adopt the sterically favorable antiparallel conformation. The most optimum bridge is benzobis(thiadiazole), as shown in Scheme 3.1. The bridge contains two extending thiadiazole units and exhibits an extremely low aromaticity as a result of its high polarity and electron-withdrawing tendency. When benzothiophene groups are attached as the aryl units, the steric interaction between the bulky bridge and the benzothiophene groups is so large that the rotation of benzothiophene rings is completely blocked. That is, the common interconversion between antiparallel and parallel conformers is extremely hindered. The two conformers, the ratio of which is close to 1 : 1, can be separated by conventional silica gel chromatography and recrystallization.

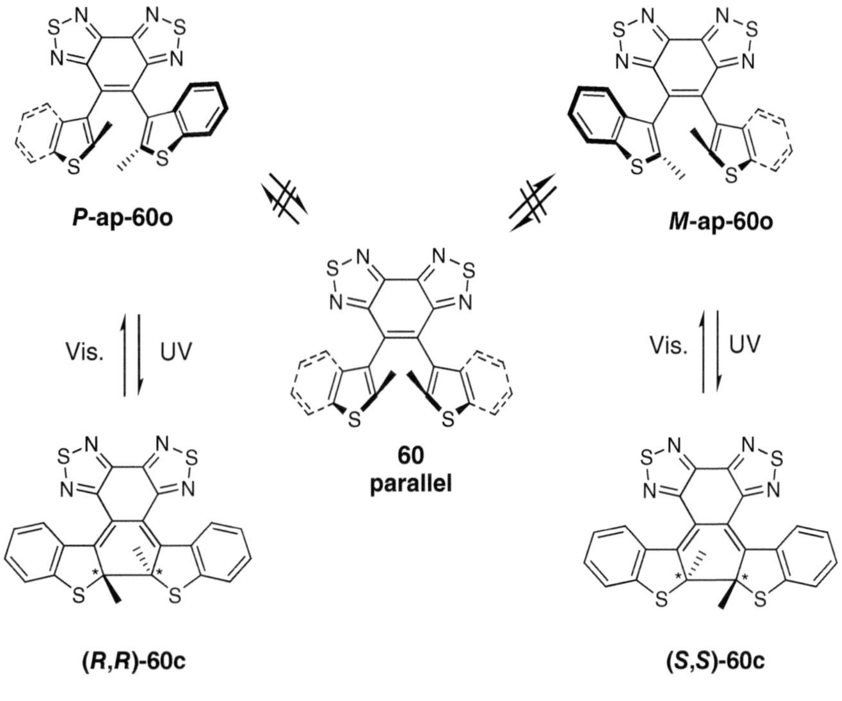

Scheme 3.1 A diarylethene derivative having a benzobis(thiadiazole) bridge. Hindered rotation leads to separate antiparallel and parallel conformers. The hindered rotation also separates *P*- and *M*-enantiomers which lead to corresponding closed-ring (*R*,*R*)- and (*S*,*S*)-enantiomers, respectively.

The photocyclization quantum yield of the isolated pure antiparallel diarylethene derivative **60o** was measured and is shown in Table 3.5. The yield was determined to be as high as 0.73 in THF. When nitro groups were introduced at 6- and 6′-positions, the yield further increased to 0.91, possibly due to suppression of deactivation processes. The bulky ethene bridge enabled extremely high photocyclization quantum yields. The molecular design makes it possible to isolate not only the antiparallel and parallel conformers, but also the pair of enantiomers of the antiparallel conformers.

Table 3.5 Photocyclization and photocycloreversion quantum yields of diarylethene derivatives having benzobis(thiadiazole) bridge in THF.

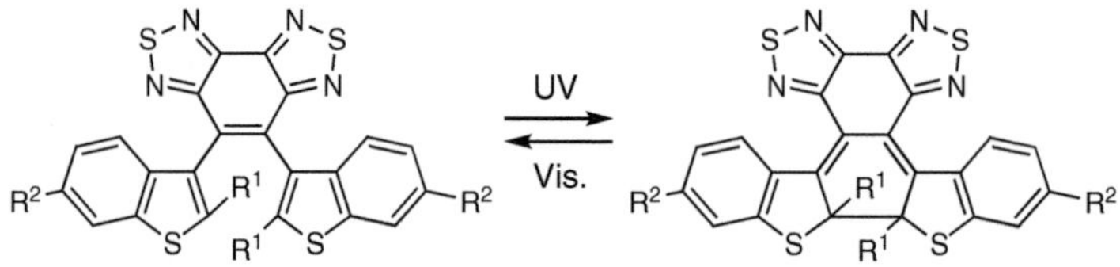

	R^1	R^2	Open-ring λ_{max} (nm)	Closed-ring λ_{max} (nm)	Φ_{OC}
60	Me	H	282	522	0.73
61	Bu	H	282	522, 548	0.67
62	Me	Acetyl	285	519, 544	0.84
63	Me	NO_2	284	554	0.91

The open-ring antiparallel conformers have two enantiomers, *P*-helix and *M*-helix. Upon irradiation with UV light at 280 nm, *P*-helix and *M*-helix enantiomers undergo enantiospecific cyclization reactions to produce closed-ring (*R*,*R*)- and (*S*,*S*)-enantiomers, respectively, as shown in Scheme 3.1. These four enantiomers can be isolated and potentially used for nondestructive optical memory media and chiral photoswitches (see Section 3.5) [33].

3.1.2 Solvent Effect on Cyclization Quantum Yield

When maleic anhydride or maleimide is used as the ethene bridge, the diarylethene derivatives undergo the photocyclization upon irradiation with visible light ($\lambda > 400$ nm). Although the visible light sensitivity is favorable for applications to biological systems, one of the drawbacks of the derivatives is that the photocyclization reactions are strongly suppressed in polar solvents. In nonpolar alkane solutions, diarylmaleic anhydride **64** and diarylmaleimides **65**, shown in Figure 3.2, undergo conventional photoswitching reactions [34, 35]. Upon irradiation with 405 nm light, the yellow solutions of **64** and **65** turned red-purple and the cyclization quantum yields were measured to be 0.13 and 0.22 in *n*-hexane, respectively. The colors disappeared upon irradiation with visible light ($\lambda > 500$ nm). The cyclization quantum yields dramatically decreased in polar solvents. The yield of **64** in THF was 0.04, which is less than 1/3 of the value in nonpolar *n*-hexane. In polar acetonitrile, it was only 0.003. Figure 3.2c shows the color changes of **65** upon irradiation with 405 nm light in various solvents. Although in nonpolar cyclohexane and less polar toluene the solution turned red–purple, any appreciable color change was not observed in polar ethanol and acetonitrile. On the other hand, the photocycloreversion quantum yields were almost the same in cyclohexane, toluene, THF, ethanol, and acetonitrile.

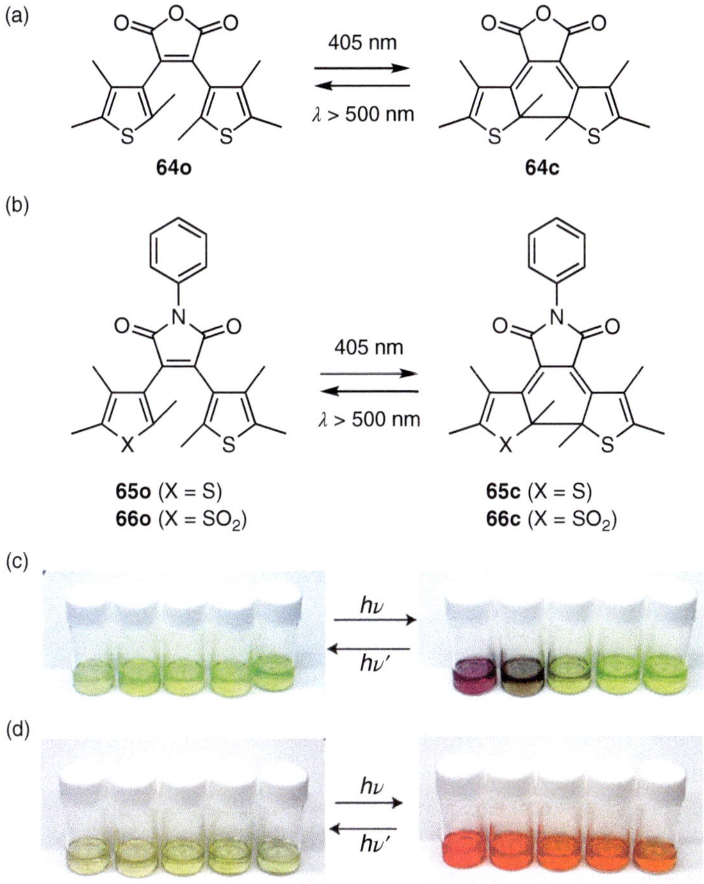

Figure 3.2 (a) Chemical structures of open- and close-ring isomers of **64**. (b) Chemical structures of open- and closed-ring isomers of **65** and **66**. (c) Photographs of color changes of **65** upon irradiation with 405 nm in various solvents (from left to right: cyclohexane, toluene, THF, ethanol, acetonitrile). Left: before photoirradiation, right: after photoirradiation. (d) Photographs of color changes of **66** upon irradiation with 405 nm in various solvents (from left to right: cyclohexane, toluene, THF, ethanol, acetonitrile). Left: before photoirradiation, right: after photoirradiation.

The solvent effect is attributed to the contribution of a twisted intramolecular charge transfer (TICT) interaction [34]. The donor acceptor interaction between the thiophene and maleic anhydride or maleimide moieties plays a role in controlling the reactivity. The cyclization reaction is considered to proceed as shown in Figure 3.3. The Franck–Condon state conformation of the molecule is in between planar and twisted geometries. In less polar solvents, the excited state stabilizes to a planar conformation with a low dipole moment and undergoes a concerted cyclization reaction in the conrotatory mode. In polar solvents, on the other hand, the Franck–Condon state converts to the energetically more stable charge-separated state. This state with a large dipole moment forces the plane of a thiophene ring

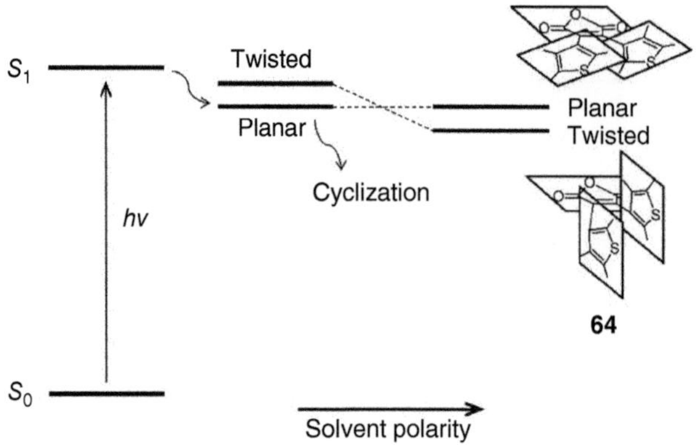

Figure 3.3 Schematic illustration of the solvent effect on the photocyclization of dithienylmaleic anhydride **64**.

to become perpendicular to that of the maleic anhydride or maleimide moiety. The twisted geometry is disadvantageous to the cyclization reaction and deactivates to the ground state without any reaction.

When one of the thiophene rings of diarylmaleimide **65** was replaced with a *S,S*-dioxide thiophene ring, such as **66**, the solvent effect disappeared [35]. Diarylmaleimide **66** underwent the photocyclization reaction even in polar ethanol and acetonitrile, as shown in Figure 3.2d. This is attributed to the loss of electron-donating property of the thiophene ring by oxidation. Upon alternate irradiation with 405 nm light and visible ($\lambda > 500$ nm) light, the nonsymmetric derivative **66** underwent the reversible photocyclization/photocycloreversion reactions even in polar solvents. The cyclization and cycloreversion quantum yields of **66** in ethanol were 0.17 and 0.030, respectively. The visible light-sensitive derivative is a potential candidate for applications to biological systems.

3.1.3 Photocycloreversion Quantum Yield

Table 3.1 shows cycloreversion quantum yields of typical diarylethene derivatives. The yields were measured upon irradiation at the absorption maxima of the closed-ring isomers. According to theoretical calculations and femtosecond laser photolysis studies, the cycloreversion reaction is considered to proceed as shown in Figure 2.8. The excited state wavepacket has to overcome the energy barrier $2ATS_1$ on surface 2A. The molecules excited with light of shorter wavelengths having vibrational excess energy go over the energy barrier more readily and reach 2A/1A conical intersection (2A/1ACI(o)). At the conical intersection, fast deactivation to the ground state takes place and thus the open-ring isomers are formed. The route suggests that thermal activation or excess kinetic energy on surface 2A is required for the ring-opening reaction to take place. Therefore, the quantum yield is considered to increases upon irradiation with light of shorter wavelengths. The wavelength dependence of the cycloreversion quantum yield was carefully examined for diarylethenes **13** and **5** [5].

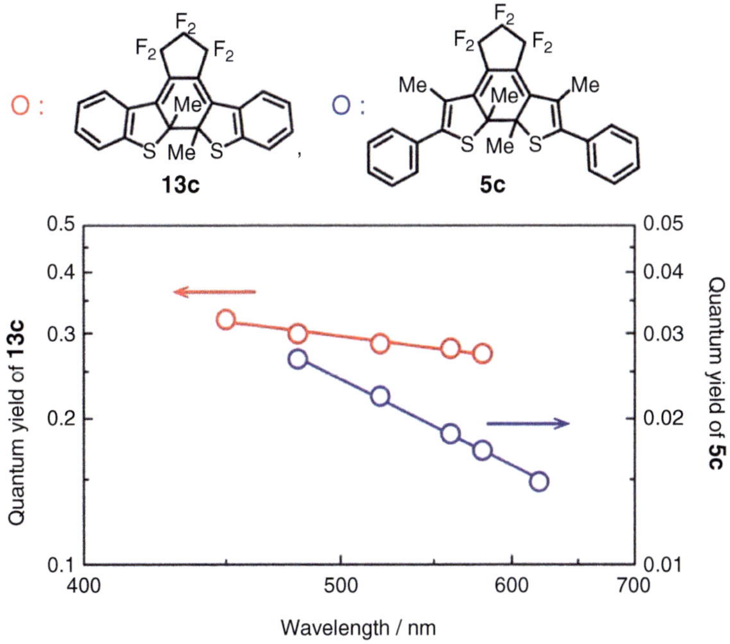

Figure 3.4 Photoirradiation wavelength dependence of the cycloreversion quantum yields of **13c** (red) and **5c** (blue) in *n*-hexane at 22 °C. *Source*: Adapted from Ref. [5] with permission from The Royal Society of Chemistry.

The cycloreversion quantum yields (ϕ_{co}) were measured at several irradiation wavelengths in the visible region. Log ϕ_{co} was plotted against wavelengths (expressed in photon energy), as shown in Figure 3.4. The quantum yields increase upon increasing the irradiation photon energy. The quantum yield of **13c** at 450 nm is 18% larger than the value at 580 nm, while the yield of **5c** at 480 nm is 75% larger than the value at 620 nm. In both compounds, linear behavior was observed between log ϕ_{co} and the photon energy, which is typical of energy-activated reactions. The wavelength dependency can be expressed as follows for compound **13c** and **5c**.

$$\text{Compound 13: } \log \Phi_{co} = -0.796 + 133/\lambda\,(\text{nm})\,(450\ \text{nm} < \lambda < 580\ \text{nm})$$

$$\text{Compound 5: } \log \Phi_{co} = -2.70 + 539/\lambda\,(\text{nm})\,(480\ \text{nm} < \lambda < 620\ \text{nm})$$

The wavelength dependence of the quantum yields can be used for visible-range chemical actinometry from 450 to 620 nm.

According to the reaction mechanism shown above, reaction temperature is also anticipated to affect the quantum yield. Temperature dependence of photocyclization and photocycloreversion reactions of **20** was measured in a 3-methylpentane solution [6]. Although no appreciable temperature dependence was observed in the cyclization reaction between 70 and −50 °C, the cycloreversion reaction efficiency was found to decrease with the decrease in temperature. The activation energy of the cycloreversion quantum yield was determined to be 16 kJ/mol.

Temperature dependence of **67** dispersed in polystyrene was measured between 5 and 300 K to know the response at extremely low temperatures [36]. The quantum yield for the cyclization reaction did not show any significant temperature dependence. It amounts on average to 0.0673 ± 0.029. On the other hand, the cycloreversion quantum yield decreased with the decrease in temperature, as shown in Figure 3.5. The temperature dependence is divided into two parts. Between 150 and 300 K, a strong decrease from 4.43×10^{-3} to 5.42×10^{-5} was observed, while below 150 K the yield showed a constant value of $2.9 \times 10^{-5} \pm 2.6 \times 10^{-5}$. The lack of any temperature dependence in the temperature range between 5 and 150 K indicates tunneling processes are effective for crossing the barrier, as shown in Figure 3.5b.

The cycloreversion quantum yields of the closed-ring isomers were found to depend on the substituents at the central reactive carbon atoms, as shown in Table 3.6. The yields are strongly suppressed by introducing electron-donating substituents, such as methoxy groups, at the reactive carbon atoms, while the yield increases by introducing electron-withdrawing substituents, such as cyano groups [37–39]. The difference in the cycloreversion quantum yields between **70** having methoxy substituents and **72** having cyano substituents was as large as 2.4×10^4. Introduction of bulky substituents also increases the cycloreversion quantum yield. The yield increased by as much as 2.7 times by replacing the methyl substituents of **5c** with isopropyl substituents, such as **75c**. The yield further increased by introducing isobutyl substituents, such as **76c**. The cycloreversion quantum yield of **76c** was 3.6 times larger than that of **5c**. Similar increase in the cycloreversion quantum yield was observed for bis(benzothienyl)perfluorocyclopentene derivatives. The yield of **80c** having isobutyl substituents was 1.3 times larger than that of **13c** having methyl substituents. The closed-ring isomer **80c** maintained the thermal stability at 80 °C for 20 h even in the presence of bulky isobutyl substituents. Introduction of isobutyl substituents at the reactive carbon atoms is highly effective to increase the cycloreversion quantum yield. The effect of isobutyl substituents will be also discussed in Section 9.3.1.

Diarylethene derivatives having long π-conjugated aryl groups, such as terthiophene groups, decreases the cycloreversion quantum yield. Diarylethenes having *S,S*-dioxide benzothiophene rings also exhibit marked low cycloreversion quantum yields (see Section 3.4.2).

As described in Figure 2.8 the cycloreversion reaction has to overcome the energy barrier $2ATS_1$ to reach the branching conical interaction 2A/1A CI (o). Theoretical calculations were carried out to estimate the energy differences between 2Ac and 2Ao and to try to correlate the difference with the cycloreversion quantum yield [40]. Figure 3.6 shows the correlation between the yields and calculated energy differences ΔE (=E(2Ao) − E(2Ac)). When the energy difference is positive, the quantum yield decreases, while it increases when ΔE is negative. The energy difference controls the cycloreversion reaction. A fairly good correlation strongly supports the proposed reaction mechanism described in Section 2.2.

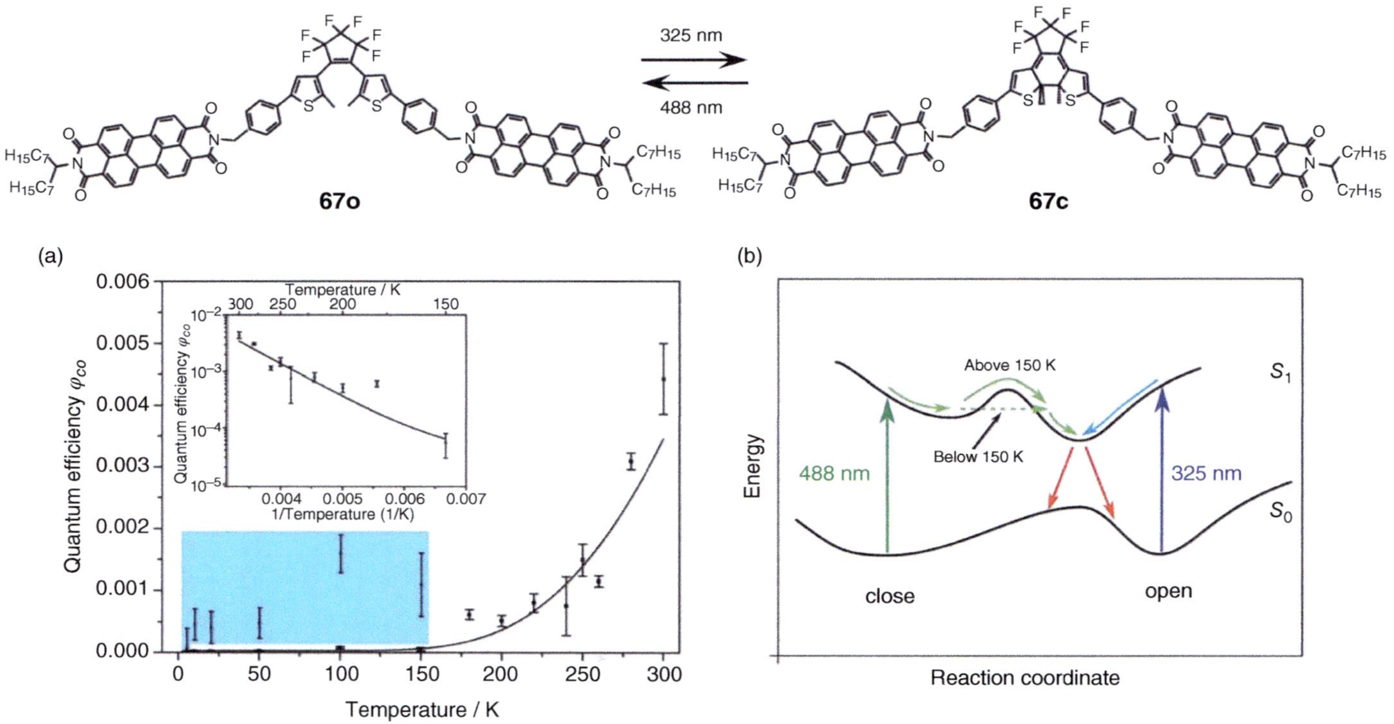

Figure 3.5 (a) Quantum yields for the ring-opening reaction of **67c** as a function of temperature. The inset shows the same dataset between 150 and 300 K, where the quantum yields are plotted against $1/T$ on a logarithmic scale. The data points below 150 K were multiplied by a factor of 20 and reproduced in the blue shaded area. (b) Simplified energy surfaces for the electronically excited **67**. Upon excitation of the open-ring isomer **67o** with 325 nm light, the system relaxes along the potential energy surface to a minimum. Similarly the system relaxes to this minimum upon excitation of the closed-ring isomer **67c** with 488 nm light. This relaxation includes thermal activation (above 150 K) or tunnelling (below 150 K) to overcome the barrier. *Source*: Adapted from Ref. [35] with permission from the PCCP Owner Societies.

Table 3.6 Photocyclization and photocycloreversion quantum yields of diarylethenes in *n*-hexane.

		R	Open-ring isomer λ_{max} (nm)	Open-ring isomer ε (M^{-1} cm^{-1})	Closed-ring isomer λ_{max} (nm)	Closed-ring isomer ε (M^{-1} cm^{-1})	Φ_{OC}	Φ_{CO} [a]
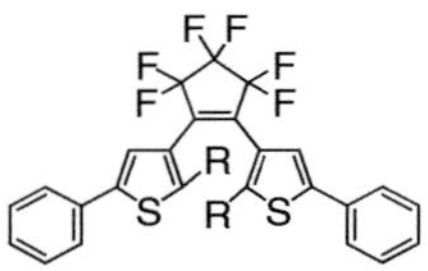	**20**	CH_3	280	35 600	575	15 600	0.59	0.013
	68	CH_2CH_3	286	40 000	600	17 000	0.52	0.0081
	69	$CH(CH_3)_2$	290	34 900	600	14 800	0.50	0.026
	70	OCH_3	309	33 000	625	15 000	0.44	~1.7×10^{-5}
	71	OCH_2CH_3	310	33 000	625	13 000	0.48	2.5×10^{-4}
	72	CN	284	39 000	454	21 000	0.42	0.41
	73	C≡CPh	309	49 000	575	16 000	0.17	0.32
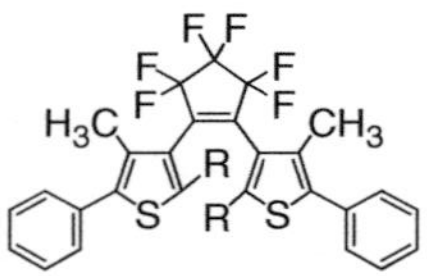	**5**	CH_3	268	28 400	562	10 900	0.45	0.017[b]
	74	CH_2CH_3	269	28 400	589	10 400	0.45	0.013[b]
	75	$CH(CH_3)_2$	270	27 100	586	9300	0.49	0.046[b]
	76	$CH_2CH(CH_3)_2$	270	28 400	598	9500	0.46	0.062[b]
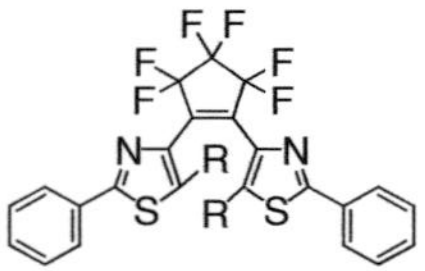	**4**	CH_3	305	37 000	535	14 500	0.42	0.017
	77	OCH_3	322	33 000	555	13 100	0.29	3.3×10^{-4}
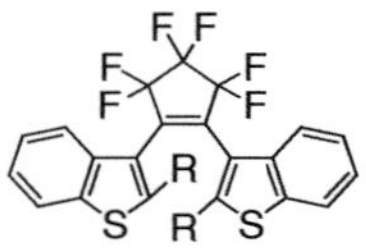	**13**	CH_3	258	15 200	516	10 000	0.31	0.29
	43	CH_2CH_3	258	16 500	536	9620	0.39	0.35
	78	OCH_3	260	20 000	547	7200	0.34	0.030
	79	CN	273	24 000	496	12 000	0.28	0.59
	80	$CH_2CH(CH_3)_2$	258	16 000	543	8400	0.47	0.38

a) Measured upon irradiation at the absorption maxima of the closed-ring isomers.
b) Measured upon irradiation at 580 nm.

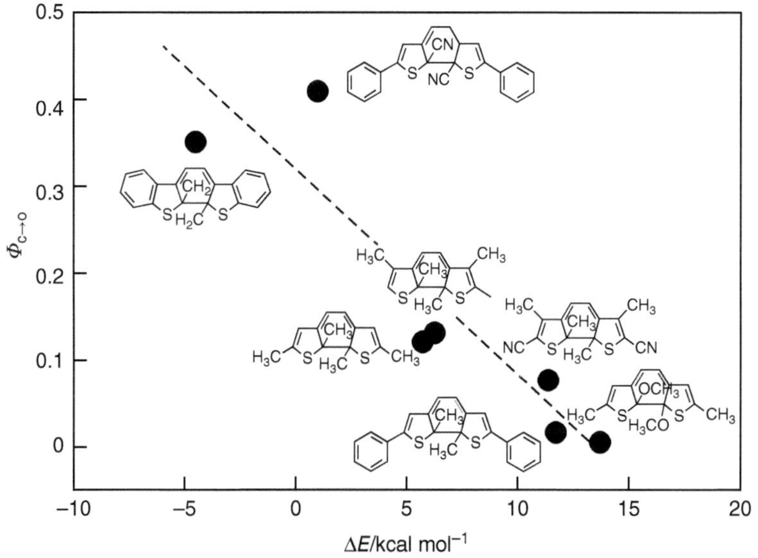

Figure 3.6 Correlation between the cycloreversion quantum yield of diarylperfluorocyclopentens and the calculated energy difference ΔE (= $E(2A_0) - E(2A_c)$). The calculation was carried out for the structures excluding perfluorocyclopentene as shown in the graph. *Source*: Adapted from Ref. [40].

3.2 Thermal Stability

Thermal stability of both isomers is an essential and indispensable property for the applications to optical memory media and photoswitching devices. All attempts carried out in the 1970–1980s to synthesize thermally irreversible molecular photoswitches failed, because there was no guiding principle on how to prepare such thermally stable photoswitchable molecules. In 1988, the serendipitous discovery of thermally irreversible diarylmaleic anhydrides paved the way for us to solve the problem [41, 42]. Inferring from experimental as well as theoretical analyses, the guiding principle for the synthesis of thermally irreversible diarylethenes was proved as follows. "The thermally irreversible photoswitching diarylethenes can be prepared by employing aryl groups with low aromatic stabilization energy." Since then, various kinds of diarylethene derivatives have been prepared based on this guiding principle.

There are three factors, which control the thermal stability of the closed-ring isomers.

1. Aromatic stabilization energy of the aryl groups.
2. Electron-withdrawing substituents at the aryl groups.
3. Steric hindrance of the substituents at the central reactive carbon atoms.

When the diarylethenes have heterocyclic aryl groups, such as thiophene, benzothiophene, thiazole, oxazole, furan, or benzofuran rings, the colored closed-ring isomers are stable and practically never revert back to the open-ring isomers at room temperature. Among them, diarylethene **4** showed extraordinarily high thermal stability [43]. The half-life time of the closed-ring isomer at 180 °C was measured to be 30 h. The activation energy of the thermal fading rate was as large as 142 kJ/mol. The half-life time at 30 °C is estimated to be 4.7×10^5 years. The large activation energy practically prohibited the thermal cycloreversion reaction at room temperature. On the other hand, the closed-ring isomers of diarylethenes having phenyl, naphthyl, pyrrolyl, or indole rings are thermally unstable. The difference in the thermal stability among the diarylethenes having heterocyclic aryl groups is well explained by the aromatic stabilization energy of the aryl groups.

When electron-withdrawing substituents are introduced at the aryl groups, the closed-ring isomers become thermally unstable, as shown in Figure 3.7a [44]. The ground state energy differences between the open- and the closed-ring isomers were calculated by the B3LYP/6-31G* program and are shown below the molecular structures. The electron-withdrawing groups are considered to weaken the central C—C bond and destabilize the closed-ring isomer, resulting in the increase in energy difference. When the energy difference becomes large, thermal stability of the closed-ring isomer decreases. The correlation between the thermal stability and the ground state energy difference between the open- and closed-ring isomer is well explained by the Horiuti–Polanyi rule as shown in Figure 2.3. The closed-ring isomers become unstable with increase in the energy difference.

Bulky alkyl substituents at the central reactive carbon atoms also affect thermal stability of the closed-ring isomers, as shown in Figure 3.7b [45]. Introduction of isopropyl substituents, such as **69**, decreased the half-life time of the closed-ring isomer to 0.33 hours at 100 °C. The lifetime is 1/1670 of that of methyl-substituted **20** (half-life time: 23 days at 100 °C). The thermal cycloreversion reaction is dependent on the corrected Taft's steric substituent constants ($E_s^{\,c}$) of the substituents at the reactive carbon atoms. This means that neither the hyperconjugation of the α-hydrogens in the substituent nor the polarity of the substituent affects the thermal cycloreversion reaction. Only the steric hindrance has an effect on the thermal cycloreversion [46].

It is worth noting that the substituent effect on the thermal stability of the closed-ring isomers is not directly correlated to that of the photocycloreversion quantum yield. The potential energy surface of the ground state is different from those of the excited states. Although the half-life time of **68c** (40 h at 100 °C) decreased to less than 1/10 of **20c** (23 days at 100 °C) by changing the substituents at the reactive carbon atoms from methyl to ethyl, the photocycloreversion quantum yield of **68c** (0.0081) is smaller than that of **20c** (0.013). There is no correlation between them.

The thermal cycloreversion reaction of **68c** was followed in the single crystalline phase by in situ X-ray crystallography [47]. According to the Woodward–Hoffmann rule, the thermal cycloreversion reaction of cyclohexadiene derivatives should proceed in a disrotatory mode. It is of interest to reveal what kind of mechanism takes place in the thermal cycloreversion reaction to understand the large activation energy in the reaction. After heating the closed-ring form crystal of **68c** at 100 °C for 256 h, four new peaks appeared around the sulfur atoms and the reactive carbon

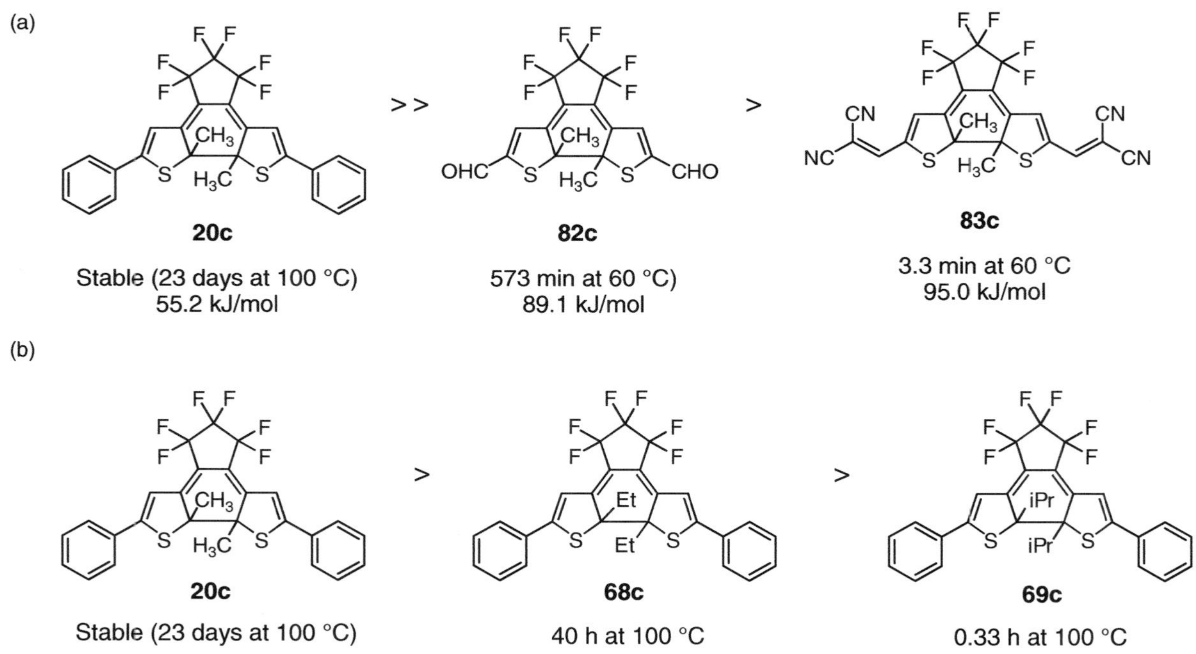

Figure 3.7 Thermal stability of closed-ring isomers. The values below the molecular structures are the half-life times of the thermal fading and ground state energy differences between open- and closed-ring isomers calculated by the B3LYP/6-31G* program. (a) Electron-withdrawing substituents and (b) steric hindrance of the substituents.

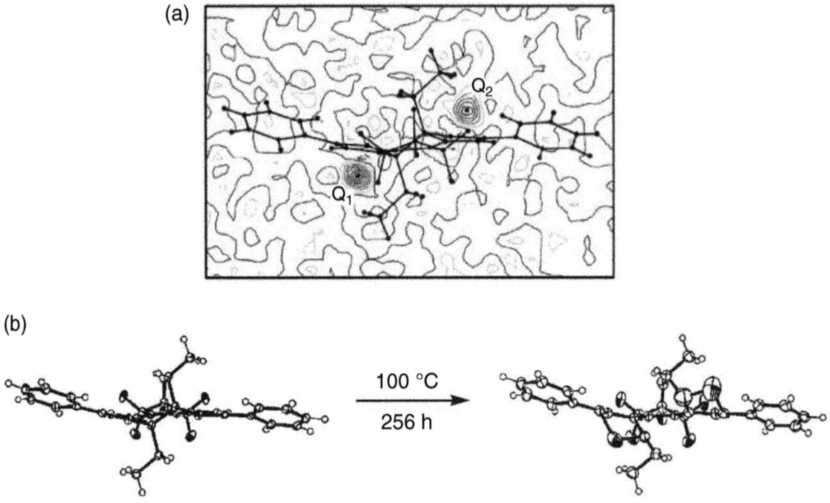

Figure 3.8 (a) $F_o - F_c$ difference Fourier electron density maps through peaks Q1 and Q2 after the thermal cycloreversion reaction of **68c**. (b) The molecular structures before and after the thermal cycloreversion reaction.

atoms, as shown in Figure 3.8a. These peaks are ascribed to the open-ring isomer and the peak intensity suggests that around 8% of the closed-ring isomers thermally converted to the open-ring isomers. The structure of the open-ring isomer clearly indicates that the thermal cycloreversion reaction proceeds in a conrotatory mode, as shown in Figure 3.8b. The conrotatory cycloreversion reaction accounts for the large activation energy in the thermal reaction.

Although typical diarylethenes with thiophene or benzothiophene aryl groups undergo P-type photoswitching reactions, the P-type photoswitches convert to T-type ones by replacing the aryl groups to pyrrole or indole rings, as described in Section 2.1. The stability of the closed-ring isomers can be controlled by changing the aryl groups. Thermally reversible photoswitching chromophores are used for light attenuation control, such as in eyeglasses. Candidates so far used for the application are spiropyrans, spironaphthoxazines, naphthopyrans, and hexaaryl-bisimidazoles. Diarylethenes with pyrrole groups are also possible candidates for the application [48].

Various derivatives **84–87** with cyano-pyrrole groups were prepared and their photoswitching performance was examined. Photogenerated closed-ring isomers **84c–86c** are unstable and they undergo thermally reversible photoswitching reactions. The half-life times of the closed-ring isomers are summarized in Table 3.7. When the methyl substituents at the reactive carbon atoms were replaced with ethyl substituents, the thermal fading rate was dramatically accelerated. The half-life time of **84c** at 25 °C is 28.2 s, while that of **85c** is as short as 7.8 seconds at −9.0 °C. The lifetime of **86c**, which has methyl substituents at 3- and 3′-positions, increased to 60.6 s at 25 °C. The activation energies of the thermal decay reactions of **84c, 85c,** and **86c** were determined to be 78.2, 65.7, and 83.7 kJ/mol, respectively. The red color of **87c** remained stable at room temperature and fading of the color was

Table 3.7 T-type molecular photoswitches. Half-life times of closed-ring isomers, activation energies of thermal cycloreversion reactions in ethyl acetate, and calculated ground state energy differences between open- and closed-ring isomers.

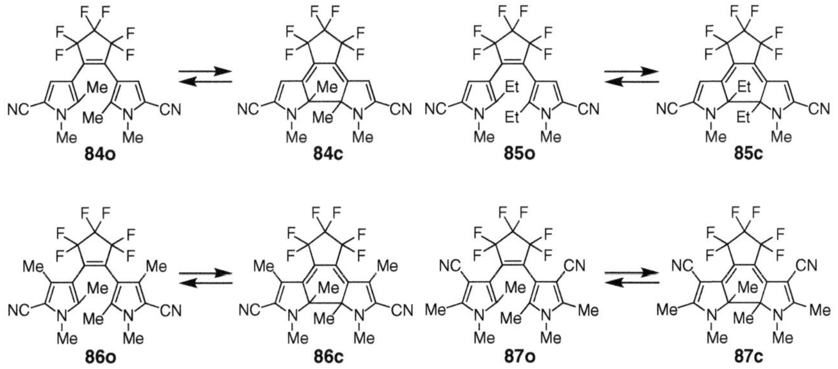

	84	85	86	87
Half-life times (s)	28.2[a]	7.8[b]	60.6[a]	—[c]
Activation energies for thermal cycloreversion reactions (kJ/mol)	78.2	65.7	83.7	—
Calculated ground state energy differences between open-ring and closed-ring isomers (kJ/mol)	174	187	159	128

a) At 25 °C.
b) At −9.0 °C.
c) Stable at 25 °C.

not observed. To know the origin of the differences in the thermal decay rates and the activation energies, the ground state energy differences between the open- and closed-ring isomers were calculated based on density functional theory (DFT, Gaussian09). As shown in Figure 2.3, thermal stability of the colored closed-ring isomer is controlled by the ground state energy difference. The difference of **84** is 174 kJ/mol, while that of **85** increases to 187 kJ/mol. The differences of **86** and **87** become small and are calculated to be 159 and 128 kJ/mol, respectively. The thermal stability differences can be well explained by the energy differences. When the difference was large as in **85**, the lifetime became short and the activation energy decreased. On the other hand, thermal fading was not observed when it was small as in **87**. The fine-tuning of thermal fading rate is useful for the application to light attenuation.

3.3 Fatigue Resistance

Fatigue resistance is an indispensable property for practical applications. Although well-designed diarylethenes were reported to undergo the coloration/decoloration reactions more than 14 000 cycles in solution [49] and 30 000 cycles in the crystalline state [50], it is strongly desired to improve fatigue resistance. To improve fatigue resistance, it is essential to understand fatigue mechanism.

The fatigue mechanism of compound **3** was studied in detail [51]. Upon irradiation with 254 nm light, the colorless *n*-hexane solution of **3** turned red, as shown in Figure 1.7a. The red color disappeared by irradiation with visible light (>400 nm). When the irradiation with 254 nm was prolonged, a stable photoproduct, which cannot be bleached, was produced. The absorption spectrum of the product is very similar to the closed-ring isomer **3c**. The by-product was isolated with a high-performance liquid chromatography (HPLC) and was found by molecular mass determination to be isomeric with **3**. X-ray crystallographic analysis was carried out for the single crystal. Figure 3.9 shows the molecular structure determined by the X-ray crystallographic analysis. It has a condensed-ring system with two 6-membered heterocyclic rings. The polyene structure is similar to **3c**. This is the reason why the spectrum is similar to **3c**. The stable photoproduct was also formed when the *n*-hexane solution of **20** was irradiated with 313 nm light. The condensed-ring structure of the by-product was the same as observed for **3** [52].

Irradiation time dependence of the by-product formation was followed. Although an induction period existed in the formation of the by-product from open-ring isomer **3o**, such a period was not observed when the closed-ring isomer **3c** was irradiated with 254 nm light. The by-product was generated from the beginning of irradiation. The by-product formation was not observed when the *n*-hexane solution of **3c** was irradiated with 546 nm light. Inferring from the above results, it is concluded that the by-product is produced by irradiation of **3c** with 254 nm light.

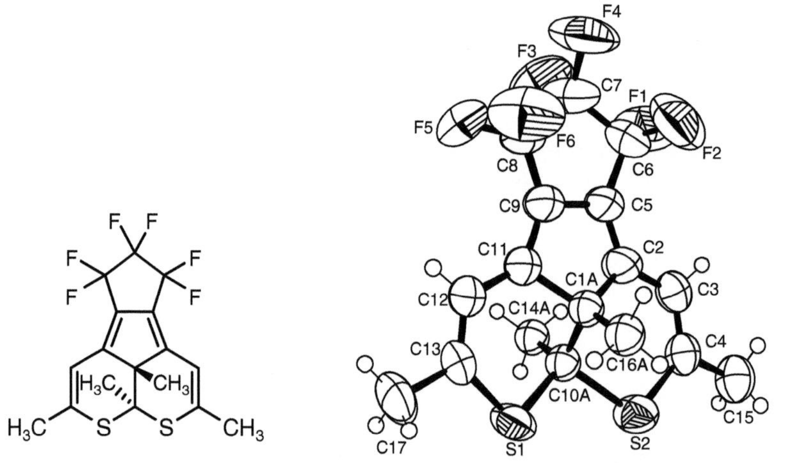

88

Figure 3.9 ORTEP drawing of by-product **88** showing 50% probability displacement ellipsoids.

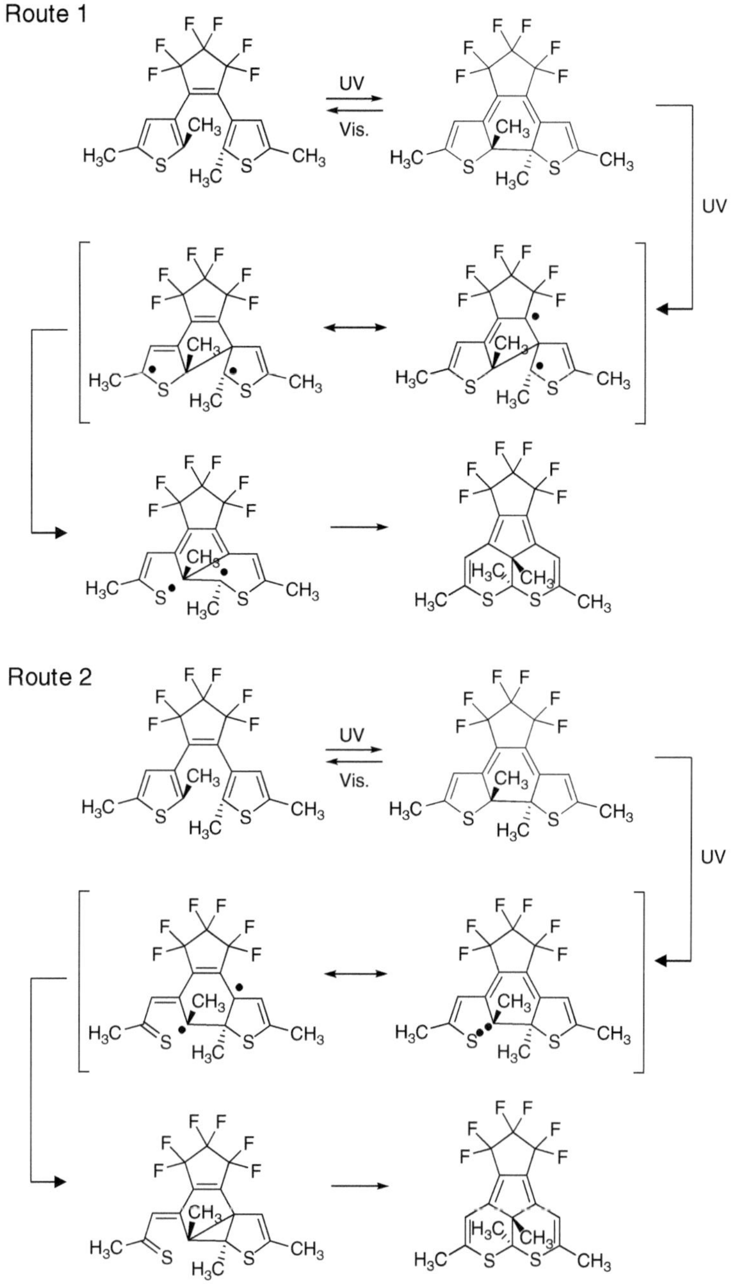

Scheme 3.2 Two routes (1 and 2) of the byproduct formation.

Scheme 3.2 shows two possible reaction routes to produce the by-product from closed-ring isomer **3c**. Route 1 goes through the conical intersection 2A/1ACI (o). It is known that photoexcited cyclohexadiene has a channel to produce a five-membered ring, methylenecyclopentene diradical. Route 1 undergoes the reaction via the five-membered ring. The radical migration in the π-conjugation system, and bond formation and cleavage finally produce the condensed-ring by-product. Another possible route 2 is initiated by photochemical C—S bond cleavage. Radical migration and bicyclohexane formation lead to the condensed-ring by-product. Similar C—S bond cleavage was reported for 2-methyl-2,3-dihydro-1-benzothiophene [53]. Although both routes can account for the formation of the by-product, route 2 is more probable, because the by-product formation is caused upon irradiation with UV light and not observed during the photocycloreversion reaction with visible light, which goes through the same conical intersection as the reaction with UV light.

The by-product formation was not observed for compound **13** even after prolonged UV irradiation [49]. Figure 3.10 shows the fatigue-resistant property of diarylethene derivatives having benzothiophene aryl groups and thiophene ones. Methylcyclohexane solutions of the derivatives were irradiated alternatively with UV and visible light in the presence of air as well as under deaerated conditions. Bis(benzothienyl)perfluorocyclopentene derivatives, **13** and **24**, can repeat the coloration/decoloration cycles more than 10^4 times without appreciable fatigue. Bis (2-ethyl-6-(5-methyl-2-thienyl)-1-benzothiophen-3-yl)perfluorocyclopentene and 2,3-bis(2-methyl-1-benzothiophen-3-yl)maleic anhydride also exhibit excellent fatigue resistance in comparison with diarylethenes having thiophene aryl groups. The superior fatigue resistance of bis(benzothienyl)ethenes is well explained by the

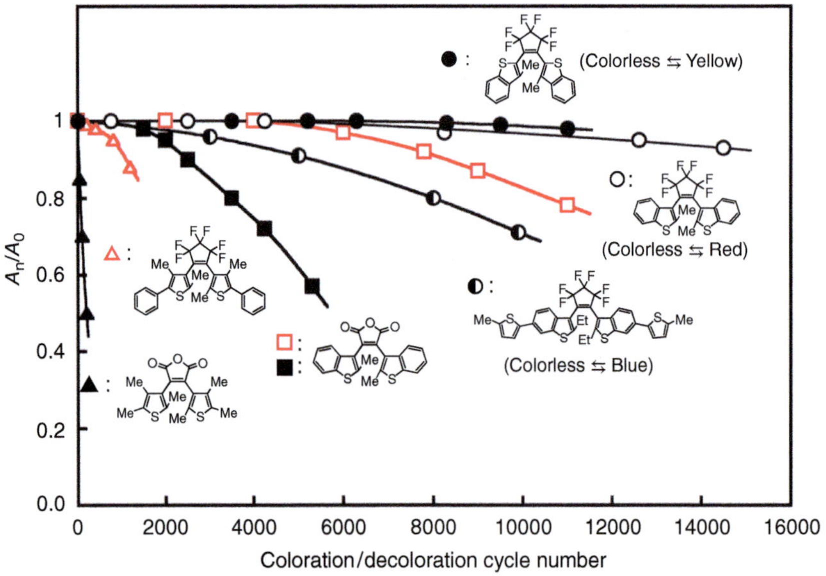

Figure 3.10 The fatigue resistant property of diarylethene derivatives. Initially, methylcyclohexane solutions of the derivatives were irradiated with UV light until the conversion from the open-ring isomers to the closed-ring isomers reached 90% of the photostationary state. Then, the colors of the closed-ring isomers were completely bleached by irradiation with visible light. The coloration/decoloration cycles were repeated (black marks and lines) in the presence of air or (red marks and lines) under deaerated conditions.

above fatigue mechanism. In the derivatives, radical formation and migration are considered to be impeded because of the condensed phenyl ring [49, 54–56].

The fatigue resistance was improved under deaerated conditions. The repeatable cycle number of 2,3-bis(2-methyl-1-benzothiophen-3-yl)maleic anhydride obviously increased by the elimination of oxygen. It is favorable to eliminate oxygen from the systems to improve the fatigue resistance.

Route 2 was also supported by theoretical calculations [57, 58]. The cycloreversion reaction is considered to take place as shown in Figure 2.8. After photoexcitation to the 1B Franck–Condon state, the excited wavepacket undergoes internal conversion to the local minimum of 2A state (2Ac). The 2A state goes over the energy barrier $2ATS_1$, and then reaches 2A/1A conical intersection (2A/1ACI(o)). At the conical intersection, fast deactivation to the ground state takes place and thus the open-ring isomer is formed. In addition to this cycloreversion process, another reaction pathway on 2A that is relevant for the by-product formation has been revealed [58].

The potential energy surfaces for the 1A and 2A electronic states were calculated for the model compound **89** at the CASSCF level and are represented in the space of two nuclear coordinates in Figure 3.11 [58]. The model compound was chosen because of computational limitation. The two independent coordinates of Figure 3.11 correspond to (1) the internuclear C—C distance connecting open- and closed-ring isomers (reaction coordinate) and (2) the internuclear C—CH_2 distance (initial fatigue coordinate). Two minima corresponding closed-ring isomer (2Ac) and open-ring isomer (2Ao), and two transition states ($2ATS_1$ and $2ATS_2$) are located on the 2A potential energy surface. $2ATS_1$ is the barrier connecting the closed- and the open-ring isomer. $2ATS_2$ barrier describes the bond-breaking of C—CH_2 bond of the penta-ring. After equilibration in 2Ac, two outcomes can be expected: (1) the system can cross $2ATS_1$ barrier (17.0 kcal/mol at the CAS(10,10)/6-31G(d)level)

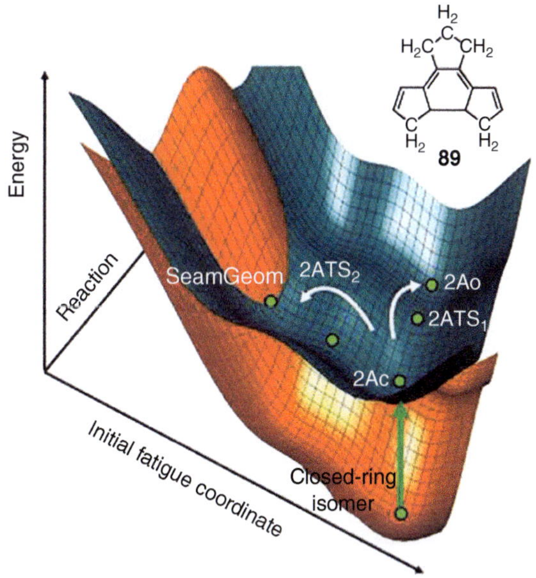

Figure 3.11 Schematic representation of structures of the S_0 (orange) and S_1 (green) potential energy surfaces corresponding to the photoisomerization between open- and closed-ring isomers of **89** and the fatigue reaction. The energies for S_1 and S_2 are represented in the space of two nuclear coordinates; (1) the internuclear C—C distance connecting the open- and closed-ring isomers and (2) the internuclear C—CH_2 distance. *Source*: Adapted from Ref. [58] with permission from the PCCP Owner Societies.

to reach 2A/1A conical intersection (2A/1ACI(o)) or (2) the system can cross the $2ATS_2$ barrier (43.7 kcal/mol) via bond breaking of C—CH_2, provided there is sufficient energy in the dissociation coordinate. The barrier height of $2ATS_2$ is substantially higher than that of $2ATS_1$. The illumination wavelength dependence of the by-product formation can be explained as follows. UV irradiation can provide enough energy for the molecule to overcome the bond breaking energy barrier and produce the by-product, while visible light cannot supply enough energy for the molecule to go through the $2ATS_2$ barrier because of its low energy. Presence of such a channel, which leads to the bond breaking, strongly supports the route 2, which is initiated by C—S bond breaking. It was confirmed by the CASSCF computations that there is no significant difference between compounds having CH_2 units and sulfur atoms.

The fatigue resistance of molecular photoswitches is evaluated by measuring number of coloration/decoloration cycles, as shown in Figure 3.10. The difficulty in obtaining fatigue-resistant photoswitches can be easily understood by the following simplified reaction sequence, in which a side reaction to produce by-product (**C**) is involved.

$$\mathbf{A} \rightleftarrows \mathbf{B} \xrightarrow{\Phi_{BC}} \mathbf{C}$$

Even if the side reaction quantum yield (ϕ_{BC}) is as small as 0.001 and **A** converts to **B** without any side reaction, 63% of the initial concentration of **A** will decompose after 1000 coloration/decoloration cycles. The occurrence of a slight side reaction that is hardly observed when the photoisomerization is performed only a few times will lead to a significant loss of photoswitching ability after many coloration/decoloration cycles. Therefore, the quantum yield of conversion to by-product should be less than 0.00001 to repeat the cycles more than 10^4 times.

It is highly important to find a molecular design approach, which improves fatigue resistance of dithienylethenes. In order to gain a systematic insight into the structure–switch relationship, a series of dithienylethene derivatives was prepared and their fatigue behavior was examined [59]. Table 3.8 shows the effect of substitution on photoswitching performance and fatigue resistance of dithienylperfluorocyclopentenes **90–92** and dithienylperhydrocyclopentenes **93–95** upon irradiation with UV (313 nm) light. The cyclization quantum yield was not influenced by the substitution in both derivatives. The cycloreversion quantum yield slightly increased by introducing electron-acceptor or donor substituents. The most striking feature is the effect of electron-acceptor substituents on the quantum yield of by-product formation. In general, the quantum yield of by-product formation of dithienylperhydrocyclopentene derivatives is one order of magnitude larger than that of dithienylperfluorocyclopentene derivatives. When electron-acceptor substituents were introduced, the by-product formation was decreased in both derivatives. In the case of dithienylperfluorocyclopentenes, the introduction of 3,5-bis(pentafluorosulfanyl)phenyl groups at 5- and 5′-positions decreased the yield to 0.00019, which is the smallest value in dithienylethene derivatives. In the case of dithienylperhydrocyclopentenes, the relatively large quantum yield of

Table 3.8 Photochemical properties of diarylethenes in acetonitrile upon irradiation with UV (313 nm) light.

		R	Φ_{OC}	Φ_{CO}	$\Phi_{C\to by\text{-}product}$
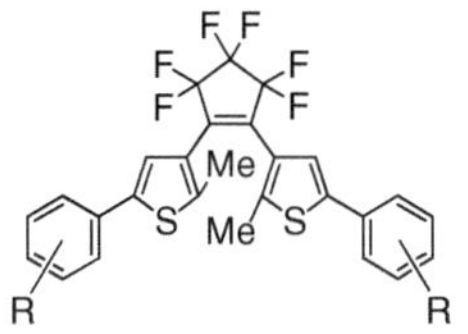	**90**	4-Me	0.62 ± 0.08	0.03 ± 0.03	0.0006 ± 0.0002
	91	3,5-$(CF_3)_2$	0.58 ± 0.06	0.05 ± 0.05	0.0004 ± 0.0001
	92	3,5-$(SF_5)_2$	0.50 ± 0.05	0.04 ± 0.05	0.00019 ± 0.00004
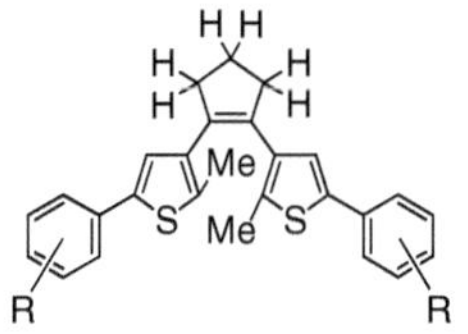	**93**	4-Me	0.43 ± 0.05	0.01 ± 0.03	0.009 ± 0.003
	94	4-MeO	0.47 ± 0.05	0.03 ± 0.03	0.007 ± 0.002
	95	3,5-$(CF_3)_2$	0.44 ± 0.04	0.02 ± 0.06	0.0004 ± 0.0001

by-product formation decreased by more than one order of magnitude by introducing 3,5-bis(trifluoromethyl)phenyl groups at 5- and 5′-positions. The fatigue resistance of the dithienylperhydrocyclopentene derivative is comparable to that of dithienylperfluorocyclopentenes. The fatigue resistance is largely improved in both dithienylperhydrocyclopentenes and dithienylperfluorocycolpenetnes by the introduction of electron-acceptor substituents, such as CF_3 and SF_5. Importantly, CF_3 and SF_5 groups are chemically inert and their introduction into diarylethenes does not alter their efficient isomerization between the open- and closed-ring isomers.

As discussed in the fatigue mechanism, the condensed-ring by-product is produced from the closed-ring isomer upon irradiation with UV light. Therefore, the dose of UV light actually absorbed by the closed-ring isomer during the ring closure process affects the fatigue [60]. The formation rate of the by-product is proportional to the number of UV photons absorbed by the closed-ring isomer times the quantum yield of the by-product formation. The fatigue resistance is dependent on the illumination process, i.e. UV wavelength, and on the absorption characteristics of the diarylethene at the exposure wavelength, besides the quantum yield of the by-product formation. The cross sections of the open- and the closed-ring isomers at the illumination wavelength play an important role. The $\varepsilon_{\text{closed-ring isomer}}/\varepsilon_{\text{open-ring isomer}}$ ratio at the illumination wavelength can be assumed as a figure of merit to determine the fatigue resistance, especially in those molecules that have similar quantum yield of by-product formation. If the ratio is large, a considerable amount of the incident light is absorbed by the closed-ring isomer, making the degradation more effective. On the other hand, if the ratio is low, the absorption by the open-ring isomer is preferred over the residual absorption of the closed-ring isomer, resulting in promotion of photocyclization and increase in the number of coloration/decoloration cycles. This relation was confirmed in both

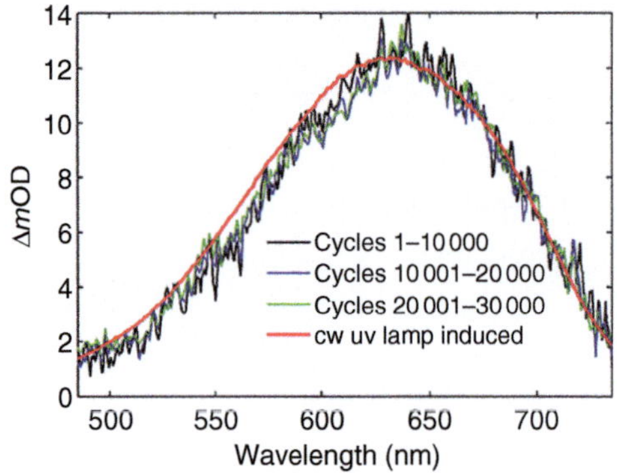

Figure 3.12 Average ΔOD of **5** for the first 3×10^4 cycles binned in 1×10^4 cycles (blue, green, and black), following excitation with 0.3 mJ/cm^2 of a 343 nm pump pulse. The red curve is the linear absorption spectrum for the closed-ring isomer after the offset correction and rescaling to match the magnitude reached with the femtosecond pump excitation. *Source*: Reprinted with permission from Ref. [50]. Copyright 2011 American Chemical Society.

dithienylethene and dithiazolylethene derivatives. This finding is important for the practical use of photoswitching molecules, because chromophores can be screened a priori on the basis of the $\varepsilon_{\text{closed-ring isomer}}/\varepsilon_{\text{open-ring isomer}}$ ratio. Alternatively, it allows for the selection of the best wavelength to induce the cyclization reaction achieving at the same time the best photofatigue resistance.

In the crystalline state, the by-product formation was strongly suppressed, because tight packing of molecules in crystal prevents C—S bond cleavage. Figure 3.12 shows the absorption spectra of the closed-ring isomer of **5** after alternate irradiation with 343- and 633 nm lasers for $1\text{–}3 \times 10^4$ cycles [50]. No detectable decrease of the absorption was observed even after 3×10^4 cycles. The molecule undergoes fatigue resistant photoswitching reactions in the crystalline phase.

3.4 Fluorescence Property

Fluorescence is a convenient tool to detect small amounts of molecules. Even a single molecule can be detected using fluorescence owing to the remarkable progress of optical detection techniques. To make use of this advantage, various types of fluorescent chromophores have been synthesized and widely applied in materials and life sciences. The applications are further extended when additional switching ability is provided to the chromophores. The modulation of fluorescence spectra and/or intensity by external stimuli, such as chemicals, electrons (or holes), or photons, is ingeniously adopted for analysis and imaging.

Diarylethenes undergo reversible photocyclization reactions. The electrocyclic reactions involve rearrangements of positions of single and double bonds in a molecule. The electronic structure change influences the absorption as well as emission property of the molecule. Although diarylethene photoswitches exhibit brilliant color change upon photoirradiation, most of them are nonfluorescent or very weakly fluorescent in both isomer forms. Therefore, it was a long-standing target to prepare photoswitchable highly fluorescent diarylethenes.

There are two types of photoswitchable fluorescent diarylethenes. One is the turn-off mode photoswitching diarylethenes. The open-ring isomer emits fluorescence, while the fluorescence is quenched when the open-ring isomer converts the closed-ring isomer upon photoirradiation by an energy transfer or an electron transfer mechanism. Another type is the turn-on mode photoswitching diarylethenes. Diarylethenes, the closed-ring isomers of which are fluorescent, belong to this type. Initially, the molecule is nonluminous under excitation with visible light ($\lambda > 420\,nm$), because the open-ring isomer has no absorption at the excitation wavelength. Upon irradiation with UV light the open-ring isomer converts to the closed-ring isomer, which has absorption in the visible wavelength region, and the closed-ring isomer emits fluorescence under excitation at the wavelength, where the closed-ring isomer has absorption. The turn-off mode photoswitching diarylethenes can be applied to optical memory media, but are hardly applicable to super-resolution fluorescence imaging, such as PALM (photoactivatable localization microscopy) or STORM (stochastic optical reconstruction microscopy), because these imaging techniques require an absolutely dark background. The turn-on mode photoswitching ability is indispensable for the application.

3.4.1 Turn-Off Mode Photoswitching

The most convenient way to prepare photoswitchable diarylethenes is to adopt fluorescent chromophores as the aryl groups. The open-ring isomers are expected to emit strong fluorescence, because the fluorescent chromophores are just substituted by the perfluorocyclopentene group. Based on this strategy, many fluorescent diarylethene derivatives having fluorescent units as the aryl groups were prepared as shown in Table 3.9.

2-Substituted indoles are known to exhibit strong fluorescence. The fluorescence quantum yield of 1-methyl-2-phenylindole is reported to be 0.85. Diarylethene derivatives having a 2-substituted indole, compounds **6** and **97**, were prepared and their photoswitching performance was examined. Both derivatives underwent efficient reversible photoswitching reactions (ϕ_{oc} = 0.44 (**6o**) and 0.26 (**97o**); ϕ_{co} = 0.075 (**6c**) and 0.21 (**97c**)). Although the open-ring isomers emitted fluorescence, the fluorescence quantum yields of **6** and **97** were limited to 0.046 and 0.063, respectively. Contrary to the expectation, the fluorescence quantum yields of the open-ring isomers were very small. The strategy to use fluorescence chromophores as the aryl groups was not successful to construct fluorescent diarylethenes. A relatively large fluorescence quantum yield was observed for 1,2-bis(2-phenyl-4-methyl-5-oxazolyl)perfluorocyclopentene (**38**), which has oxazole rings as the aryl groups. The quantum yield of the open-ring isomer is as large as 0.45 in *n*-hexane, while the closed-ring isomer is nonluminous.

An approach to increase the fluorescence quantum yield of the open-ring isomers is to integrate a fluorescent chromophore and a diarylethene unit in a molecule.

Table 3.9 Photophysical properties of fluorescent diarylethenes (λ_{max}: absorption maximum, λ_f: fluorescence maximum, Φ_f: fluorescence quantum yield).

Molecular structure	Open-ring isomer			Closed-ring isomer			References
	λ_{max} (nm)	λ_f (nm)	Φ_f	λ_{max} (nm)	λ_f (nm)	Φ_f	
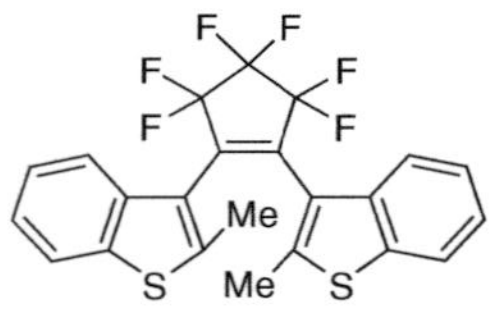 **13**	258 (EtOAc[a])	436 (EtOAc)	0.012 (EtOAc)	523 (EtOAc)	630 (*n*-hexane)	~0.0001	[54, 61, 62]
 **96**	293 (*n*-hexane)	440 (*n*-hexane)	0.011 (*n*-hexane)	540 (*n*-hexane)	640 (*n*-hexane)	—	[63]
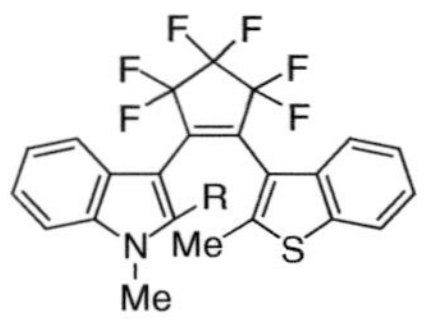 **6**: R = Me **97**: R = Ph	**6**: 260. **97**: 296 (*n*-hexane)	— **97**: 433 (*n*-hexane)	**6**: 0.046 **97**: 0.063 (*n*-hexane)	**6**: 556 **97**: 586 (*n*-hexane)	—	—	[64]

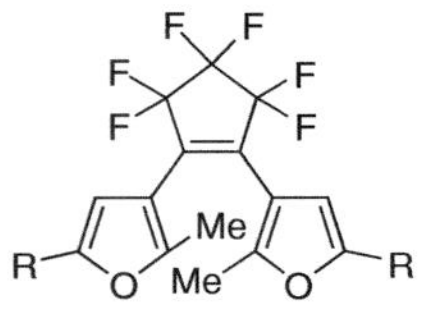 **98**: R = Me **26**: R = Ph	**98**: 302 **26**: 285 (*n*-hexane)	**98**: 382 **26**: 438 (*n*-hexane)	**98**: 0.03 **26**: 0.03 (*n*-hexane)	**98**: 449 **26**: 525 (*n*-hexane)	—	—	[11]
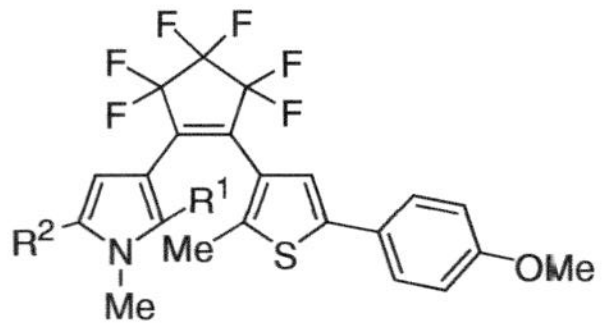 **99**: R^1= CN, R^2 = Me **100**: R^1= Me, R^2 = CN	**99**: 288 **100**: 296 (*n*-hexane)	**99**: 423 **100**: 421 (*n*-hexane)	**99**: 0.019 **100**: 0.016 (*n*-hexane)	**99**: 537 **100**: 592 (*n*-hexane)	—	—	[65]
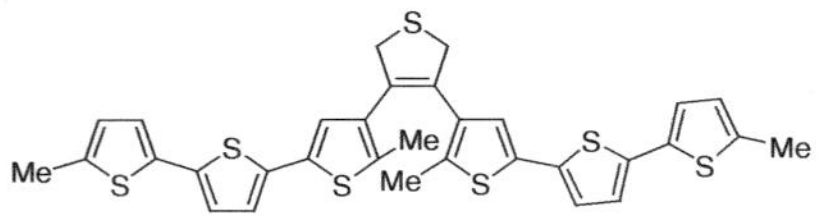 **101**	365 (CH_3CN)	457 (CH_3CN)	0.028 (CH_3CN)	595 (CH_3CN)	—	—	[66]

(Continued)

Table 3.9 (Continued)

Molecular structure	Open-ring isomer λ_{max} (nm)	Open-ring isomer λ_f (nm)	Open-ring isomer Φ_f	Closed-ring isomer λ_{max} (nm)	Closed-ring isomer λ_f (nm)	Closed-ring isomer Φ_f	References
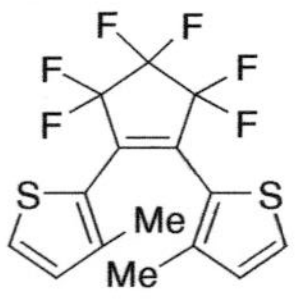 **1**	316 (3-MP[b])	500 (3-MP)	0.01 (3-MP)	432 (3-MP)	—	—	[67]
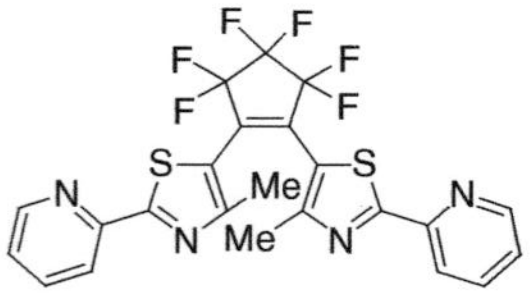 **102**	358 (CH_3CN)	500 (CH_3CN)	0.005 (CH_3CN)	414 (CH_3CN)	—	—	[68]
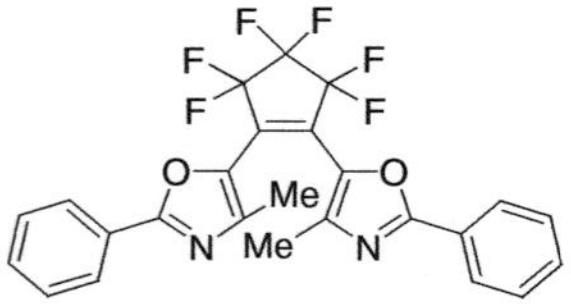 **38**	287 373 (*n*-hexane)	470 (*n*-hexane)	0.45 (*n*-hexane)	252 353 (*n*-hexane)	—	—	[24]

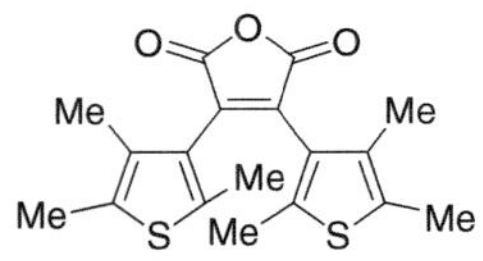 **64**	313 (*n*-hexane)	488 (*n*-hexane)	Relative yield 1.0 (*n*-hexane)	552 (*n*-hexane)	—	—	[35]
	340 (benzene)	541 (benzene)	0.35 (benzene)	564 (benzene)			
	335 (THF)	560 (THF)	0.088 (THF)	560 (THF)			
	336 (CH_3CN)	—	<0.01 (CH_3CN)	563 (CH_3CN)			
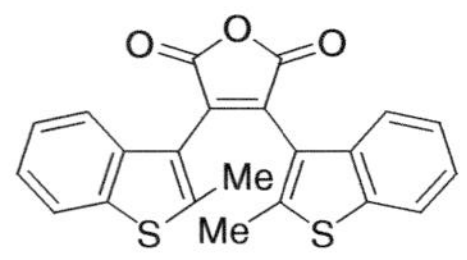 **103**	407 (n-hexane)	487 (n-hexane)	Relative yield 1.0 (n-hexane)	538 (n-hexane)	—	—	[35]
	417 (benzene)	527 (benzene)	1.1 (benzene)	541 (benzene)			
	404 (THF)	540 (THF)	0.92 (THF)	533 (THF)			
	404 (CH_3CN)	572 (CH_3CN)	0.058 (CH_3CN)	535 (CH_3CN)			
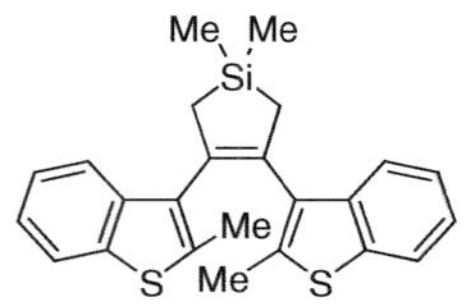 **104**	267 (3-MP)	—	—	447 (3-MP)	575 (3-MP)	0.003 (3-MP)	[69]

(Continued)

Table 3.9 (Continued)

Molecular structure	Open-ring isomer			Closed-ring isomer			References
	λ_{max} (nm)	λ_f (nm)	Φ_f	λ_{max} (nm)	λ_f (nm)	Φ_f	
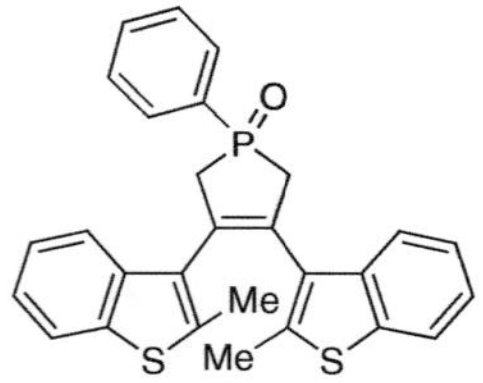 **105**	264 (3-MP)	—	—	455 (3-MP)	570 (3-MP)	0.02 (3-MP)	[69]
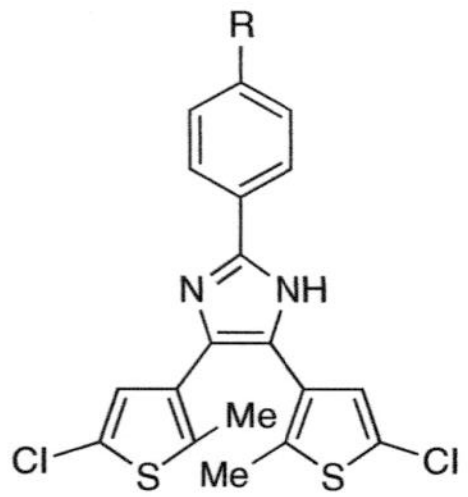 **106**: R = H **107**: R = Cl **108**: R = OMe	**106**: 298 **107**: 309 **108**: 295 (CH_3CN)	**106**: 381 **107**: 390 **108**: 378 (CH_3CN)	**106**: 0.017 **107**: 0.114 **108**: 0.002 (CH_3CN)	**106**: 520 **107**: 516 **108**: 533 (CH_3CN)	**106**: 452 **107**: 449 **108**: 489 (CH_3CN)	**106**: 0.014 **107**: 0.009 **108**: 0.13 (CH_3CN)	[70]

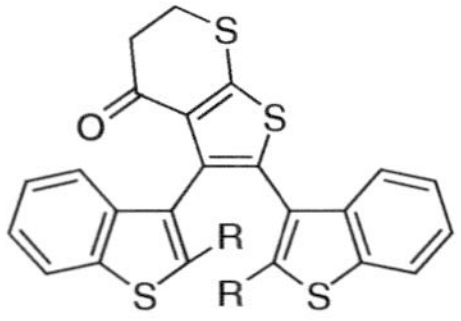 **109**: R = Me **110**: R = Et	**109**: 252 **110**: 252 (EtOAc)	—	—	**109**: 520 **110**: 524 (EtOAc)	**109**: 580 **110**: 594 (EtOAc)	**109**: 0.11 **110**: 0.05 (EtOAc)	[71]
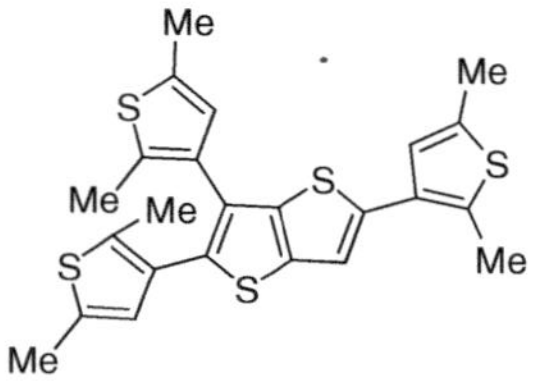 **111**	334 (benzene)	405 (benzene)	0.06 (benzene)	560 (benzene)	—	—	[72]
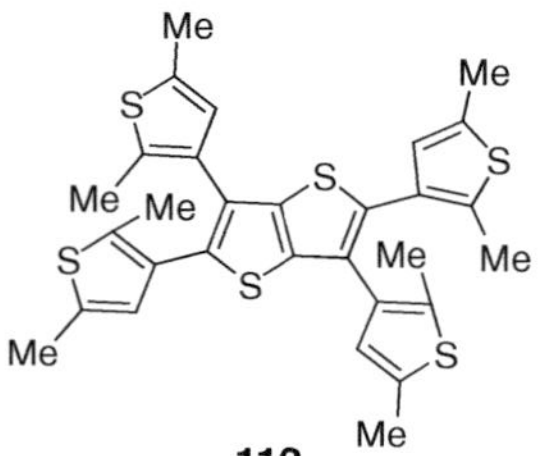 **112**	332 (benzene)	405 (benzene)	0.10 (benzene)	560 (benzene)	—	—	[72]

(Continued)

Table 3.9 (Continued)

Molecular structure	Open-ring isomer			Closed-ring isomer			References
	λ_{max} (nm)	λ_f (nm)	Φ_f	λ_{max} (nm)	λ_f (nm)	Φ_f	
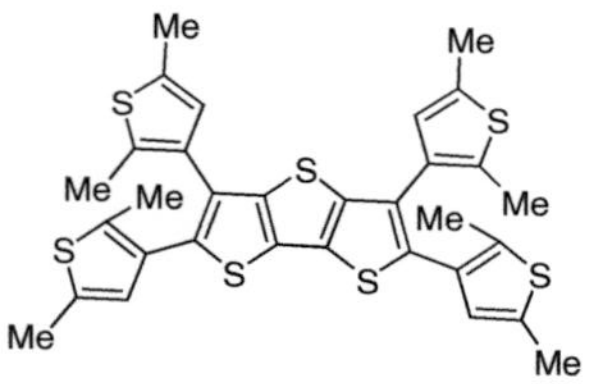 **113**	350 (benzene)	420 (benzene)	0.14 (benzene)	564 (benzene)	—	—	[72]
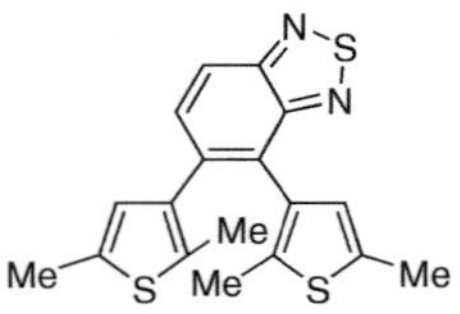 **114**	235 310 368(sh) (cyclohexane)	470 (cyclohexane)	0.077 (cyclohexane)	230 368 486 (cyclohexane)	—	—	[73]
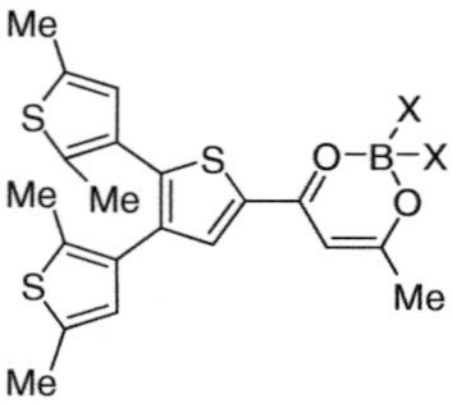 **115**: R = F **116**: R = C_6F_5 **117**: R = C_6H_5	**115**: 417 **116**: 434 **117**: 415 (benzene)	**115**: 502 **116**: 517 **117**: 484 (benzene)	**115**: 0.167 **116**: 0.160 **117**: 0.007 (benzene)	**115**: 758 **116**: 810	—	—	[74]

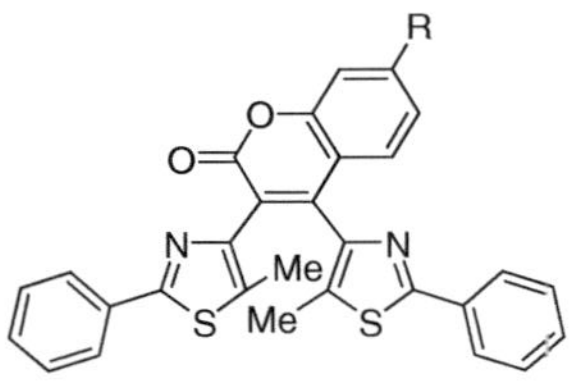 **118**: R = H **119**: R = OMe **120**: R = OH **121**: R = O⁻	**118**: ~310 (*n*-hexane) **119**: ~300 (*n*-hexane) **120**: ~310 (CH_2Cl_2) **121**: ~400 (MeOH/water buffer sol.)	— — — **121**: 490 (MeOH/water buffer sol.)	**118**: 0.0025 (*n*-hexane) **119**: 0.0043 (*n*-hexane) **120**: 0.019 (CH_2Cl_2) **121**: 0.25 (MeOH/water buffer sol.)	**118**: ~560 (*n*-hexane) **119**: ~580 (*n*-hexane) **120**: ~570 (CH_2Cl_2) **121**: ~610 (MeOH/water buffer sol.)	—	—	[75]
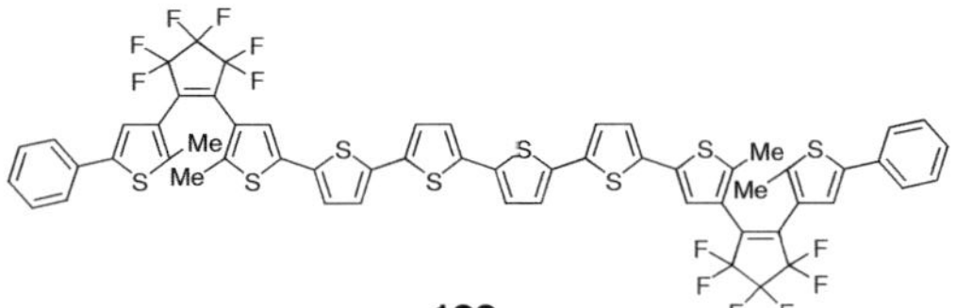 **122**	270 435 (heptane)	500 (isopentane)	0.2 (isopentane)	468 634 (heptane)	—	—	[76]
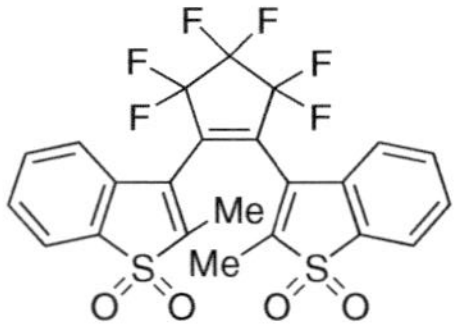 **123**	276 308 (EtOAc)	464 (EtOAc)	0.025 (EtOAc)	262 398 (EtOAc)	492 (EtOAc)	0.011 (EtOAc)	[54]

(Continued)

Table 3.9 (Continued)

Molecular structure	Open-ring isomer			Closed-ring isomer			References
	λ_{max} (nm)	λ_f (nm)	Φ_f	λ_{max} (nm)	λ_f (nm)	Φ_f	
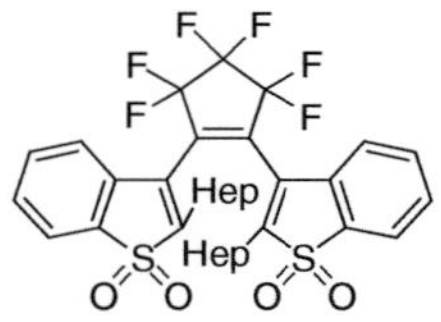 **124**	278 (EtOAc)	—	—	414 (EtOAc)	476 (EtOAc)	0.19 (EtOAc)	[55]
 **8**	298 336 (1,4-dioxane)	—	—	456 (1,4-dioxane)	595 (1,4-dioxane)	0.87 (1,4-dioxane)	[77]
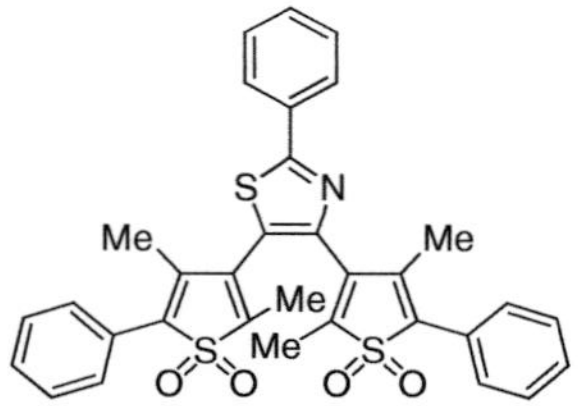 **125**	307 (2-MTHF[c])	—	—	324, 490 (2-MTHF)	577 (2-MTHF)	0.35 (2-MTHF)	[78]

a) EtOAc: ethyl acetate.
b) 3-MP: 3-methylpentane.
c) 2-MTF: 2-methyltetrahydrofuran.

A Förster-type resonance energy transfer (FRET) or a photoinduced intramolecular electron transfer (IET) mechanism is adopted to reversely photoswitch the fluorescence property. Figure 3.13 shows typical dyad **126**, in which a fluorescent bis(phenylethynyl)anthracene is connected to a dithienylethene unit through an adamantyl spacer [79]. When the diarylethene unit is in the open-ring form, the fluorescence quantum yield is 0.73, while it decreases to less than 0.001 when the open-ring form converts to the closed-ring form upon UV irradiation. The fluorescence spectrum of the bis(phenylethynyl)anthracene unit matches the absorption spectrum of the closed-ring form of the diarylethene unit. Therefore, the fluorescence is efficiently quenched by the closed-ring form according to the FRET mechanism. By measuring the fluorescence lifetimes (τ (open-ring form) = 4.9 ns, τ (closed-ring form) = 4.5 ps), the intramolecular energy transfer rate was estimated to be $2.2 \times 10^{11}\ \mathrm{s}^{-1}$ (see Section 5.1).

Although at the single-molecule level, discrete *on/off* switching between two states is observed, in the ensemble solution fluorescence change is gradual and the fluorescence quenching is incomplete at the photostationary state (PPS). At the PPS, fluorescent open-ring isomers still remain and they contribute to remnant fluorescence. To minimize the remnant fluorescence at the PPS, perylenemonoimide derivatives having multiple dithienylethene (DTE) units were synthesized, as shown in Scheme 3.3 [80]. The fluorescence quantum yields of the derivatives **127**, **128,** and **129** having one, two, and three DTE units were measured to be 0.85, ~1.0, and 0.97 in toluene, respectively. Upon irradiation with 302 nm light, the open-ring DTEs in the derivatives were converted to the closed-ring DTEs with the yield of 95% at the PPS in toluene. The fluorescence *on/off* ratio of **127** at the PPS was 30. For **128,** the ratio increased to 865. The fluorescence on/off ratio further increased to 3026 for **129**. Introduction of multiple DTEs into one fluorophore is a useful design approach to minimize the remnant fluorescence in turn-off mode photoswitching diarylethenes. Based on similar design strategy, four tetradithienylethene-substituted porphyrins **130–133** were prepared and photoresponsive performance of these derivatives was examined [81]. Although efficient fluorescence quenching was observed, at the PPS the fluorescence quantum yields were still in the order of ~0.05% and the *on/off* ratio was 120–160. The residual fluorescence is considered to result from the presence of low concentration of all-open and partially closed-ring isomers.

Another approach to increase the *on/off* photoswitching ratio is to use aggregate nanoparticles. In the solid state where molecular diffusion is prohibited, a single acceptor molecule quenches fluorescence of surrounding multiple donor molecules. According to the Perrin formulation [82], donor molecules within a quenching sphere are quenched with unity efficiency by the acceptor molecule. Molecules outside of the quenching sphere are not quenched at all. The Perrin model predicts the following relationship:

$$\ln(\Phi o/\Phi_{\mathrm{D}}) = VN[Q]$$

where ϕ_{o} and ϕ_{D} are the fluorescence quantum yields of the donor molecule in the absence and presence of quencher, respectively. V is the volume of the active sphere of quenching, N is Avogadro's number, and $[Q]$ is the concentration of the

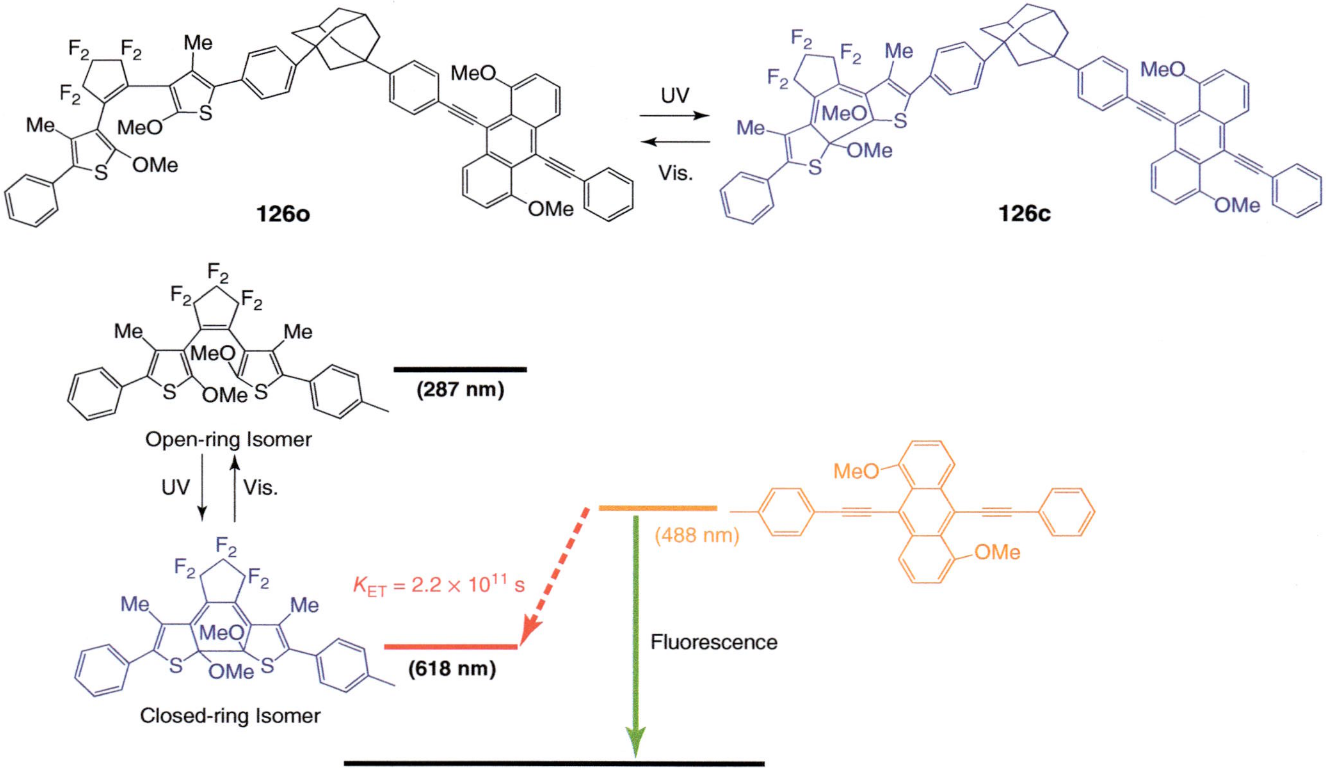

Figure 3.13 Fluorescence quenching of **126** based on a Förster-type intramolecular energy transfer. The fluorescence of the anthracene unit was reversibly quenched by the closed-ring isomer unit upon alternate irradiation with UV and visible light.

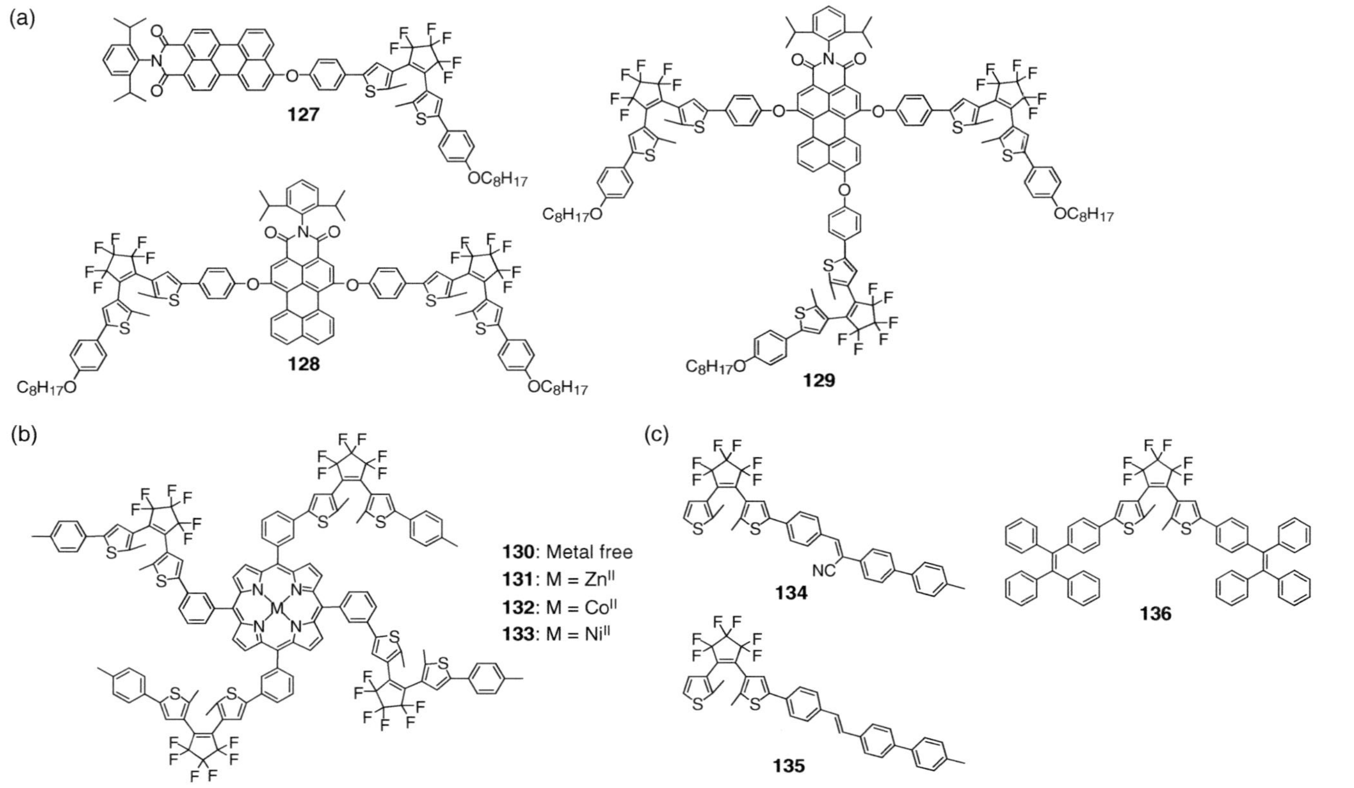

Scheme 3.3 Chemical structures of (a) dithienylethenes–perylenemonoimide composites **127–129**, (b) porphyrin–tetradithienylethenes **130–133** and (c) dithienylethenes having 1-cyano-1-(4′- methylbiphenyl)-2-phenylethene side group **134**, 1-(4′-methylbiphenyl)-2-phenylethene side group **135** and tetraphenylethene side groups **136**.

acceptor molecule. According to this Perrin model, donor fluorescence *exponentially* decreases with increasing the acceptor concentration in high efficiency.

The highly efficient quenching in the condensed state was demonstrated using aggregate nanoparticles of compound **137**, in which a fluorescent benzothiadiazole (BTD) unit is covalently attached to DTE, as shown in Figure 3.14 [83]. The DTE unit is able to undergo reversible photoswitching reactions in solution as well as in the solid state. Nanoparticles of **137**, diameter of which is 25 ± 10 nm, were prepared by the precipitation method in H_2O/THF mixture. The fluorescence of BTD was quenched in both solution and nanoparticles upon UV irradiation. The quenching behavior in the nanoparticles was quite different from that observed in a THF solution. In the THF solution the fluorescence *linearly* decreased with increase in conversion from **137o** to **137c**, as shown in Figure 3.14b. On the other hand, in the suspension of the nanoparticles the fluorescence *exponentially* decreased at the early conversion range, as shown in Figure 3.14c. The fluorescence reached almost zero (99% fluorescence quenching) after only 5% conversion. The fluorescence was switched off in a few second of UV (313 nm) irradiation (50 μW cm^{-1}). This rapid turn-off photoswitching is ascribed to a very efficient intermolecular FRET process between the BTE unit and the closed-ring form of DTE unit within the quenching sphere. Fluorescence of the donor molecules in the sphere, which is estimated to be 4.9 nm, is quenched by the closed-ring form of DTE acceptor molecule. The Perrin model well explains the efficient quenching.

Similar efficient photoswitching was observed in aggregates of diarylethenes having 1-cyano-1-(4′-methylbiphenyl)-2-phenylethylene or tetraphenylethene side groups, such as **134** and **136** (Scheme 3.3) [84, 85]. 1-Cyano-*trans*-bis(4′-methylbiphenyl)ethylene (CN-MBE) is well known to exhibit aggregate-induced enhanced emission (AIEE). CN-MBE quenches its fluorescence in a THF solution, while it exhibits blue fluorescence in the solid state. **134** was designed and synthesized by replacing one of the end tolyl groups of CN-MBE with a dithienylethene. Diarylethene **135**, in which the CN group is lacking, does not show any AIEE effect. In the THF solution of **135,** only a small amount of fluorescence was reduced upon irradiation with UV light, which causes the photoisomerization of DTE unit from the open- to the closed-ring isomers (29% reduction at the PSS). On the other hand, the fluorescence of nanoparticles and PMMA films containing very high concentration of **134** (20 wt%) was efficiently quenched upon irradiation with UV light. The fluorescence of suspension of the nanoparticles was reduced by 170 times (from $\phi_f = 3.4\%$ to $\phi_f = 0.02\%$ at the PSS).

The PMMA film containing **136** emitted brilliant green fluorescence when the DTE unit was in the open-ring form [85]. The fluorescence was also efficiently quenched when the open-ring form converts to the closed-ring form upon UV irradiation. These efficient fluorescence quenching in nanoparticles and PMMA films suggests that the Perrin model mechanism also plays an important role.

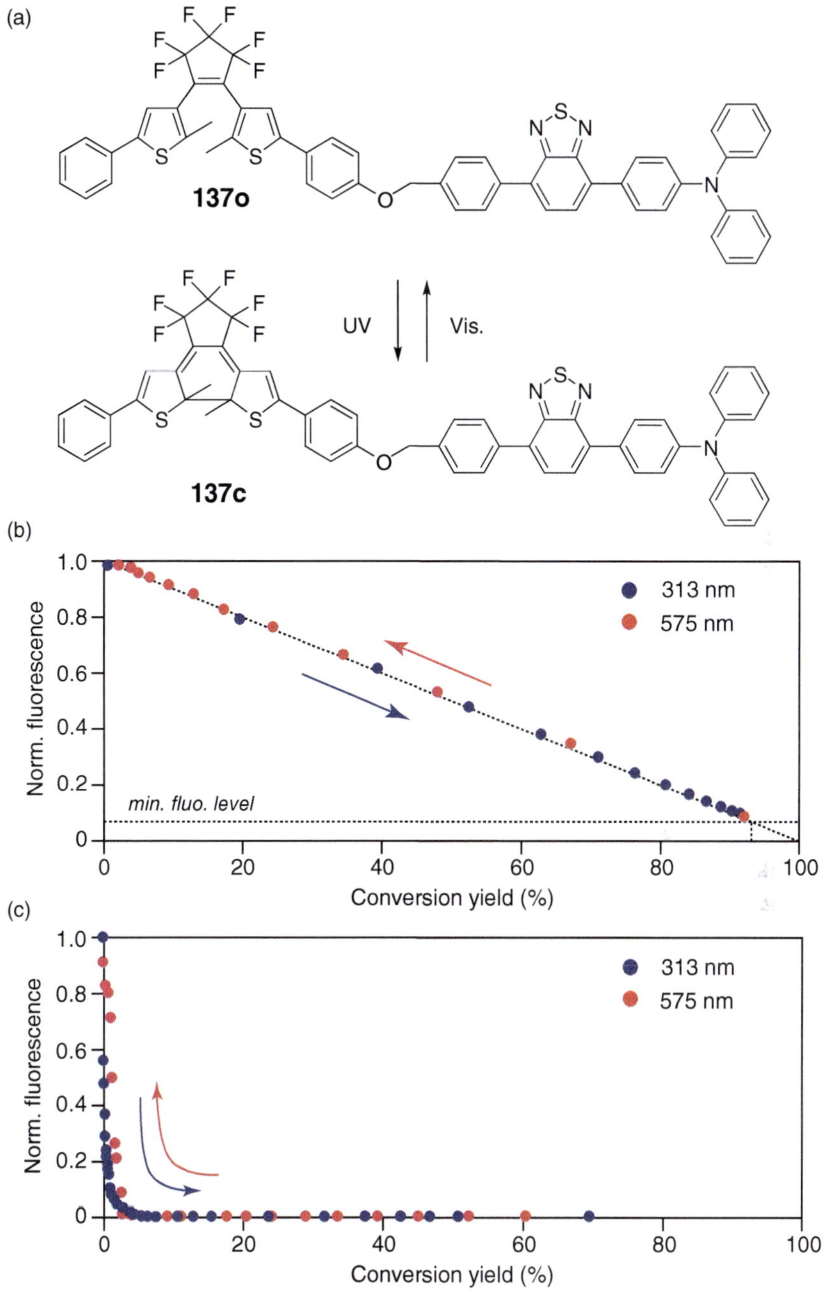

Figure 3.14 (a) Chemical structures of **137o** and **137c**. (b) Fluorescence intensity versus conversion yield correlation plots in THF solution. (c) The correlation in the nanoparticle suspension in H_2O/THF (80/20) mixture.

3.4.2 Turn-On Mode Photoswitching

When the closed-ring isomer emits fluorescence, the diarylethene belongs to a turn-on mode photoswitching molecule. The molecule is initially nonluminous, while it emits fluorescence when the open-ring isomer converts to the closed-ring isomer upon UV irradiation. The fluorescence disappears upon visible light irradiation. Although vast numbers of diarylethenes have been reported, most of them are nonfluorescent or weakly fluorescent in both isomers. Diarylethene, which emits appreciable fluorescence in the closed-ring isomer, is very rare. Turn-on mode photoswitching molecules are indispensable for super-resolution microscopy. Therefore, it is strongly desired to develop highly fluorescent turn-on mode photoswitching diarylethenes.

As shown in Table 3.9, **13** emits fluorescence in both open- and closed-ring isomer forms, but the fluorescence yields are very low. The yields were found to increase by oxidation of the sulfur atoms to sulfones, such as **123** [54]. Although **123** possesses the turn-on mode fluorescence switching property, the fluorescence quantum yield of the closed-ring isomer ($\phi_f = 0.011$) is still too low to be applied to fluorescence imaging. For the practical application to super-resolution fluorescent microscopy, it is indispensable to accomplish the following requirements.

1. Increase in the absorption coefficient of the closed-ring isomer.
2. Bathochromic shift of absorption and fluorescence spectra.
3. Increase in the fluorescence quantum yield of the closed-ring isomer.

Sulfone derivatives of various bis(benzothienyl)ethenes were prepared, as shown in Scheme 3.4. Introduction of both short alkyl substituents at 2- and 2′-positions and aryl substituents, such as phenyl or thiophene rings, at 6- and 6′-positions of the benzothiophene-1,1-dioxide groups was found to dramatically improve their fluorescence as well as photoswitching performance [77, 86–89].

Figure 3.15a shows the absorption and fluorescence spectra of diarylethene **8** having ethyl substituents at 2- and 2′-positions and phenyl substituents at 6- and 6′-positions in 1,4-dioxane. The open-ring isomer **8o** is colorless and has no absorption in the visible wavelength region. Upon irradiation with UV light, the colorless 1,4-dioxane solution turned yellow and a new absorption appeared at 456 nm. The absorption is ascribed to the closed-ring isomer **8c**. The absorption coefficient of **8c** was 4.6×10^4 M^{-1} cm^{-1}, which is twice larger than that of **123c**. The closed-ring isomer exhibited brilliant green fluorescence around 550 nm under excitation with 488 nm light. The fluorescence quantum yield of **8c** was 0.87 in 1,4-dioxane, which is ~80 times higher than that of **123c**. It is worth noting that the yield of **8c** in ethanol ($\phi_f = 0.76$) is 3 times higher than that of **138c** ($\phi_f = 0.26$), which has methyl substituents at 2- and 2′-positions. Both the absorption and fluorescence spectra showed significant bathochromic shifts in comparison with those of **123c**. In addition, **8c** exhibited excellent thermal stability. No absorption spectral change was observed even after 100 hours in the dark at 80 °C. The closed-ring isomer **8c** returned to the open-ring isomer **8o** upon irradiation with visible light ($\lambda > 470$ nm). The cycloreversion quantum yield in 1,4-dioxane was measured to be 5.9×10^{-4},

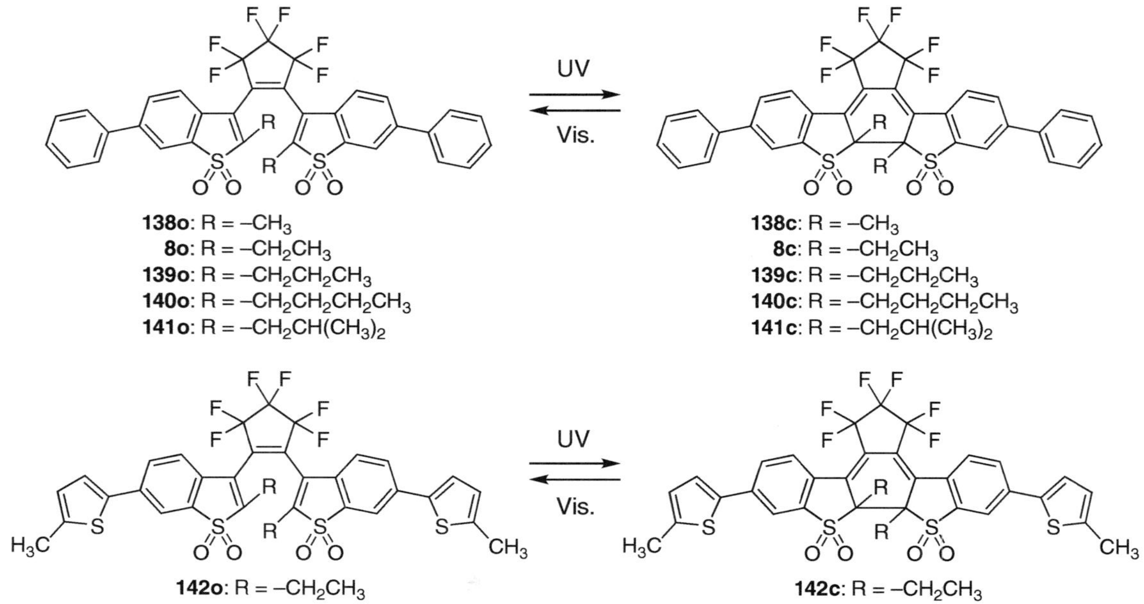

Scheme 3.4 Photoisomerization of fluorescent diarylethenes **138–142** and **8**.

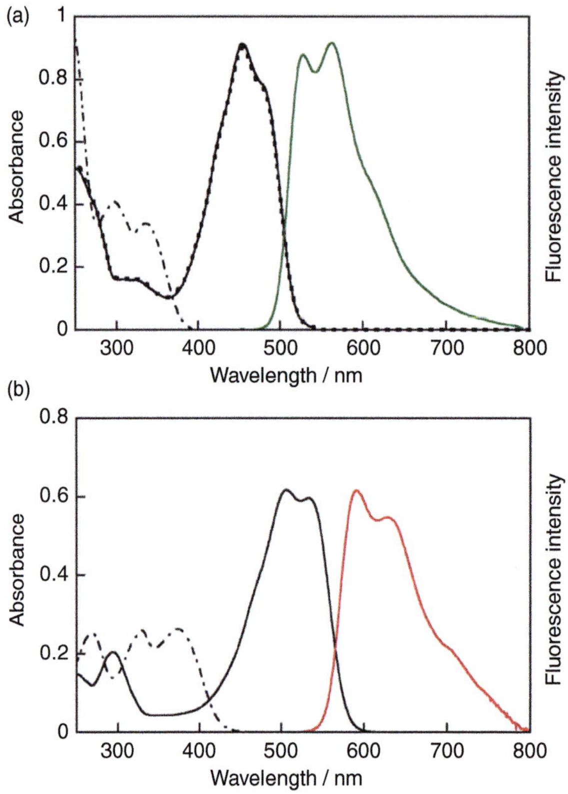

Figure 3.15 (a) Absorption spectra of **8o** (black dashed line), **8c** (black solid line) and photostationary state under irradiation with 313 nm light (black dotted line) in 1,4-dioxane (2.0×10^{-5} M) and fluorescence spectrum of **8c** (green solid line) under excitation with 488 nm light. (b) Absorption spectra of **142o** (black dashed line), **142c** (black solid line) and photostationary state under irradiation with 365 nm light in 1,4-dioxane (1.0×10^{-5} M) and fluorescence spectrum of **146c** (red solid line) under excitation with 532 nm light. The photostationary spectrum overlaps with the spectrum of **142c** due to ~100% photoconversion.

which is much lower than the value of **123c** (ϕ_{co} = 0.061 in ethyl acetate). The above requirements 1–3 have been fulfilled by introducing both ethyl and phenyl substituents.

Replacing the phenyl substituents at 6- and 6′-positions with thiophene rings shifted the absorption and fluorescence bands to longer wavelengths. Figure 3.15b shows the absorption and fluorescence spectra of **142** having thiophene rings at 6- and 6′-positions in 1,4-dioxane. The absorption maximum of **142c** is located at 506 nm ($\varepsilon = 5.8 \times 10^4$ M^{-1} cm^{-1}) and can be excited with 532 nm laser light. The fluorescence quantum yield was 0.78 in 1,4-dioxane and the cycloreversion quantum yield was less than 10^{-5}. Table 3.10 summarizes photophysical and photochemical properties of the sulfone derivatives. The absorption and fluorescence maxima of the closed-ring isomers having aryl substituents at 6- and 6′-positions of benzothiophene-1,1-dioxide groups show a significant bathochromic shift in comparison with non-substituted derivative **123** and the absorption coefficients of the closed-ring isomers increase by 2 or 3 times.

It is worth noting the effect of substitution of isobutyl groups (see also Section 9.3.1). Introduction of isobutyl substituents at the central reactive carbon atoms dramatically increased the cycloreversion quantum yield. The yield of **141c** was as large as 1.0×10^{-2}, which is 17 times larger than that of **8c** having ethyl substituents at the reactive carbon atoms.

Table 3.10 Photophysical and photochemical properties of **123**, **8**, and **138–142**.

	Open-ring isomer		Closed-ring isomer		
	λ_{max} (nm) (ε/10^4 M^{-1} cm^{-1})	Φ_{oc}	λ_{max} (nm) (ε/10^4 M^{-1} cm^{-1})	Φ_{co}	Φ_f
123	276 (0.37), 308 (0.41)	0.22[a)]	398 (2.10)	0.061[d)]	0.011
138	296 (1.8), 335 (1.4)	0.61[b,c,g)]	443 (5.1)	1.2×10^{-3}[e,g)]	0.64
8	298 (1.9), 336 (1.5)	0.62[b,c,g)]	456 (4.6)	5.9×10^{-4}[e,g)]	0.87
139	297 (2.0), 336 (1.6)	0.58[c,g)]	456 (4.5)	9.0×10^{-4}[e,g)]	0.89
140	298 (2.0), 336 (1.6)	0.56[c,g)]	456 (4.5)	1.2×10^{-3}[e,g)]	0.85
141	300 (1.9), 336 (1.5)	0.53[b)]	460 (4.2)	1.0×10^{-2}[e)]	0.80
142	330 (2.4), 374 (2.5)	0.23[c,g)]	506 (5.8)	$<1.0 \times 10^{-5}$[e,f,g)]	0.78

λ_{max}, Absorption maximum; ε, molar absorption coefficient; Φ_{oc}, cyclization quantum yield; Φ_{co}, cycloreversion quantum yield; Φ_f, fluorescence quantum yield. 1,4-Dioxane were used as solvents except **123**, for which ethyl acetate was used as the solvent.

a) Under irradiation with 312 nm light.
b) Under irradiation with 313 nm light.
c) Under irradiation with 330 nm light.
d) Under irradiation at λ_{max} of the closed-ring isomer.
e) Under irradiation with 450 nm light.
f) Under irradiation with 488 nm light.
g) Measured using a photoreaction quantum yield measuring system (Shimadzu QYM-01) with a corrected power meter.

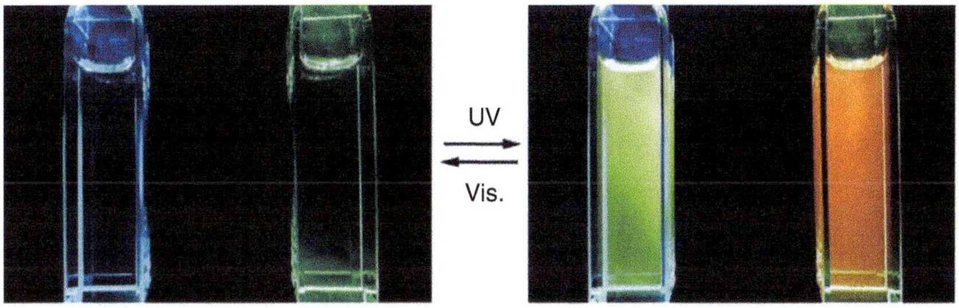

Figure 3.16 Photographs of 1,4-dioxane solutions containing **8** and **142** before and after irradiation with 365 nm light under excitation with 488 nm blue light (left-side solution containing **8**) and 532 nm green light (right-side solution containing **142**). *Source*: Adapted with permission from Ref. [77]. Copyright 2011 American Chemical Society.

Figure 3.16 shows the photographs of the fluorescence switching of 1,4-dioxne solutions containing **8** and **142** upon irradiation with 365 nm light under excitation with 488 nm (left-side solution containing **8**) and 532 nm light (right-side solution containing **142**). Before UV irradiation, both solutions were nonluminous, while brilliant green and orange fluorescence emissions instantaneously appeared upon UV irradiation. The emissions are attributed to the formation of fluorescent closed-ring isomers **8c** and **142c**.

Above turn-on mode fluorescent diarylethenes show very low cycloreversion quantum yields. For real-time cell imaging, it is favorable for the probe to have a relevant large cycloreversion quantum yield. Ahn and co-workers [90] developed a turn-on mode fluorescent probe **109** for live cell imaging, which has relatively large cycloreversion as well as cyclization quantum yield. The derivative **109** underwent a cyclization reaction upon UV irradiation (ϕ_{oc} = 0.25) and the closed-ring isomer gave a fairly high fluorescence at 580 nm (ϕ_f = 0.11). The closed-ring isomer returned to the open-ring isomer upon irradiation with visible light (ϕ_{co} = 0.17). The high fluorescence contrast and excellent reversibility makes the compound suitable for live cell imaging. In addition, the compound showed no toxicity to the cells even in the millimolar concentration range.

3.5 Chiral Property

As shown in Scheme 2.4, the open-ring isomer has two conformations, parallel and antiparallel ones, and the conrotatory photocyclization reaction takes place only from the antiparallel conformer. The open-ring antiparallel conformer has C_2 symmetry. The C_2 symmetry conformer can be further subdivided into a pair of enantiomers with *M*- and *P*-helicity. It is worth noting that in solution these two antiparallel enantiomers undergo rapid racemization reactions through fast rotation of the aryl groups. Photochemical conrotatory cyclization reactions of the *P*-helical (right-handed) and *M*-helical (left-handed) conformers yield two enantiomers of closed-ring isomers with (*R*,*R*) and (*S*,*S*) configuration, respectively, as shown in Scheme 3.5. The photogenerated equal amounts of closed-ring enantiomers are stable and can be separated using chiral HPLC. It is of interest from both fundamental as well as application points of view to develop diarylethene derivatives

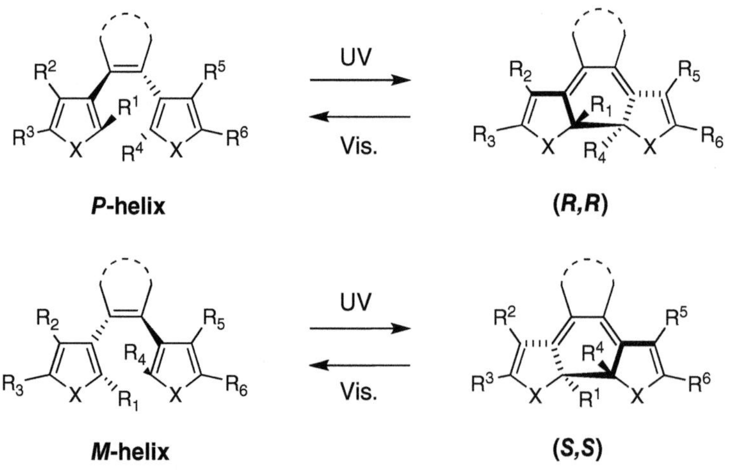

Scheme 3.5 Photocyclization reactions of ***P***- and ***M***-helical open-ring conformers yield (*R*,*R*) and (*S*,*S*) closed-ring enantiomers, respectively.

which undergo stereoselective photoisomerization reactions to induce enrichment of one of the enantiomers or diastereomers of the closed-ring isomers. When such diarylethene derivatives are used for optical memory media, the photochemically generated chirality can be read out by optical rotation at the wavelength, where both isomers have no absorption, without destruction.

One of the convenient approaches to induce the diastereo-differentiation in a photocyclization reaction is to introduce optically active substituents to diarylethenes. In general, however, the photocyclization of diarylethenes in solution results in the formation of the two diastereomers in equal amounts. Special conditions are required to induce diastereo-differentiation. The photochemical asymmetric induction was examined for diarylmaleimides having an optically active *l*- or *d*-menthyl group at the 2-position of benzothiophene ring (**143o** and **144o**), as shown in Scheme 3.6 [91]. Although no asymmetric induction was observed in *n*-hexane, the diastereoselective photocyclization reaction took place in slightly polar (or polarizable) solvents, such as in toluene or hexane/THF mixture. In toluene at −40 °C, diastereomeric excess (*de*) as high as 86.6% was observed. To reveal the asymmetric induction mechanism, CD spectra were measured in *n*-hexane, *n*-hexane/THF mixture, and toluene. No detectable difference in the CD spectra was observed. This indicates that in the ground state, the solvent polarity does not affect the conformation of the open-ring isomers. The asymmetric induction is ascribed to the energy difference between *P*- and *M*-helical open-ring isomers in the excited state. One of the conformers is stabilized in slightly polar solvents in the excited state and efficiently converts to the closed-ring isomer, resulting in the enrichment of one of the closed-ring diastereomers.

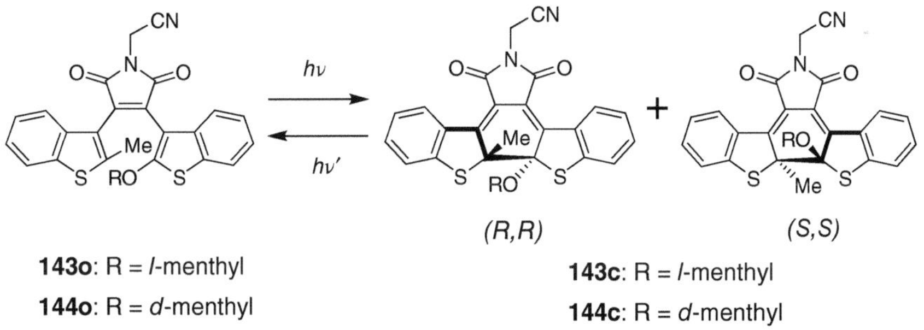

Scheme 3.6 Chiral diarylethenes **143** and **144**.

The diastereoselective photocyclization reaction can take place in crystal. A diarylethene having (*S*)-3-methyl-1-penten-1-yl substituents **145** (Scheme 3.7) was synthesized and crystalized to form chiral crystals [92]. The open-ring isomers **145o** are packed in the crystal in only *P*-helical conformer of the possible two atropisomers. The diarylethene can undergo a cyclization reaction both in solution and in the single crystalline phase. In crystal, only one closed-ring diastereomer was generated, while the diastereoselection was not observed in solution. (*S*)-**145o** underwent the photocyclization reaction to produce diastereomer (*S*,*R*,*R*)-**145c** in crystal. The diastereometric excess (*de*) value was over 95%.

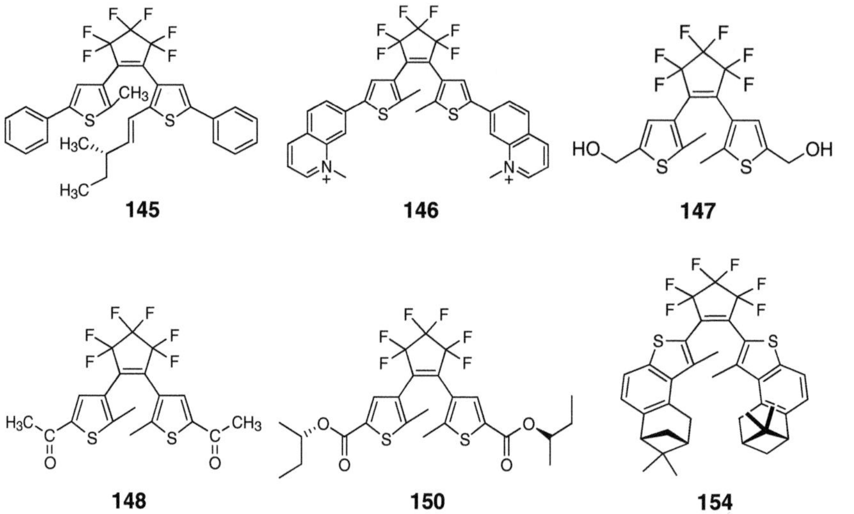

Scheme 3.7 Chiral (**145, 150, 154**) and achiral (**146, 147, 148**) diarylethenes employed for stereoselective photocyclization reactions.

Achiral diarylethene molecules placed in chiral environments undergo enantioselective photocyclization reactions. Andréasson and co-workers [92] found that both open- and closed-ring isomers of methylquinolium-appended diarylethene **146** interact with calf thymus DNA through intercalation and a significant enantioselective photocyclization reaction takes place upon UV irradiation. This enantiomeric enhancement demonstrates direct transfer of chirality from DNA to the photoproduct.

Yokoyama and co-workers [94] reported enantioselective photoisomerizations of diarylethenes incorporated in human serum albumin (HAS). HAS can incorporate a remarkably wide range of chemical species within its hydrophobic pockets. Since HAS consists of optically active amino acids, compounds incorporated into the HAS pockets are surrounded by chiral environments. Diarylethene derivative **147** having hydroxymethyl groups was irradiated with UV (313 nm) light in the presence of HAS in a buffered aqueous solution. When the concentration of HAS was increased, the enantiomeric excess (*ee*) value of the closed-ring isomers increased. A value as high as 63% *ee* was observed for **147** when the HAS/**147o** ratio was 10. The *ee* value dramatically increased with decreasing the temperature and 71% *ee* value was observed at −4 °C. Finally, in order to further increase the enantioselectivity, the addition of acetonitrile to the buffer aqueous solution was examined [95]. The *ee* value of the photogenerated closed-ring isomers of methyl ester substituted diarylethene **148** increased up to 94% in the 15% acetonitrile buffer. When the incubation temperature of the diarylethene in HAS in 15% acetonitrile-buffer solution was decreased to −4 °C, the value was further increased and achieved >99% *ee*. The chirality of HAS was effectively transferred to diarylethenes.

Supramolecular chirality of gel aggregates is able to be transferred to molecular chirality and locked by the photochemical switching of diarylethenes. Feringa and co-workers [95] reported a diastereoselective photocyclization reaction of chiral diarylethene **149** in the gel, as shown in Figure 3.17. The self-assembly property of **149** due to multiple hydrogen-bonding interactions between the amide groups leads it to gel in toluene. The gel showed strong circular dichroism around 320 nm ascribed to exciton coupling within the gel aggregates. Transmission electron microscopy revealed the formation of helical fibers. The photocyclization reactions in the helical fibers produced closed-ring isomers with a large *de* value (>96%). The chirality of the fibers is locked in the closed-ring isomers.

Not only in the gel state but also in a bulk amorphous state the diastereoselective cyclization reaction was observed [97]. Diarylethene **150** having chiral (*S*)-*sec*-butyl groups formed a stable amorphous thin film at room temperature. The closed-ring isomers generated upon UV irradiation in the amorphous state showed 25% *ee*. The result suggests that in the amorphous state there exist microcrystals or short-distance regularity and the chiral environment enriches one of the helical conformers of the open-ring isomers.

Diastereo- and enantioselective reactions described above are strongly dependent on environmental conditions. For applications to optical memory media and optical switches, it is required to develop asymmetric induction systems, which are not influenced by environmental conditions. Yokoyama and co-workers [98] employed facial chirality to achieve ultimate diastereoselectivity (100% *de*) in the photocyclization reaction of diarylethene **151**. The facial chiral diarylethene does not have any asymmetric carbon atoms. When one of the two surfaces of the thiophene ring (ring A) is occupied by bulky triethylene glycol, photocyclization can take place only when another thiophene ring (ring B) approaches from the back side of the ring A, as shown in Figure 3.18. The facial chirality resolves the antiparallel open-ring isomers to equal amounts of two enantiomers. When one of enantiomers was excited by irradiation with UV light, the open-ring isomer underwent photocyclization reaction to produce exclusively one diastereomer of the closed-ring isomer. The diastereoselectivity was 100% in *n*-hexane, toluene, and ethyl acetate. Diarylethene **151** undergoes ultimate diastereoselective photocyclization irrespective of environmental conditions.

Takeshita and co-workers [99] prepared optically active thiophenophan-1-enes **152** and **153** to achieve highly stereoselective photocyclization reactions. Both **152** and **153** have a dithienylethene unit and an additional chiral bridge between the two thiophene rings. The short chain prevents the rotation of the thiophene rings (diastereomerization). Upon irradiation with UV light, the open-ring isomers of **152o** and **153o** underwent photocyclization reactions to produce chiral closed-ring isomers, as shown in Scheme 3.8. The photogenerated closed-ring isomers **152c** and **153c** are enantiomers. The cyclization reactions may be called "diastereospecific photoisomerization reactions."

Branda and co-workers [100] prepared diarylethene **154**, which has chiral substituents in the benzothiophene aryl groups. The molecule also underwent a stereoselective photocyclization reaction to produce only a single closed-ring stereoisomer. The closed-ring isomer showed very large specific rotation at 422 nm ($[\alpha] = 5725°$)

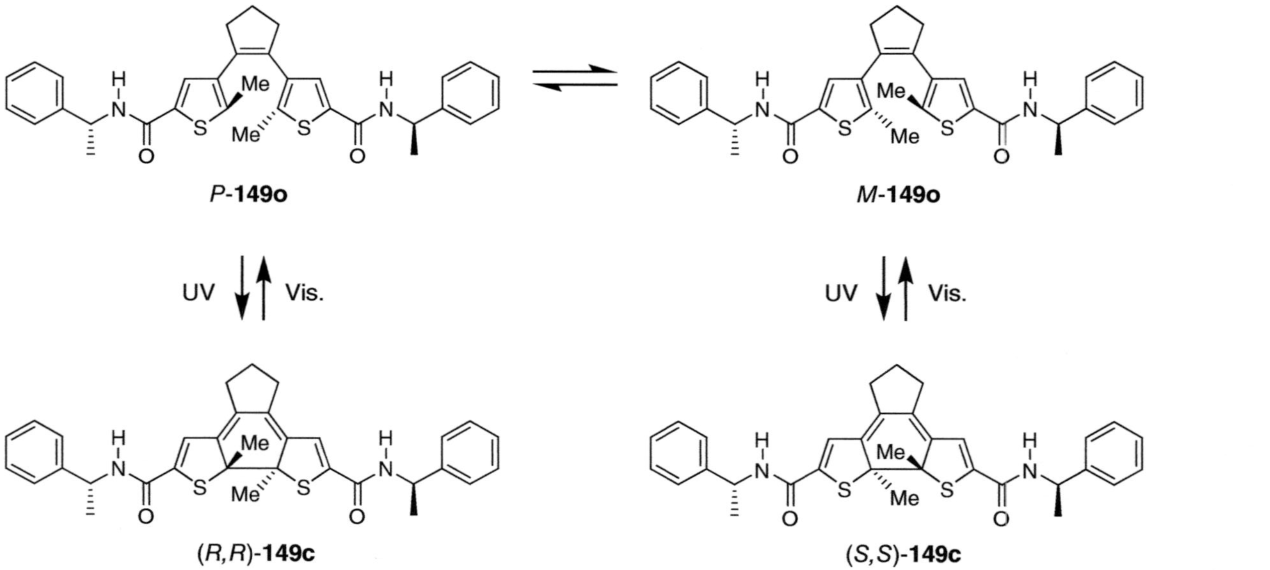

Figure 3.17 Chiral switch **149**, which has (*R*)-1-phenylethylamine-derived amide side groups, exists in the open form in two helical conformations *P*-**149o** and *M*-**149o** in rapid equilibrium. These can be transformed by irradiation into closed forms **149c**. (*R*,*R*)-**149c** can only be obtained from *P*-**149o**, and (*S*,*S*)-**149c** is exclusively formed from *M*-**149o**.

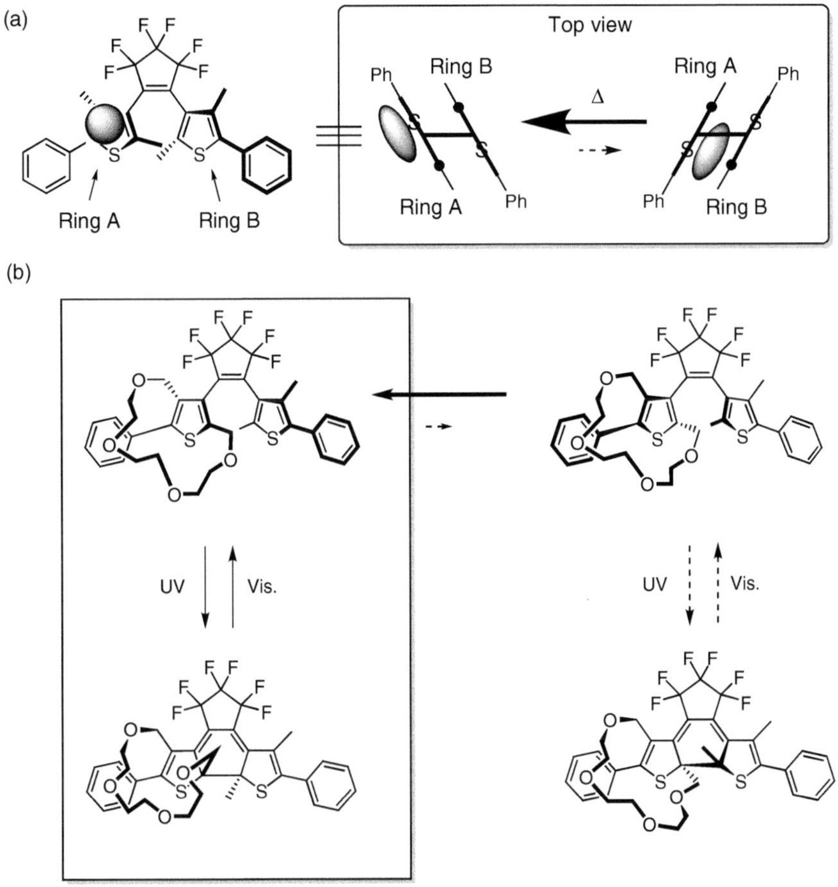

151

Figure 3.18 Molecular modeling concept and the molecular design of chiral diarylethene **151**. *Source*: Adapted from Ref. [98] with permission from The Royal Society of Chemistry.

and the most significant difference between the signal for the PSS and that for the open-ring form was observed at 373 nm as large as $\Delta[\alpha] = 8698°$.

In most cases, aryl groups are chemically modified to restrict their rotation. The rotation of aryl groups can also be restricted by employing an appropriate ring structure as the ethene bridge. When the ethene bridge is bulky, such as phenanthrene or benzobis(thiadiazole), the rotation of the aryl groups is strongly hindered and the open-ring isomers are resolved to *M*- and *P*-helix enantiomers, as shown in Scheme 3.1 [32, 33]. The interconversion (or racemization) of the two enantiomers is controlled by the steric hindrance between the ethene bridge and the aryl groups. The stabilized enantiomers undergo the photocyclization reactions to produce two enantiomers of closed-ring isomers with (*R,R*) and (*S,S*) configuration originating from the central two asymmetric carbon atoms. Zhu and co-workers [33] isolated all the five sterically hindered and thermally stable isomers of **60** and

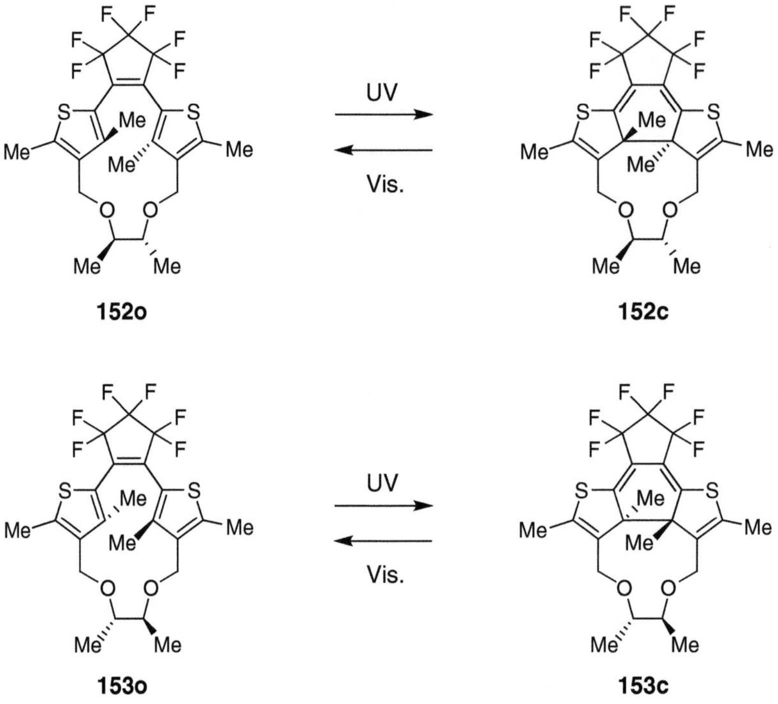

Scheme 3.8 Diastereospecific photoisomerization reactions of **152** and **153**.

fully confirmed the geometrical structures by X-ray crystallographic analysis. They are one parallel conformer $\mathbf{60}_{\mathbf{parallel}}$, one pair of antiparallel open-ring enantiomers ***P*-ap-60o** and ***M*-ap-60o,** and another pair of closed-ring enantiomers **(*R,R*)-60c** and **(*S,S*)-60c**, as shown in Scheme 3.1. These separated open-ring isomers showed absolute enantiospecific photocyclization and photocycloreversion reactions in both solution and polymer matrices. The two pairs of enantiomers ***P*-ap-60o** and ***M*-ap-60o** were thermally stable enough without showing any racemization. The compound **60** offers a unimolecular enantiospecific platform for potential applications as bistable chiroptical switches and all-photonic optical memory media with optical rotation as nondestructive readout signals.

References

1 Kobatake, S., Yamada, T., Uchida, K. et al. (1999). Photochromism of 1,2-bis(2,5-dimethyl-3-thienyl)perfluorocyclopentene in a single crystalline phase. *J. Am. Chem. Soc.* 121 (11): 2380–2386.

2 Shibata, K., Muto, K., Kobatake, S., and Irie, M. (2002). Photocyclization/cycloreversion quantum yields of diarylethenes in single crystals. *J. Phys. Chem. A* 106 (1): 209–214.

3 Irie, M., Sakemura, K., Okinaka, M., and Uchida, K. (1995). Photochromism of dithienylethenes with electron-donating substituents. *J. Org. Chem.* 60 (25): 8305–8309.

4 Pu, S., Yan, L., Wen, Z. et al. (2008). Synthesis and chlorine atom position effect on the properties of unsymmetrical photochromic diarylethenes. *J. Photochem. Photobiol. A: Chem.* 196: 84–93.

5 Sumi, T., Takagi, Y., Yagi, A. et al. (2014). Photoirradiation wavelength dependence of cycloreversion quantum yields of diarylethenes. *Chem. Commun.* 50: 3928–3930.

6 Irie, M., Lifka, T., Kobatake, S., and Kato, N. (2000). Photochromism of 1,2-bis(2-methyl-5-phenyl-3-thienyl)perfluorocyclopentene in a single-crystalline phase. *J. Am. Chem. Soc.* 122 (20): 4871–4876.

7 Uchida, K., Tsuchida, E., Aoi, Y. et al. (1999). Substituent effect on the coloration quantum yield of a photochromic bisbenzothienylethene. *Chem. Lett.* 28: 63–64.

8 Yamaguchi, T., Uchida, K., and Irie, M. (2007). Photochromism of 1,2-bis(3-*n*-alkyl-1-benzothiophen-2-yl)perfluorocyclopentene. *Mol. Cryst. Liq. Cryst.* 474: 111–118.

9 Takami, S., Kawai, T., and Irie, M. (2002). Photochromism of dithiazolylethenes having methoxy groups at reaction centers. *Eur. J. Org. Chem.*: 3796–3800.

10 Liu, G., Pu, S., and Wang, X. (2010). Photochromism of new 3,5-position hybrid diarylethene derivatives bearing both thiophene and thiazole moieties. *Tetrahedron* 66: 8862–8871.

11 Yamaguchi, T. and Irie, M. (2006). Photochromic and fluorescent properties of bisfurylethene derivatives. *J. Mater. Chem.* 16: 4690–4694.

12 Yamaguchi, T. and Irie, M. (2005). Photochromism of bis(2-alkyl-1-benzofuran-3-yl)perfluorocyclopentene derivatives. *J. Org. Chem.* 70 (25): 10323–10328.

13 Yamaguchi, T., Hosaka, M., Ozeki, M. et al. (2011). Photochromism of diarylethene derivatives bearing a benzo[*b*]silole unit. *Tetrahedron Lett.* 52: 5601–5604.

14 Yamaguchi, T. and Irie, M. (2006). Photochromism of diarylethene derivatives having *n*-alkylbenzothiophene and *n*-alkylbenzofuran units. *Bull. Chem. Soc. Jpn.* 79 (7): 1100–1105.

15 Yamaguchi, T., Uchida, K., and Irie, M. (2008). Photochromic properties of diarylethene derivatives benzofuran and benzothiophene rings based on regioisomers. *Bull. Chem. Soc. Jpn.* 81 (5): 644–652.

16 Liu, G., Pu, S., Wang, X. et al. (2011). Effects of substitution on the optoelectronic properties of photochromic diarylethenes bearing a pyrrole moiety. *Dyes Pig.* 90: 71–81.

17 Liu, G., Pu, S., Wang, X. et al. (2011). The effect of the cyano group position upon unsymmetrical isomeric diarylethenes bearing a pyrrole unit. *Dyes Pig.* 90: 89–99.

18 Pu, S., Fan, C., Miao, W., and Liu, G. (2008). New photochromic diarylethenes with a six-membered aryl unit. *Tetrahedron* 64: 9464–9470.

19 Pu, S., Miao, W., Cui, S. et al. (2010). The synthesis of novel photochromic diarylethenes bearing a biphenyl moiety and the effects of substitution on their properties. *Dyes Pig.* 87: 257–267.

20 Wang, R., Pu, S., Liu, G. et al. (2011). New photochromic diarylethenes bearing a condensed aromatics moiety. *Tetrahedron Lett.* 52: 3306–3310.

21 Pu, S., Yan, P., Liu, G. et al. (2011). New photochromic diarylethenes bearing a pyridine moiety. *Tetrahedron Lett.* 52: 143–147.

22 Pu, S., Yang, T., Xu, J., and Chen, B. (2006). Syntheses and properties of new photochromic diarylethene derivatives having a pyrazole unit. *Tetrahedron Lett.* 47: 6473–6477.

23 Pu, S., Li, H., Liu, G. et al. (2011). Synthesis and the effects of substitution upon photochromic diarylethenes bearing an isoxazole moiety. *Tetrahedron* 67: 1438–1447.

24 Shibata, K., Kuroki, L., Fukaminato, T., and Irie, M. (2008). Fluorescence switching of a diarylethene derivative having oxazole rings. *Chem. Let.* 37 (8): 832–833.

25 Yamaguchi, T., Taniguchi, W., Ozeki, T. et al. (2009). Photochromism of diarylethene oxazole derivatives in a single-crystalline phase. *J. Photochem. Photobiol. A: Chem.* 207: 282–287.

26 Takeshita, M., Kato, N., Kawauchi, S. et al. (1998). Photochromism of diarylethenes included in cyclodextrins. *J. Org. Chem.* 63 (25): 9306–9313.

27 Matsuda, K., Shinkai, Y., Yamaguchi, T. et al. (2003). Very high cyclization quantum yields of diarylethene having two *N*-methylpyridinium ions. *Chem. Lett.* 32 (12): 1178–1179.

28 Yamaguchi, T. and Irie, M. (2006). Photochromism of bis(2-alkyl-1-benzothiophen 3-yl)perfluoroethene derivatives. *J. Photochem. Photobiol. A: Chem.* 178: 162–169.

29 Fukumoto, S., Nakashima, T., and Kawai, T. (2011). Photon-quantitative reaction of a dithiazolylarylene in solution. *Angew. Chem. Int. Ed.* 50: 1565–1568.

30 Li, R., Nakashima, T., Galangau, O. et al. (2015). Photon-quantitative 6π-electrocyclization of a diarylbenzo[*b*]thiophene in polar medium. *Chem. Asian J.* 10: 1725–1730.

31 Göstl, R., Kobin, B., Grubert, L. et al. (2012). Sterically crowding the bridge of dithienylcyclopentenes for enhanced photoswitching performance. *Chem. Eur. J.* 18: 14282–14285.

32 Li, W., Jiao, C., Li, X. et al. (2014). Separation of photoactive conformers based on hindered diarylethenes: Efficient modulation in photocyclization quantum yields. *Angew. Chem. Int. Ed.* 53: 4603–4607.

33 Li, W., Li, X., Xie, Y. et al. Enantiospecific photoresponse of sterically hindered diarylethenes for chiroptical switches and photomemories. *Sci. Rep.* 5: 9186.

34 Irie, M. and Sayo, K. (1992). Solvent effects on the photochromic reactions of diarylethene derivatives. *J. Phys. Chem.* 96 (19): 7671–7674.

35 Ohsumi, M., Hazama, M., Fukaminato, T., and Irie, M. (2008). Photocyclization reaction of a diarylmaleimide derivative in polar solvents. *Chem. Commun.*: 3281–3283.

36 Thurn, J., Maier, J., Pärs, M. et al. (2017). Temperature dependence of the conversion efficiency of photochromic perylene bisimide dithienylcyclpentene triads embedded in a polymer. *Phys. Chem. Chem. Phys.* 19: 26065–26071.

37 Shibata, K., Kobatake, S., and Irie, M. (2001). Extraordinarily low cycloreversion quantum yields of photochromic diarylethenes with methoxy substituents. *Chem. Lett.*: 618–619.

38 Morimitsu, K., Shibata, K., Kobatake, S., and Irie, M. Dithienylethenes with a novel photochromic performance. *J. Org. Chem.* 67 (13): 4574–4578.

39 Morimitsu, K., Kobatake, S., Nakamura, S., and Irie, M. (2003). Efficient photocycloreversion reaction of diarylethenes by introduction of cyano substituents to the reactive carbons. *Chem. Lett.* 32 (9): 858–859.

40 Nakamura, S., Kobayashi, T., Takata, A. et al. (2007). Quantum yields and potential energy surfaces: a theoretical study. *J. Phys. Org. Chem.* 20: 821–829.

41 Irie, M. and Mohri, M. (1988). Thermally irreversible photochromic systems. Reversible photocyclization of diarylethene derivatives. *J. Org. Chem.* 53 (4): 803–808.

42 Nakamura, S. and Irie, M. (1988). Thermally irreversible photochromic systems. A theoretical study. *J. Org. Chem.* 53 (26): 6136–6138.

43 Takami, S., Kobatake, S., Kawai, T., and Irie, M. (2003). Extraordinarily high thermal stability of the closed-ring isomer of 1,2-bis(5-methyl-2-phenylthiazol-4-yl)perfluorocyclopentene. *Chem. Lett.* 32 (10): 892–893.

44 Gilat, S.L., Kawai, S.H., and Lehn, J.-M. (1995). Light-triggered molecular devices: photochemical switching of optical and electrochemical properties in molecular wire type diarylethene species. *Chem. Eur. J.* 1 (5): 275–284.

45 Shoji, H., Kitagawa, D., and Kobatake, S. (2014). Alkyl substituents effects in photochemical and thermal reactions of photochromic thiophene-*S*,*S*-dioxidized diarylethenes. *New J. Chem.* 38: 933–941.

46 Kitagawa, D., Sasaki, K., and Kobatake, S. (2011). Correlation between steric substituents constants and thermal cycloreversion reactivity of diarylethene closed-ring isomers. *Bull. Chem. Soc. Jpn.* 84 (2): 141–147.

47 Kobatake, S., Shibata, K., Uchida, K., and Irie, M. (2000). Photochromism of 1,2-bis(2-ethyl-5-phenyl-3-thienyl)perfluorocyclopentene in a single-crystalline phase. Conrotatory thermal cycloreversion of the closed-ring isomer. *J. Am. Chem. Soc.* 122 (49): 12135–12141.

48 Inaba, K., Iwai, R., Morimoto, M., and Irie, M. (2019). Thermally reversible photochromism of dipyrrolylethenes. *Photochem. Photobiol. Sci.* 18: 2136–2141.

49 Hanazawa, M., Sumiya, R., Horikawa, Y., and Irie, M. (1992). Thermally irreversible photochromic systems. Reversible photocyclization of 1,2-bis(2-methylbenzo[*b*]thiophene-3-yl)perfluorocycloalkane derivatives. *J. Chem. Soc., Chem. Commun.* 3: 206–207.

50 Jean-Ruel, H., Cooney, R.R., Gao, M. et al. (2011). Femtosecond dynamics of ring closing process diarylethene: a case study of electrocyclic reactions in photochromic single crystals. *J. Phys. Chem. A* 115: 13158–13168.

51 Higashiguchi, K., Matsuda, K., Kobatake, S. et al. (2000). Fatigue mechanism of photochromic 1,2-bis(2,5-dimethyl-3-thienyl)perfluorocyclopentene. *Bull. Chem. Soc. Jpn.* 73 (10): 2389–2394.

52 Irie, M., Lifka, T., Uchida, K. et al. (1999). Fatigue resistant properties of photochromic dithienylethenes: by-product formation. *Chem. Commun.*: 747–750.

53 Neckers, D.C. and DeZwaan, J. (1969). Photochemical reactions of cyclic sulphides. *Chem. Commun.*: 813–814.

54 Jeong, Y.-C., Yang, S.I., Kim, E., and Ahn, K.-H. (2006). Development of highly fluorescent photochromic material with high fatigue resistance. *Tetrahedron* 62: 5855–5861.

55 Jeong, Y.-C., Park, D.G., Lee, I.S. et al. (2009). Highly fluorescent photochromic diarylethene with an excellent fatigue property. *J. Mater. Chem.* 19: 97–103.

56 Jeong, Y.-C., Gao, C., Lee, G.I. et al. (2009). The considerable photostability improvement of photochromic terarylene by sulfone group. *Tetrahedron Lett.* 50: 5288–5290.

57 Patel, P.D., Mikhailov, I.A., Belfield, K.D., and Masunov, A.E. (2009). Theoretical study of photochromic compounds, part 2: thermal mechanism for byproduct formation and fatigue resistance of diarylethenes used as data storage materials. *Int. J. Quantum Chem.* 109 (815): 3711–3722.

58 Mendive-Tapia, D., Perrier, A., Bearpark, M.J. et al. (2014). New insights into the by-product fatigue mechanism of the photo-induced ring-opening in diarylethenes. *Phys. Chem. Chem. Phys.* 16: 18463–18471.

59 Herder, M., Schmidt, B.M., Grubert, L. et al. (2015). Improving the fatigue resistance of diarylethene switches. *J. Am. Chem. Soc.* 137: 2738–2747.

60 Pariani, G., Quintavalla, M., Colella, L. et al. (2017). New insight into the fatigue resistance of photochromic 1,2-diarylethenes. *J. Phys. Chem. C* 121: 23592–23598.

61 Shim, S., Joo, T., Bae, S.C. et al. (2003). Ring opening dynamics of a photochromic diarylethene derivative in solution. *J. Phys. Chem. A* 107 (40): 8106–8110.

62 Shim, S., Eom, I., Joo, T. et al. (2007). Ring closure reaction dynamics of diarylethene derivatives in solution. *J. Phys. Chem. A* 111 (37): 8910–8917.

63 Kim, M.-S., Kawai, T., and Irie, M. (2002). Fluorescence modulation in photochromic amorphous diarylethenes. *Opt. Mater.* 21: 271–274.

64 Yagi, K. and Irie, M. (2003). Fluorescence property of photochromic diarylethenes with indole groups. *Bull. Chem. Soc. Jpn.* 76 (8): 1625–1628.

65 Pu, S., Liu, G., Shen, L., and Xu, J. (2007). Efficient synthesis and properties of isomeric photochromic diarylethenes having a pyrrole unit. *J. Org. Lett.* 9 (11): 2139–2142.

66 Xie, N. and Chen, Y. (2006). Combination of fluorescent switch and electrochemical switch based on a photochromic diarylethene. *New J. Chem.* 30: 1595–1598.

67 Fukaminato, T., Kawai, T., Kobatake, S., and Irie, M. (2003). Fluorescence of photochromic 1,2-bis(3-methyl-2-thienyl)ethene. *J. Phys. Chem. B* 107 (33): 8372–8377.

68 Girard, M., Léaustic, A., Charlot, M.-F. et al. (2005). Synthesis and photochromism of two new 1,2-bis(thiazolyl)perfluorocyclopentene with chelating sites. *New J. Chem.* 29: 439–446.

69 Cipolloni, M., Ortica, F., Bougdid, L. et al. (2008). New thermally irreversible and fluorescent photochromic diarylethenes. *J. Phys. Chem. A* 112 (21): 4765–4771.

70 Liu, M. and Chen, Y. (2009). The photochromism and fluorescence of diarylethenes with an imidazole bridge unit: a strategy for the design of turn-on fluorescent diarylethene system. *J. Phys. Chem. A* 113 (19): 5550–5553.

71 Ai, Q., Pang, S., and Ahn, K.-H. (2016). Photoswitchable "turn-on" fluorescent diarylethenes: substituent effects on photochemical properties and electrochromism. *Chem. Eur. J.* 22: 656–662.

72 Ko, C.-C., Lam, W.H., and Yam, V.W.-W. (2008). Photochromic oligothienoacene derivatives with photo-switchable luminescene properties and computational studies. *Chem. Commun.*: 5203–5205.

73 Zhu, W., Meng, X., Yang, Y. et al. (2010). Bisthienylethenes containing a benzothiadiazole unit as a bridge: photochromic performance dependence on substitution position. *Chem. Eur. J.* 16: 899–110.

74 Poon, C.-T., Lam, W.H., Wong, H.-L., and Yam, V.W.-W. (2010). A versatile photochromic dithienylethenes-containing β-diketonate ligand: near-infrared photochromic behavior and photoswitchable luminescence properties upon incorporation of a boron(III) center. *J. Am. Chem. Soc.* 132 (40): 13992–13993.

75 Suzuki, K., Ubukata, T., and Yokoyama, Y. (2012). Dual-mode fluorescence switching of photochromic bisthiazolylcoumarin. *Chem. Commun.* 48: 765–767.

76 Milder, M.T.W. and Herek, J.L. (2009). Tunable aggregation and luminescence of bis(diarylethene)sexithiophenes. *J. Phys. Chem. A* 113 (27): 7717–7724.

77 Uno, K., Niikura, H., Morimoto, M. et al. (2011). In situ preparation of highly fluorescent dyes upon photoirradiation. *J. Am. Chem. Soc.* 133: 13558–13564.

78 Taguchi, M., Nakagawa, T., Nakashima, T., and Kawai, T. (2011). Photochromic and fluorescence switching properties of oxidized triangle terarylenes in solution and in amorphous solid states. *J. Mater. Chem.* 21: 17425–17432.

79 Irie, M., Fukaminato, T., Sasaki, T. et al. (2001). A digital fluorescent molecular photoswitch. *Nature* 420: 759–760.

80 Li, C., Yan, H., Zhao, L.-X. et al. (2014). A trident dithienylethenes-perylenemonoimide dyad with super fluorescent switching speed and ratio. *Nat. Commun.* 5: 5709.

81 Biellmann, T., Galanti, A., Boixel, J. et al. (2018). Fluorescence commutation and surface photopatterning with porphyrin tetradithienylethene switches. *Chem. Eur. J.* 24: 1631–1639.

82 (a) Perrin, F. (1924). *CR Acad. Sci.* 178: 1978. (b) Turro, N.J. (1978). *Modern Molecular Photochemistry*, p. 317. Menlo Park, CA: Benjamin/Cummings Pub.

83 Su, J., Fukaminato, T., Placial, J.-P. et al. (2016). Giant amplification of photoswitching by a few photons in fluorescent photochromic organic nanoparticles. *Angew. Chem. Int. Ed.* 55: 3662–3666.

84 Lim, S.-J., An, B.-K., Jung, S.D. et al. (2004). Photoswitchable organic nanoparticles and polymer film employing multifunctional molecules with enhanced fluorescence emission and bistable photochromism. *Angew. Chem. Int. Ed.* 43: 6346–6350.

85 Li, C., Gong, W.-L., Hu, Z. et al. (2013). Photoswitchable aggregation-induced emission of a dithienylethenes-tetraphenylene conjugate for optical memory and super-resolution imaging. *RSC Adv.* 3: 8967–8972.

86 Irie, M. and Morimoto, M. (2018). Photoswitchable turn-on mode fluorescent diarylethenes: strategies for controlling the switching response. *Bull. Chem. Soc. Jpn.* 91 (2): 237–250.

87 Takagi, Y., Kunishi, T., Katayama, T. et al. (2012). Photoswitchable fluorescent diarylethene derivatives with short alkyl chain substituents. *Photochem. Photobiol. Sci.* 11: 1661–1665.

88 Takagi, Y., Morimoto, M., Kashihara, R. et al. (2017). Turn-on mode fluorescent diarylethenes: control of the cycloreversion quantum yield. *Tetrahedron* 73: 4918–4924.

89 Morimoto, M., Sumi, T., and Irie, M. (2017). Photoswitchable fluorescent diarylethene derivatives having thiophene 1,1-dioxide groups: effect of alkyl substituents at the reactive carbons. *Materials* 10: 1021–1028.

90 Pang, S.-C., Hyun, H., Lee, S. et al. (2012). Photoswitchable fluorescent diarylethene in a *turn-on* mode for live cell imaging. *Chem. Commun.* 48: 3745–3747.

91 Yamaguchi, T., Uchida, K., and Irie, M. (1997). Asymmetric photocyclization of diarylethene derivatives. *J. Am. Chem. Soc.* 119 (26): 6066–6071.

92 Kodani, T., Matsuda, K., Yamada, T. et al. (2000). Reversible diastereoselective photocyclization of a diarylethene in a single-crystalline phase. *J. Am. Chem. Soc.* 122 (40): 9631–9637.

93 Pace, T.C.S., Müller, V., Li, S. et al. (2013). Enantioselective cyclization of photochromic dithienylethenes bound to DNA. *Angew. Chem. Int. Ed.* 52: 4393–4396.

94 Fukagawa, M., Kawamura, I., Ubukata, T., and Yokoyama, Y. (2013). Enantioselective Photochromism of diarylethenes in human serum albumin. *Chem. Eur. J.* 19: 9434–9437.

95 Kawamura, K., Osawa, K., Watanobe, Y. et al. (2017). Photocyclization of photoswitches with high enantioselectivity inhuman serum albumin in an artificial environment. *Chem. Commun.* 53: 3181–3184.

96 de Jong, J.J.D., Lucas, L.N., Kellogg, R.M. et al. (2004). Reversible optical transcription of supramolecular chirality into molecular chirality. *Science* 304: 278–281.

97 Yamaguchi, T., Nomiyama, K., Isayama, M., and Irie, M. (2004). Reversible diastereoselective photocyclization of diarylethenes in a bulk amorphous state. *Adv. Mater.* 16 (7): 643–645.

98 Shiozawa, T., Hossain, M.K., Ubukata, T., and Yokoyama, Y. (2010). Ultimate diastereoselectivity in the ring closure of photochromic diarylethene possessing facial chirality. *Chem. Commun.* 46: 4785–4787.

99 Takeshita, M. and Jin-nouchi, H. (2010). Fully diastereospecific photochromic reaction of a thiophenophan-1-ene. *Chem. Commun.* 46: 3994–3995.

100 Wigglesworth, T.J., Sud, D., Norsten, T.B. et al. (2005). Chiral discrimination in photochromic helicenes. *J. Am. Chem. Soc.* 127 (20): 7272–7273.

4
Photoswitchable Crystals

In general, compounds that exhibit reversible photoswitching reactions in the crystalline phase are very rare. Only a limited number of compounds, such as paracyclophanes [1], *N*-salicylideneanilines [2], aziridines [3], triarylimidazole dimers [4], diphenylmaleonitriles [5], triazines [6], and 2-(2′,4′-dinitrobenzyl) pyridine [7], are reported to undergo the reactions in the crystalline phase. The photoswitching of diarylethene crystals was serendipitously discovered during the course of synthesis of diarylethene derivatives having thiophene oligomers **156**, as shown in Figure 4.1 [8]. The diiodo-derivative **155** was prepared to introduce oligothiophenes in the side groups. Microcrystals of **155** were accidentally found to change the color from colorless to red under irradiation with UV light, which is used daily for thin-layer chromatography. At first, iodine released by the action of UV light was thought to be the origin of the red color. But, the assumption was proved not to be true, because microcrystals of **22**, which has no iodine, also exhibited the change from colorless to red under UV light. This is the course of finding of photoswitchable diarylethene crystals.

Upon irradiation with UV light, the colorless crystals change to yellow, red, blue, or green depending on the molecular structures, as shown in Figure 4.2. The coloration takes place not only on the surface layer but also deeply in the crystal bulk. UV light can penetrate a depth of millimeters and induce the photoswitching reactions. The colors of the crystals indicate the formation of the closed-ring isomers. The colors remained stable in the dark, while they disappeared upon irradiation with visible light. Scheme 4.1 shows examples of diarylethene derivatives, which undergo the photoswitching reactions in the crystalline phase. The photoinduced coloration/decoloration cycles of the crystals can be repeated many times, typically more than 10^4 times, as shown in Figure 3.12. The photoswitching reactions in the crystalline phase were explicitly confirmed by the measurement of dichroism and in situ X-ray crystallographic analysis, as described in Sections 4.1 and 4.2.

4.1 Dichroism

A convenient way to confirm the photoswitching reactions in crystal is to observe the dichroism under linearly polarized light. In crystal, molecules are regularly oriented.

Diarylethene Molecular Photoswitches: Concepts and Functionalities, First Edition. Masahiro Irie.
© 2021 WILEY-VCH GmbH. Published 2021 by WILEY-VCH GmbH.

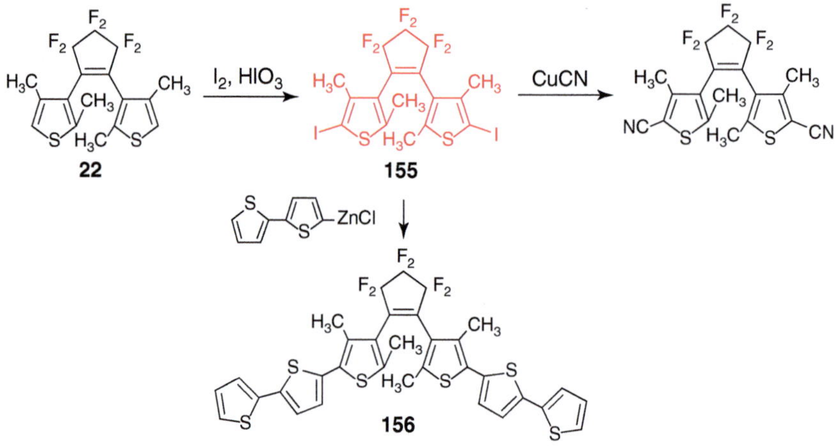

Figure 4.1 A synthetic route of diarylethene **156** having oligothiophene aryl groups. During the course of the synthesis, diarylethene crystal **155** was serendipitously discovered to undergo color change from colorless to red upon irradiation with UV light.

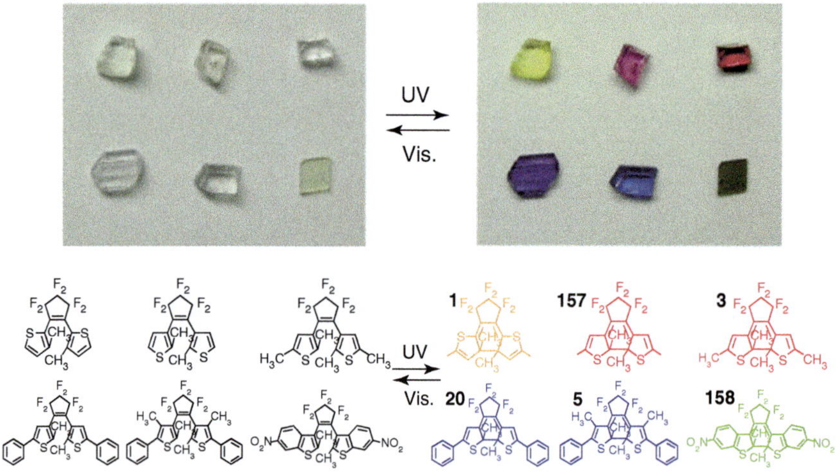

Figure 4.2 Color changes of several diarylethene single crystals upon photoirradiation.

Therefore, when the reactions take place in crystal bulk maintaining the regular orientation, the absorption spectrum exhibits dichroism.

A rhombus-shaped single crystal, as shown in Figure 4.3a, was obtained by recrystallization of diarylethene **3** from a *n*-hexane solution. Figure 4.3c–f shows photos of surface A observed under polarized light before and after UV irradiation [9]. Upon irradiation with 366 nm light, the crystal turned red. At a certain angle ($\theta = 30°$), the crystal showed a deep red color (Figure 4.3e). When the crystal was rotated as much as 90°, the color almost disappeared (Figure 4.3f). Figure 4.4 shows the polarized absorption spectra of the photogenerated closed-ring isomer at 30° and 120° and polar plots of the absorbance at 535 nm. The closed-ring isomer has the electronic

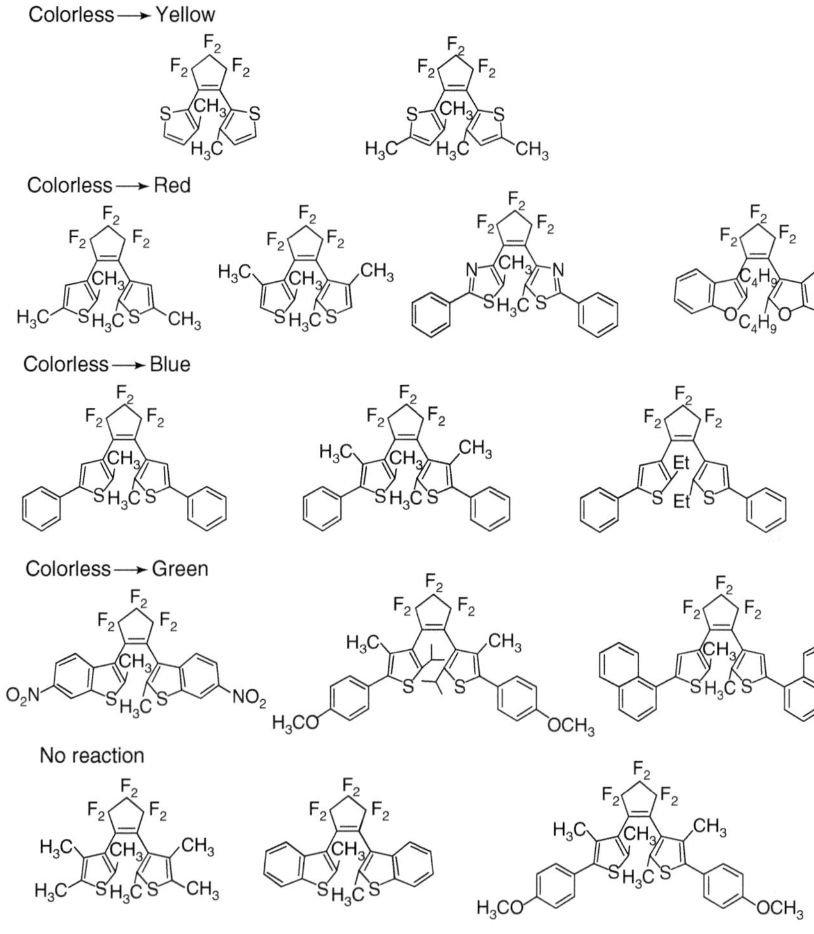

Scheme 4.1 Examples of diarylethene derivatives, which undergo photoswitching reactions in the crystalline phase. The last three diarylethenes did not show any color change upon UV irradiation.

transition moment vector along the *x*-axis (Figure 4.3b). The absorption band at 535 nm is ascribed to the electronic transition. The orientation of the closed-ring isomer can be evaluated by an order parameter, $S = (A_{\parallel} - A_{\perp})/(A_{\parallel} + 2A_{\perp})$. The S value for the single crystal of **3c** was determined to be 0.84 at 535 nm. Such a high order parameter indicates that the diarylethene molecules undergo the cyclization reaction in the regularly oriented crystal lattice.

Two-color dichroism was observed in the crystal of 1,2-bis(2-methyl-6-nitro-1-benzothiophen-3-yl) perfluorocyclopentene (**158**) [10]. The rhombus-shaped crystal turned green upon irradiation with UV light. The green color disappeared upon irradiation with visible light. When the colored crystal was observed under polarized light, the crystal exhibited yellow color at a certain angle (Figure 4.5c). When the crystal was rotated up to 90°, the color turned blue, as shown in Figure 4.5d. The yellow color reappeared at 180°. Figure 4.6b shows the polarized absorption spectra

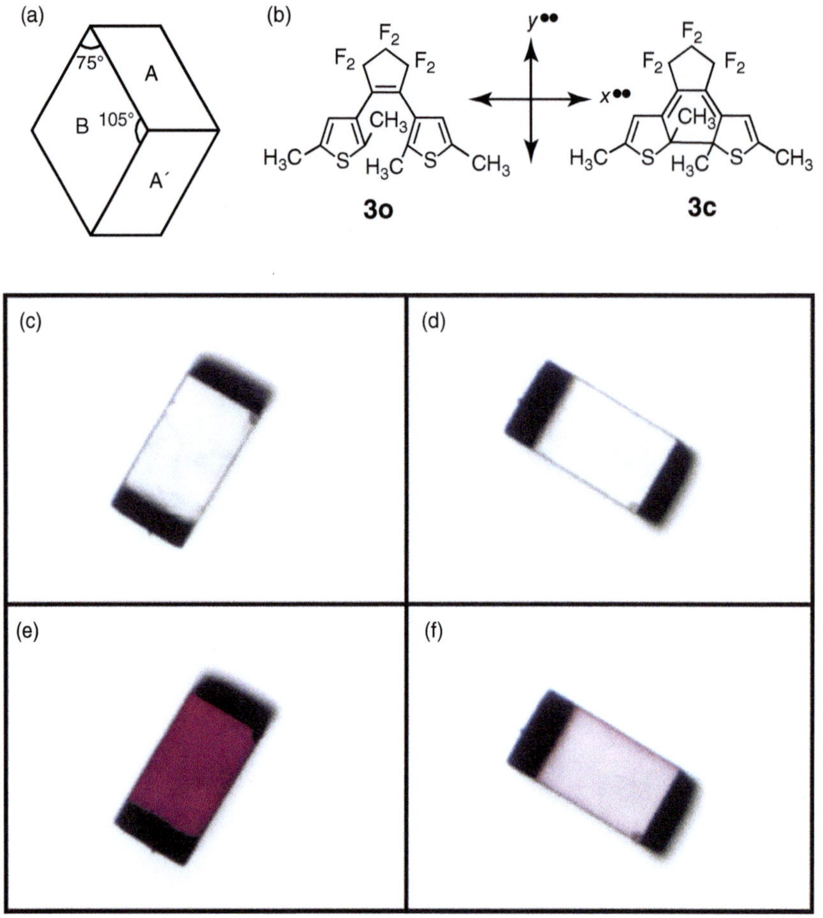

Figure 4.3 (a) Shape of single crystal **3** and (b) directions of long (x) and short (y) axes of the molecule. Photographs of single crystal **3** (face A) under polarized light before ((c) $\theta = 30°$, (d) $\theta = 120°$) and after ((e) $\theta = 30°$, (f) $\theta = 120°$) irradiation with 366 nm light. θ is defined in Figure 4.4.

of the closed-ring isomer at 0° and 90°. The order parameter was calculated to be 0.90 at 600 nm. The high order parameter indicates that the cyclization reaction takes place in the crystal lattice. According to the molecular orbital calculation of the closed-ring isomer, the transition moment vector of the longest absorption band (600 nm) is in the long axis of the closed-ring isomer and the second band (465 nm) is perpendicular to that. The molecular packing viewed from surface A (Figure 4.6c) is shown in Figure 4.6d. When the polarized light coincides with the short axis of the closed-ring isomer, the crystal shows yellow color, while it turns blue when the light coincides with the long axis of the molecule. The clear two-color dichroism indicates that each closed-ring isomer is regularly oriented in the crystal and the photoinduced cyclization reaction takes place in the crystal lattice of the open-ring isomers.

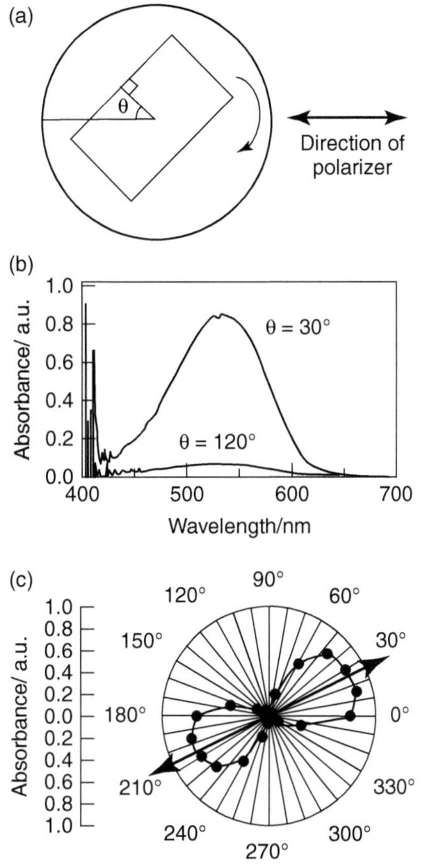

Figure 4.4 Polarized absorption spectra of the red color in single crystal **3**. (a) Direction of polarizer, (b) polarized absorption spectra, and (c) polar plots of the absorption at 535 nm. *Source*: Reprinted with permission from Ref. [9]. Copyright 1999 American Chemical Society.

4.2 X-Ray Crystallographic Analysis

Although the dichroism shown in Section 4.1 provides certain evidence for the photoswitching in the crystalline phase, the evidence is indirect. To unambiguously confirm the photoswitching in crystal, it is indispensable to determine the structures of photogenerated closed-ring isomers by X-ray crystallographic analysis. Both photocyclization and photocycloreversion reactions of diarylethene **3** were followed by in situ X-ray crystallographic analysis.

It is no easy matter to follow the cyclization reaction by X-ray analysis. The conversion ratio from the open-ring isomer **3o** to the closed-ring isomer **3c** upon irradiation with UV light is limited, because photogenerated **3c** also absorbs the UV light and induces the cycloreversion reaction. A conventional way to minimize the inner filter effect is to irradiate the crystal face, which has the least absorption coefficient of **3c** in the UV region, with linearly polarized UV light. Based on this guiding idea, crystal face B (Figure 4.3a) was selected and irradiated with linearly polarized 360 nm light, the polarization plane of which was set parallel to the *c*-axis to minimize the absorption of the closed-ring isomer in the UV region [11].

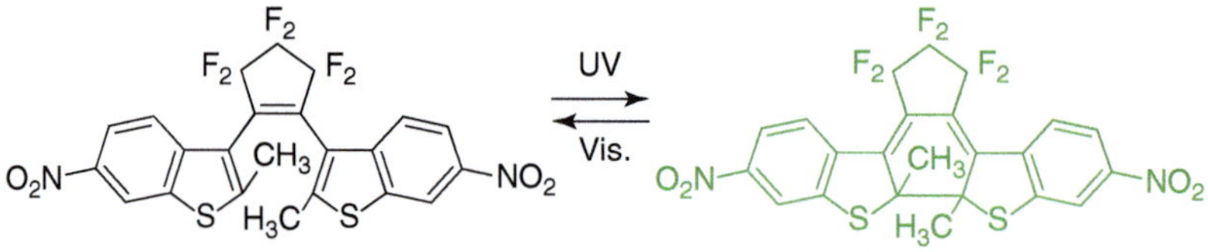

158o **158c**

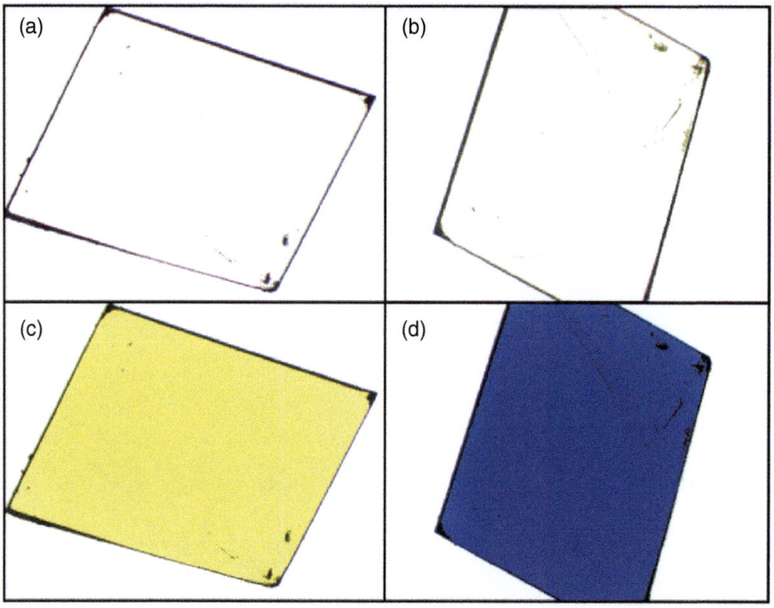

Figure 4.5 Photographs of single crystal **158** under polarized light before ((a) $\theta = 0°$, (b) $\theta = 90°$) and after ((c) $\theta = 0°$, (d) $\theta = 90°$) irradiation with 366 nm light. θ is the rotation angle of the crystal, as shown in Figure 4.6. *Source*: Adapted with permission from Ref. [10]. Copyright 1999 American Chemical Society.

The difference Fourier electron density map for crystal **3** upon UV irradiation is shown in Figure 4.7a. After the first least-square refinement, the difference Fourier electron density map showed the existence of two fairly high electron density peaks Q1 (2.24 e $Å^{-3}$) and Q2 (2.22 e $Å^{-3}$) at the central part of the molecule ascribed to the sulfur atoms of the photogenerated closed-ring isomer. In the open-ring isomer, the sulfur atoms are located at upper and lower parts of the molecule. The atoms moved their positions to the central part by the rotation of thiophene rings in a conrotatory mode. The distance between the two peaks was 3.69 Å, which is slightly longer than the length of the closed-ring isomer in **3c** crystal, 3.56 Å. Electron density peaks corresponding to two carbon atoms at 2- and 2′-positions and six new peaks corresponding to fluorine atoms of the photogenerated **3c** also appeared.

According to these peaks, the structure of the photogenerated closed-ring isomer was constructed and is shown in Figure 4.8 along with that of the open-ring isomer. The occupancy factor for the closed-ring isomer was 0.084(2), which indicates that

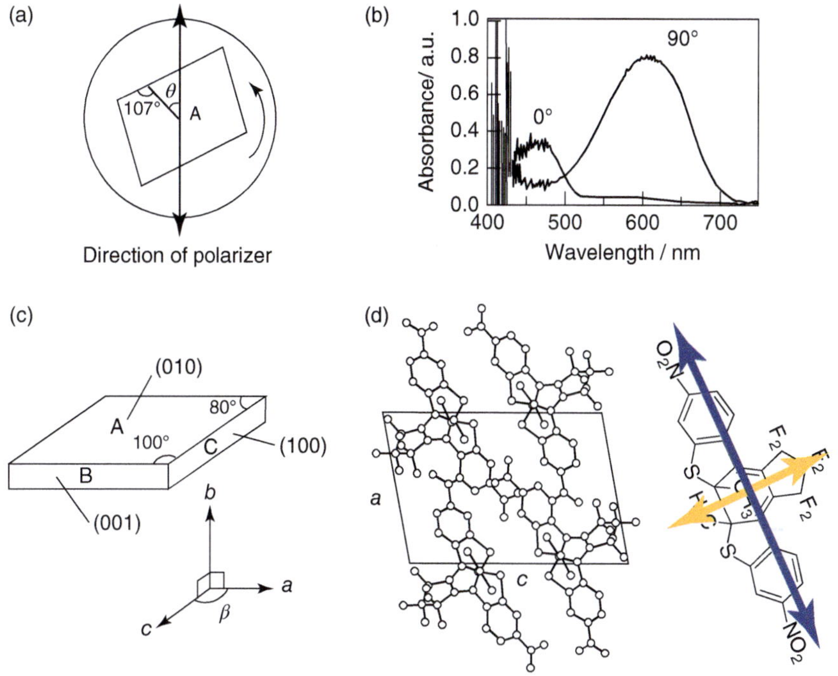

Figure 4.6 Polarized absorption spectra of colored single crystal **158** (surface A). (a) Direction of polarizer, (b) polarized absorption spectra, (c) shape of the crystal, and (d) molecular packing viewed from surface A.

~8% of molecules in the crystal underwent the photocyclization reaction in **3o** crystal.

The molecular structure of the photogenerated closed-ring isomer in **3o** crystal was found not to be the same as that of the closed-ring isomer in **3c** crystal [12]. Figure 4.9a shows the structural difference between the two closed-ring isomers. The central C1—C10 bond lengths of the two isomers were the same, while the distance between S1 and S2 of the closed-ring isomer photogenerated in **3o** crystal was larger than that of the closed-ring isomer in **3c** crystal. The structure of the photogenerated closed-ring isomer is distorted. The structural difference was reflected in the absorption spectrum. The maximum of the photogenerated isomer shifted to longer wavelengths in comparison with that observed in the closed-ring form crystal, as shown in Figure 4.9c. The red shift is ascribed to the strained structure of the closed-ring isomer. The theoretical DFT calculations based on the strained structure confirmed the red shift.

The cycloreversion reaction was also followed by in situ X-ray crystallographic analysis [13]. In the visible wavelength region, only the closed-ring isomer has absorption. Therefore, the cycloreversion reaction can proceed without any perturbation of the photogenerated open-ring isomer. The single crystal of the closed-ring isomer was irradiated with 680 nm light and formation of the open-ring isomer was followed. After the differential Fourier electron density analysis, two new peaks,

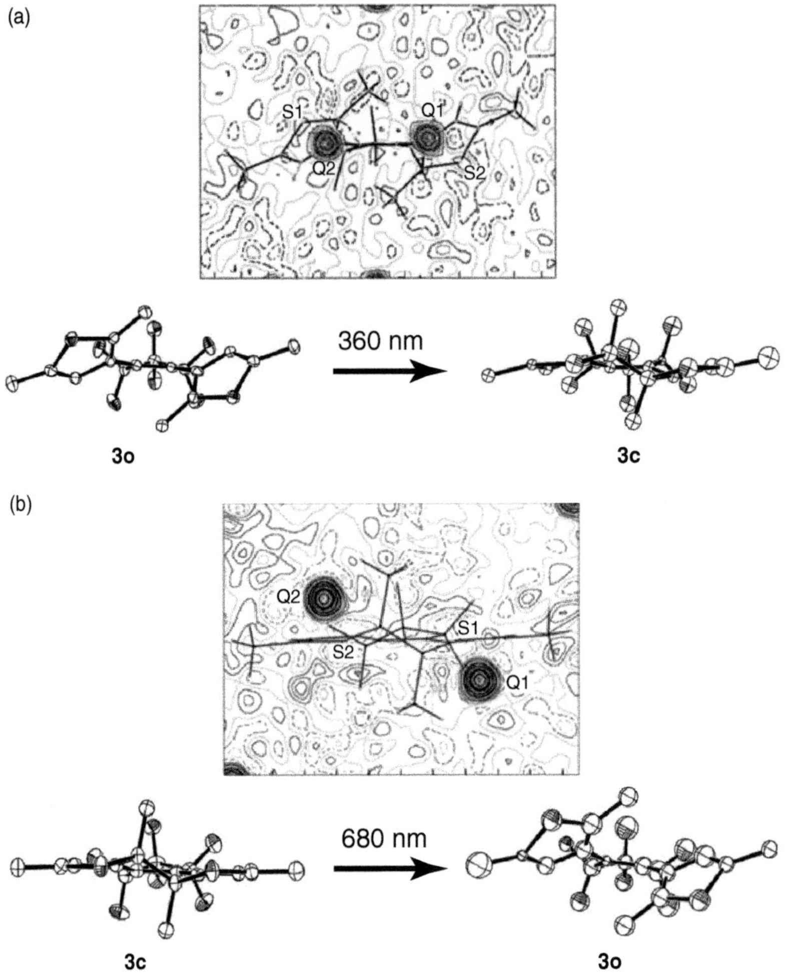

Figure 4.7 $F_o - F_c$ difference Fourier electron density maps for single crystal **3** upon irradiation with (a) UV light and (b) visible light. (a) shows the cyclization reaction from **3o** to **3c**. (b) shows the cycloreversion reaction from **3c** to **3o**.

Q1 and Q2, appeared, as shown in Figure 4.7b. At the same time, negative counter regions were observed at S1 and S2 of the closed-ring isomer. The result indicates that the S1 and S2 atoms of the closed-ring isomer move to Q1 and Q2 upon irradiation with 680 nm light. The open-ring isomer is generated in a conrotatory mode. The distance between Q1 and Q2 was 5.42 Å, which is shorter than that of the open-ring isomer, 5.96 Å. The open-ring isomer in the closed-ring form crystal has a conformation more planar than that in the most stable conformation. The occupancy factor of the open-ring isomer converged to 0.094 (2), which indicates ~9% of the closed-ring isomers converted to the open-ring isomers upon irradiation with 680 nm light.

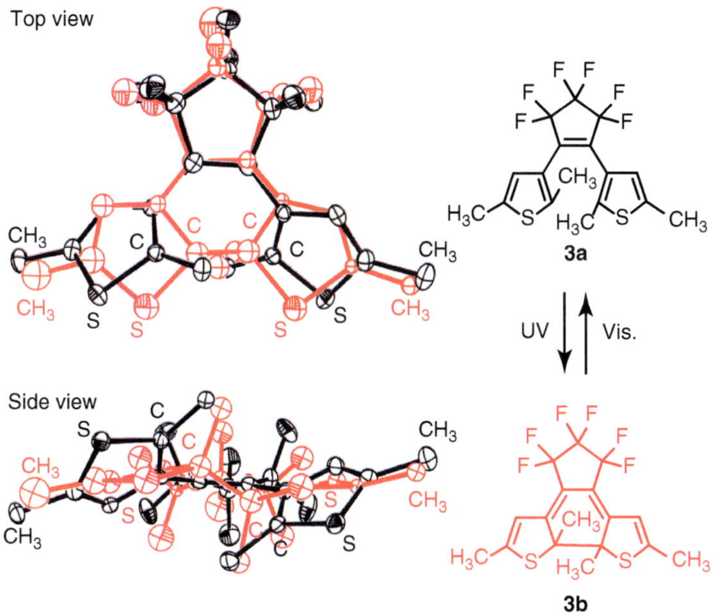

Figure 4.8 ORTEP drawings of photoirradiated crystal **3**. Black and red molecules indicate the open- and the closed-ring isomers, respectively. *Source*: Adapted with permission from Ref. [11]. Copyright 2000 The Chemical Society of Japan.

4.3 Quantum Yield

As shown in Scheme 2.4, in solution an open-ring isomer has two conformations, reactive antiparallel and nonreactive parallel ones, in equal amounts. These two conformers are interconvertible. Therefore, in general, the maximum cyclization quantum yield in solution is less than 0.5. On the other hand, molecules in crystals are fixed in either of these two conformations. When the molecules are fixed in a parallel conformation, they are not able to undergo the cyclization reaction, while the cyclization reaction readily takes place when fixed in a reactive antiparallel conformation.

The quantum yields of the photocyclization/photocycloreversion reactions in single crystals were measured using a polarizing microscope connected with a photodetector [14]. Table 4.1 shows the quantum yields in crystal and in solution. The cyclization quantum yields in crystal are almost twice as large as those in solution, and close to 1.0 (100%). The high cyclization quantum yields are attributed to high population of the reactive antiparallel conformers in single crystals. Figure 4.10 shows the correlation between the cyclization quantum yields of diarylethene crystals and the distances between the reactive carbon atoms in the crystals [24]. When the distance is shorter than 0.40 nm, the molecule readily undergoes the cyclization reaction with the quantum yield of 1.0 (100%), while the cyclization reaction does not take place when the distance is longer than 0.42 nm.

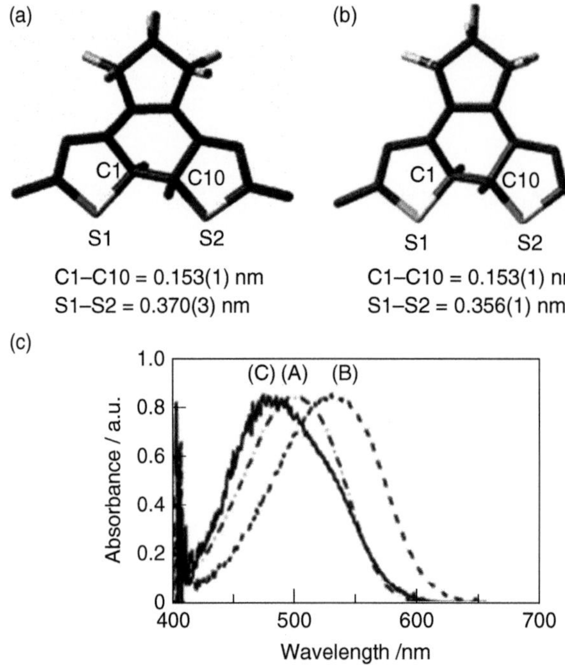

Figure 4.9 Molecular structures of the closed-ring isomer **3c** (a) generated in the open-ring form crystal and (b) in the closed-ring form crystal. (c) Absorption spectra of the closed-ring isomer **3c** (A) in *n*-hexane, (B) generated in the open-ring form crystal and (c) in the closed-ring form crystal.

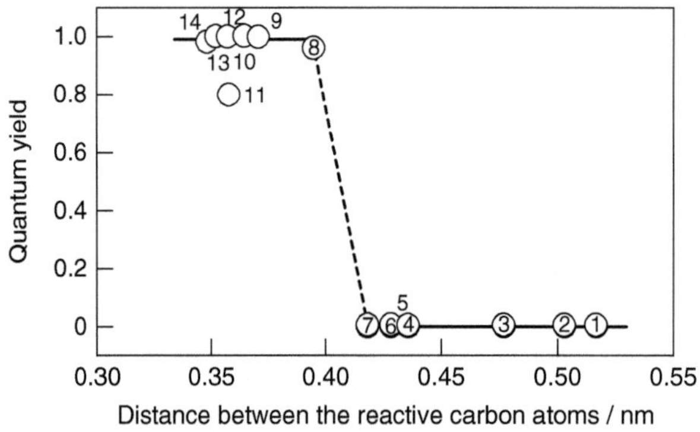

Figure 4.10 Relationship between the photocyclization quantum yield and the distance between the central reactive carbon atoms. (1) 1,2-bis(2,4,5-trimethyl-3-thienyl)perfluorocyclopentene, (2) **162**, (3) 1,2-bis(2-methyl-6-formylbenzothiophen-3-yl)perfluorocyclopentene, (4) **163**, (5) **13**, (6) 1-(2-methylbenzothiophen-3-yl)-2-(2,6-dimethylbenzothiophen-3-yl)perfluorocyclopentene, (7) 1,2-bis(2,6-dimethylbenzothiophen-3-yl)perfluorocyclopentene, (8) **5**, (9) **68**, (10) **159**, (11) **3**, (12) 1,2-bis(2-methyl-6-(5-*tert*-butylphenyl)-3-thienyl)perfluorocyclopentene, (13) **20**, and (14) **160**. *Source*: Reproduced from Ref. [24] with permission from The Royal Society of Chemistry.

Table 4.1 Photocyclization and photocycloreversion quantum yields in *n*-hexane and in crystal.

	Distance between the central reactive carbon atoms (nm)	In *n*-hexane			In crystal		
		Φ_{oc}	Φ_{co}	References	Φ_{OC}	Φ_{CO}	References
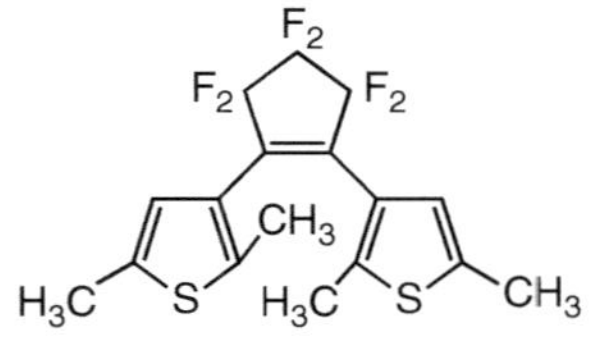 **3**	0.358	0.40	0.12	[14, 15]	0.80, 0.83[a)]	0.10, 0.11[a)]	[14]
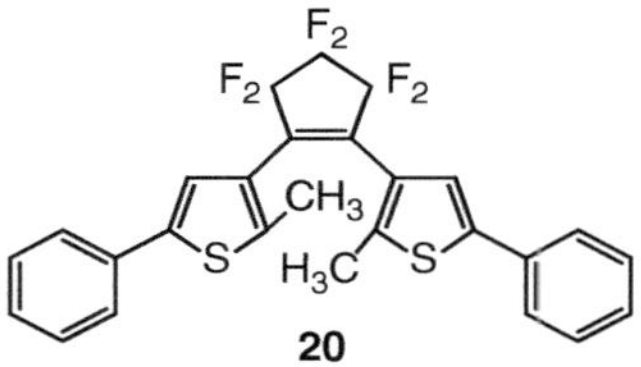 **20**	0.353, 0.361	0.59	0.013	[16]	1.0_1	0.017	[14, 17]
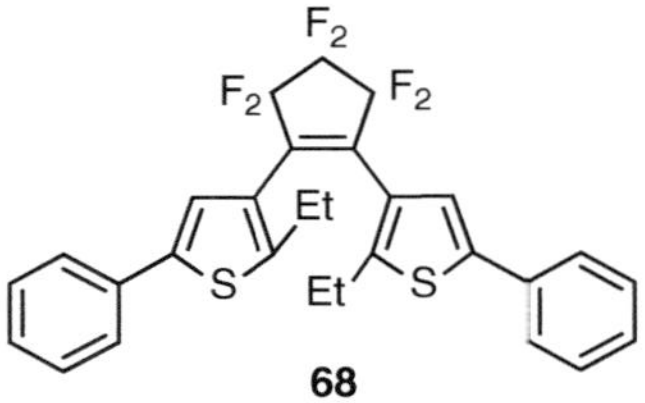 **68**	0.371	0.52	0.0081	[18]	1.0_3	0.027	[14]

(Continued)

Table 4.1 (Continued)

	Distance between the central reactive carbon atoms (nm)	In *n*-hexane			In crystal		
		Φ_{oc}	Φ_{co}	References	Φ_{OC}	Φ_{CO}	References
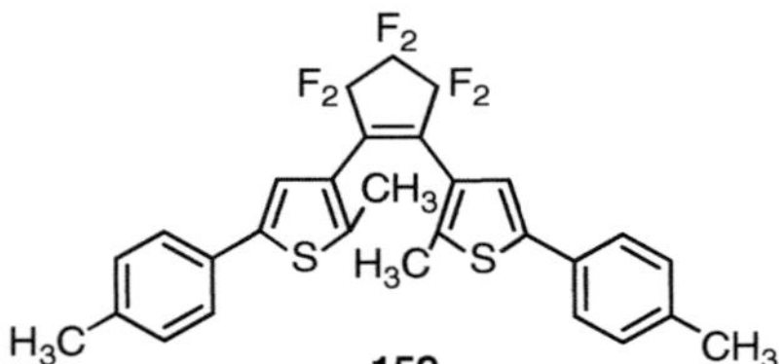 **159**	0.364	0.53	0.0097	[16]	0.98	0.029	[16]
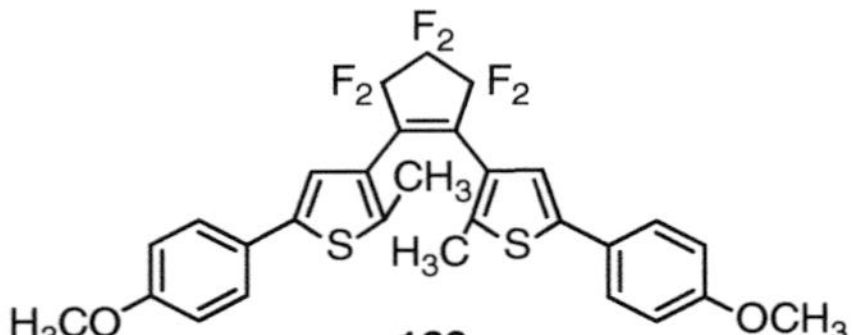 **160**	0.348–0.370	0.56	0.0052	[19]	0.95–1.00[b)]	0.0061–0.027[b)]	[19]
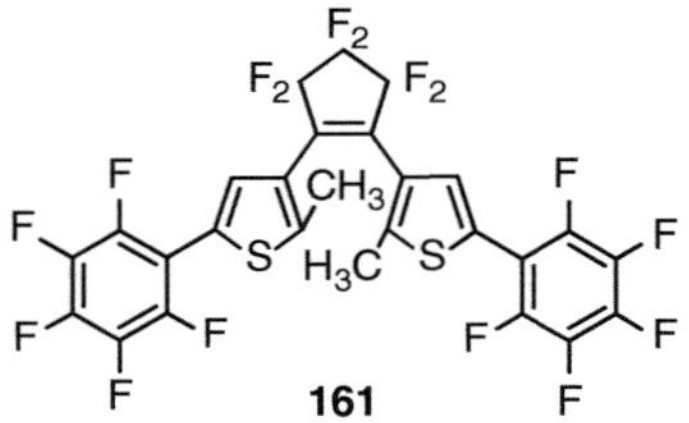 **161**	0.349	0.47	0.022	[20]	0.97	0.044	[20]

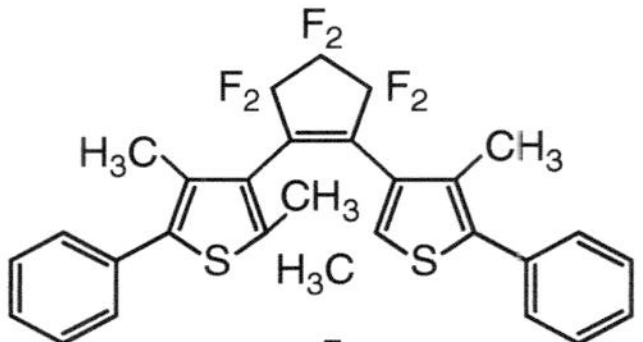 5	0.396	0.45	0.018	[21]	0.96	0.027	[14]
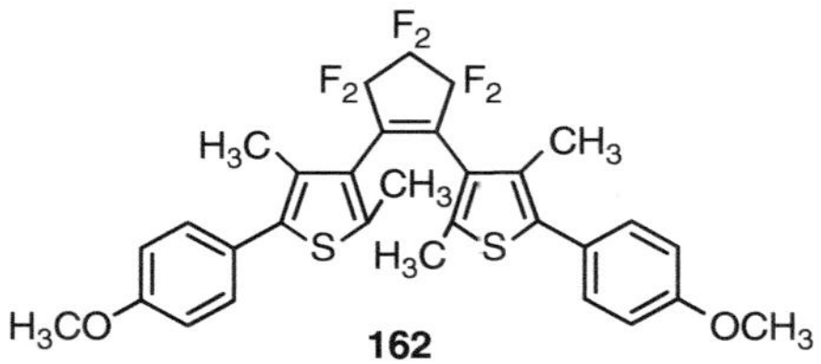 162	0.503	0.48	0.0080	[22]	0	—	[24]
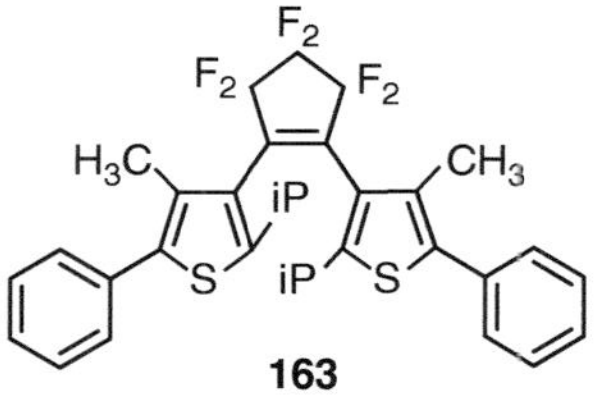 163	0.436	0.49	0.046	[23]	0	—	[24]
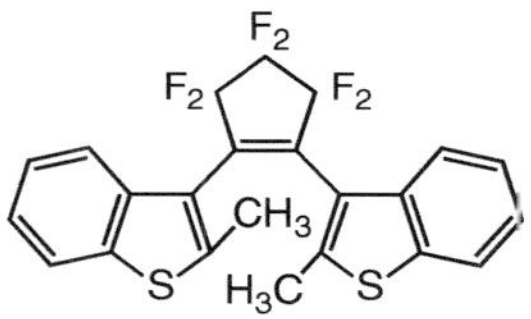 13	0.435	0.31	0.29	[21]	0	—	[24]

a) Determined by irradiation to two different surfaces.
b) Four polymorphic forms gave the different values.

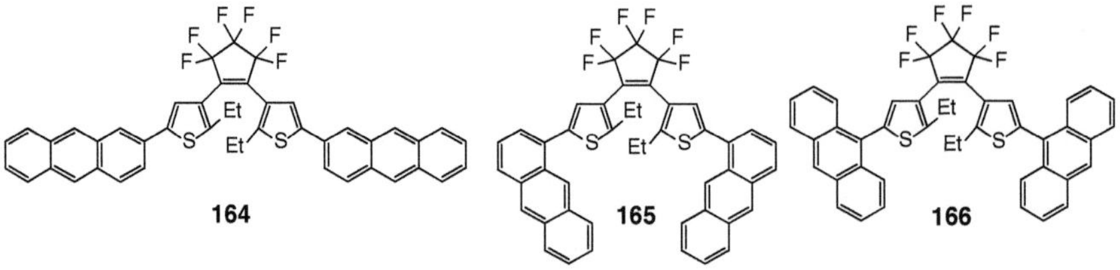

Scheme 4.2 Dithienylethenes having anthracene substituents.

An exceptional example, which does not obey the above general rule, was found in diarylethene derivatives having anthracene substituents, as shown in Scheme 4.2 [25]. These three derivatives are fixed in antiparallel conformation in crystal and the distances between the central reactive carbon atoms are shorter than 0.40 nm. Although the derivative **164** underwent the photocyclization reaction in crystal, both **165** and **166** did not show any color change upon UV irradiation. In order to understand why the reactivity is different among these three derivatives, their molecular structures and conformations were examined in detail by X-ray crystallographic analysis. The dihedral angles between the anthracene substituent and the thiophene ring of **165** and **166** were measured to be 77.09°, and 61.12° (left side) and 47.98° (right side), respectively, while the angles of **164** were 19.54° (left side) and 23.84° (right side). The dihedral angles indicate that anthracene substituents are isolated and π-conjugation is lost in **165** and **166**. On the other hand, in **164**, π-conjugation can extend throughout the molecule. The difference in the π-conjugation is considered to affect the photocyclization reactivity in crystal. Upon irradiation with UV light, the excited energy of the anthracene substituents in **164** can extend to the central part and induce the photocyclization reaction, while the excitation energy is localized in the anthracene substituents in **165** and **166**, and cannot induce the photocyclization reaction of the central part. In contrast to the different reactivity in crystal, all three derivatives underwent the photocyclization reactions in solution. This is due to free rotation of the anthracene rings in solution. The conformation of the substituents plays an important role in the single-crystalline photoreactions.

4.4 Multicolored Systems and Nano-Layered Periodic Structures

In one-component crystals, the color of crystals changes only between two states, colorless and colored ones, upon alternate irradiation with UV and visible light. On the other hand, in multicomponent crystals composed of more than two kinds of diarylethene derivatives, reversible color switching between multicolors takes place. Although it is not easy to prepare a molecule composed of three kinds of diarylethenes with different colors, yellow, red, and blue, a mixed crystal composed of three different kinds of diarylethenes can be readily prepared by choosing appropriate diarylethene derivatives. Figure 4.11 shows an example of three such derivatives, which have similar molecular size but exhibit different colors in the

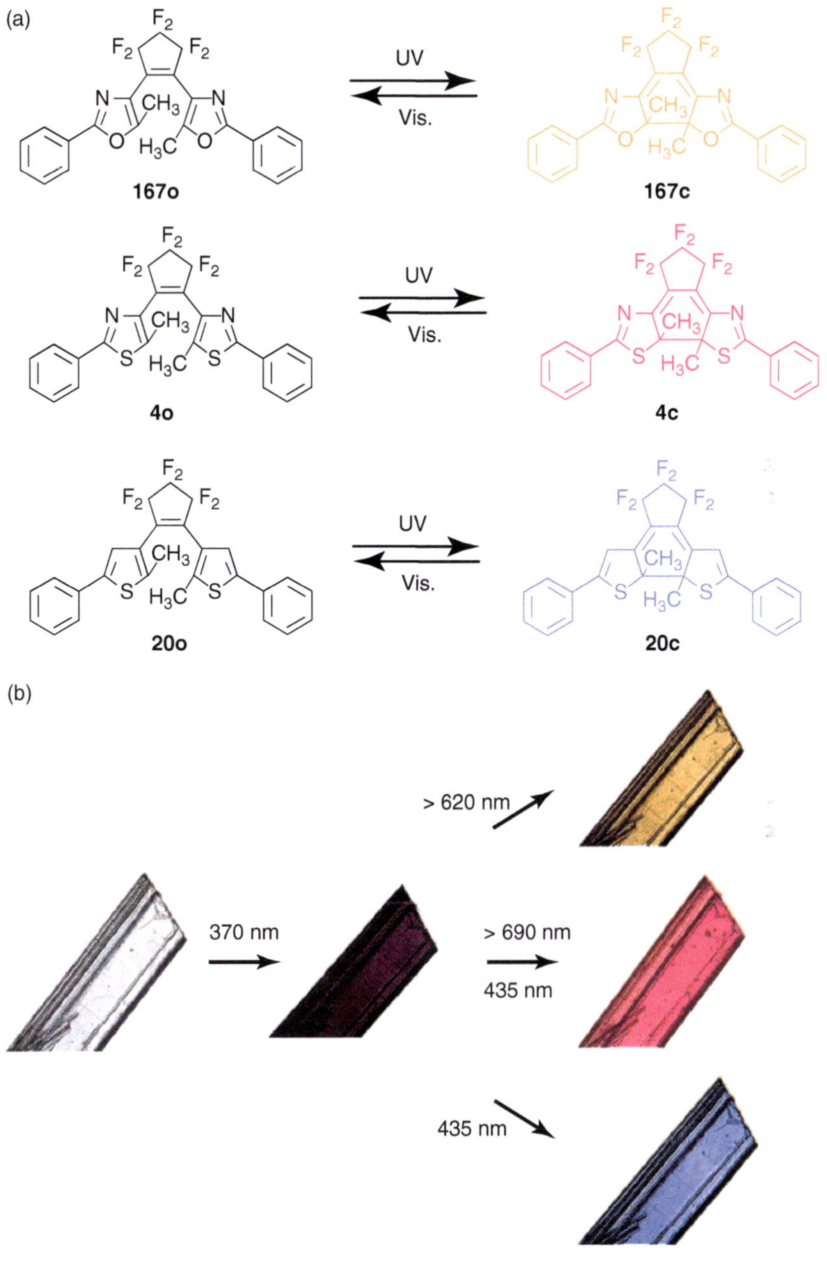

Figure 4.11 (a) Three dithienylethenes **167**, **4**, and **20**, which have similar size, but exhibit different colors in the closed-ring isomers. (b) Color changes of a three-component crystal containing **167**, **4**, and **20** upon irradiation with the appropriate wavelengths of light.

closed-ring isomers [26]. These three diarylethenes **167**, **4**, and **20** can be readily mixed with each other in a single crystal and their single component crystals exhibit yellow, red, and blue colors upon irradiation with UV light. Mixed crystals composed of the three derivatives in similar amounts were prepared by recrystallization from a *n*-hexane or an ethanol solution. The colorless crystal turned purplish black upon irradiation with 370 nm light. Upon subsequent irradiation with $\lambda > 670$ nm light, the color of the crystal turned yellow. When the purplish black crystal was irradiated with 435 nm and $\lambda > 690$ nm light, it turned red. The crystal turned blue upon irradiation with 435 nm light. All the colors were thermally stable and completely bleached by irradiation with visible light. The multicolored crystals have the potential for applications to multifrequency three-dimensional optical memory media and photoswitchable full-color displays.

A single crystal of **168** has a unique molecular packing structure in which photoactive and photoinactive conformer layers are alternatively stacked [27]. Both conformer A and B adopted an antiparallel conformation, but the distances between the reactive carbon atoms were different, as shown in Figure 4.12b. The distances of conformers A and B were 0.365 and 0.493 nm, respectively. Only A satisfied the requirement for the cyclization reaction and underwent the reaction, while conformer B did not undergo the reaction. Therefore, closed-ring isomers and open-ring isomers are alternatively packed one by one, as shown in Figure 4.12c.

4.5 Fluorescent Crystals

Molecular crystals that change luminescent properties upon application of external stimuli, such as heat, light, vapor, and mechanical force, are potentially applicable in sensors, electroluminescent display, and bioimaging. Light is a powerful tool for controlling luminescence because it can control the switching response in a wide range of spatiotemporal resolutions. Although various types of molecular crystals, such as metal complexes and metal–organic frameworks, have been reported to change the luminescence upon photoirradiation, in most cases the luminescence changes are irreversible, or another physical/chemical treatment is required to recover the original state. Reversible photoswitching of luminescence was achieved using photoswitchable diarylethenes.

Most of diarylethene crystals are nonfluorescent. An exceptional example is 1,2-bis(3-methyl-2-thienyl)perfluorocyclopentene (**1**), which undergoes reversible photoinduced cyclization/cycloreversion reactions in the single-crystalline phase and the fluorescence reversibly switches *off* and *on* [28]. Upon irradiation with UV light, the fluorescent open-ring isomer converted to the nonfluorescent closed-ring isomer and the fluorescence was bleached, while the fluorescence was recovered by irradiation with visible light. By using the bleaching of component diarylethene molecules three-dimensional erasable optical recording in the single crystal was demonstrated [29]. Each spot was written by irradiation with an Ar ion laser (336–363 nm multi lines, 10 mW) for 20 ms. The recorded spots were detected as dark spots. The spot size read by a fluorescence-detection method

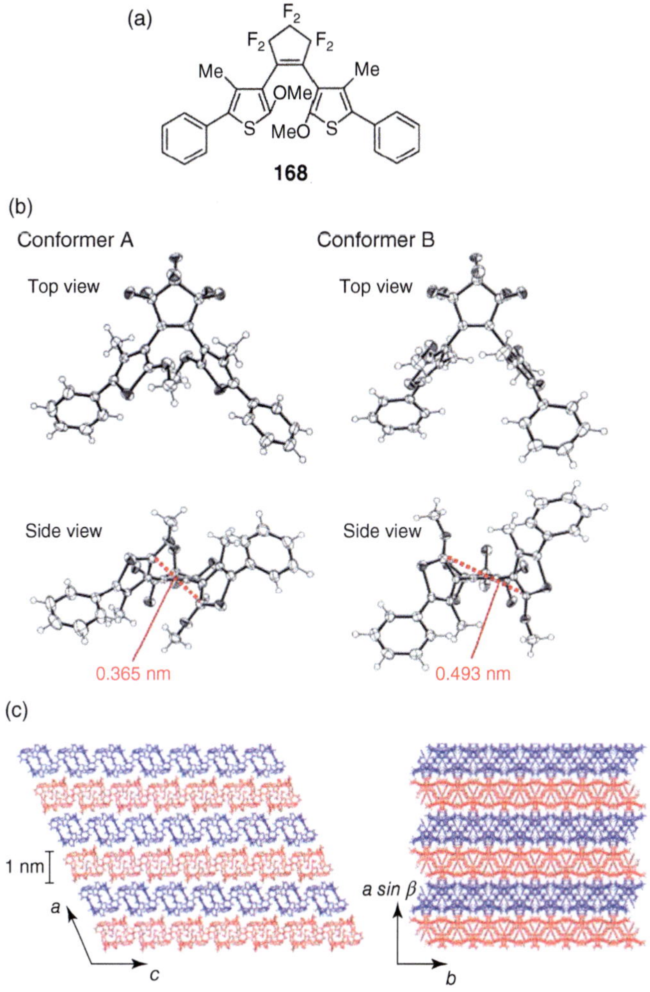

Figure 4.12 (a) Molecular structure of **168**. (b) ORTEP drawings and (c) molecular packing diagrams of crystal **168**. Blue and red molecules in the packing diagrams indicate photoreactive and photoinactive conformers A and B, respectively.

was approximately 200 nm in plane and 1.5 μm in depth, which corresponds to the recording density higher than 5 Tbit/cm^3. The recorded spots were erased by irradiation with visible light. The writing/recording/erasing cycles were able to be repeated more than 10^4 times.

As described in Section 3.4.2, sulfone derivatives of 1,2-bis(2-alkyl-6-aryl-1-benzothiophen-3-yl)perfluorocyclopentene undergo cyclization/cycloreversion reactions upon alternate irradiation with UV and visible light and the UV-generated closed-ring isomers emit fluorescence with high quantum yields. Among various sulfone derivatives, **169** and **170** were found to undergo the cyclization/cycloreversion reactions even in crystal, as shown in Figure 4.13 [30]. X-ray crystallographic analysis indicates that both compounds are fixed in an antiparallel conformation in crystal and the distances between the reactive carbon atoms are shorter than 0.42 nm (**169o**:0.402 nm, **170o**:0.399 nm, and 0.394 nm). The conformation and the distances fulfill the requirements for the diarylethenes to undergo photocyclization reactions in the single-crystalline phase.

Initially, the single crystals of **169o** and **170o** were not luminous (off state) under excitation with 450 and 480 nm light, respectively, because both crystals have no absorption band at these wavelengths. Upon irradiation with UV light, the crystals became fluorescent (on state) due to the formation of fluorescent closed-ring isomers. The UV-irradiated crystals of **169** and **170** emitted green (λem: 510 nm) and yellow–green (λem: 540 nm) fluorescence under excitation with 450 and 480 nm light, respectively, as shown in Figure 4.13a and b. Photoirradiation with visible ($\lambda > 460$ nm) light bleached the green and yellow–green fluorescence. Thus, these single crystals underwent turn-on mode fluorescence photoswitching with high on/off contrast.

Regular packing and orientation of molecules in single crystals lead to anisotropy of their optical properties. Dichroism of UV-irradiated single crystal of **170** was examined with a polarizing microscope. The crystal was illuminated with 480 nm light and the fluorescence was detected through a polarizing plate. At a certain angle ($\theta = 0°$), the crystal exhibited yellow color and gave strong yellow–green fluorescence, while the yellow color became weak and the fluorescence almost disappeared when the crystal was rotated as much as 90° ($\theta = 90°$). This indicates that the fluorescence emission from the single crystal is highly polarized because of the regular orientation of the diarylethene molecules in the crystal.

Irradiating a single crystal with spatially patterned light enables area-selective switching of the fluorescence. Figure 4.14 shows fluorescence images of the single crystal of **170**. Upon irradiation with patterned 405 nm light through an objective lens, the irradiated area in the single crystal was activated to be fluorescent and bright letter "F" was recorded on the dark background. The letter was erased by irradiation with 480 nm light. After that, upon irradiation with 405 nm light, a letter "ϕ" was recorded in the same area, indicating the rewritable recording of the fluorescence images.

4.6 Photomechanical Response

It is of specific interest from both scientific as well as technological points of view to have molecular systems make mechanical motion based on geometrical structure changes of individual molecules induced by external stimuli, such as chemicals, photons, or electrons (holes), and link the motion to macroscopic mechanical work. In particular, direct conversion of light energy into mechanical work has been a long-standing challenge since the first report by Merian in 1966 [31]. Although elaborate molecular machines that mimic biological muscles have been extensively studied [32–36], an attempt to link the molecular-scale movement of man-made devices to macroscopic motion is limited. These efforts are just like attempts to construct airplanes by mimicking the action of birds. It is indispensable to conceive

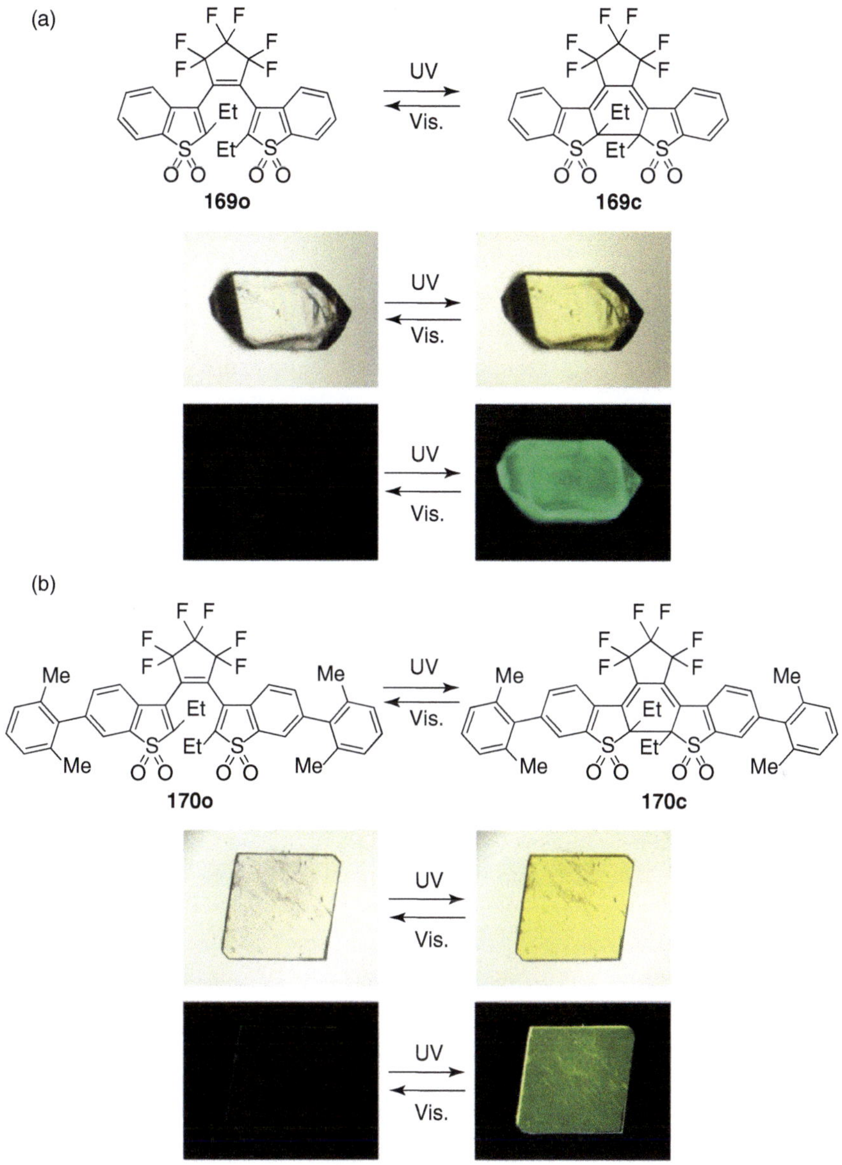

Figure 4.13 Photographs of single crystals of (a) **169o** and (b) **170o**. Top: photoinduced color changes by irradiation with UV ($\lambda = 365$ nm) and visible ($\lambda > 460$ nm) light, bottom: photoinduced fluorescence changes under dark-field observation. The excitation wavelengths for the fluorescence detection were 450 and 480 nm for **169o** and **170o**, respectively. *Source*: Adapted from Ref. [30] with permission from The Royal Society of Chemistry.

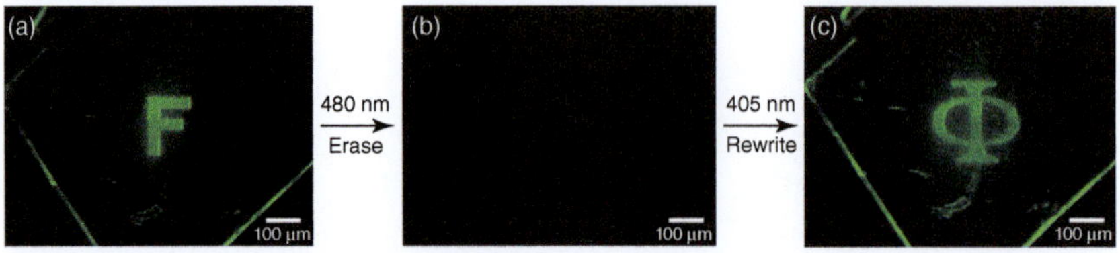

Figure 4.14 Patterned fluorescence photoswitching of single crystal **170o**. (a) "F" patterned 405 nm light (pulse width: 80 ms) was irradiated, (b) erased by irradiation with 480 nm light, and (c) "Φ" patterned 405 nm light (pulse width: 80 ms) was irradiated. *Source*: Reproduced from Ref. [30] with permission from The Royal Society of Chemistry.

a new concept to construct realistic molecular machines. In 2007, a prominent idea was proposed to use photoswitchable molecular crystals for linking the molecular-scale geometrical structure changes of component molecules to macroscopic movement of crystals [37]. Molecular crystals are densely packed organized systems. In such dense systems, strain energy produced by the photoisomerization of individual photoswitchable molecules is considered to directly induce the bulk crystal shape change.

4.6.1 Surface Morphology Change

As shown in Figure 4.8, diarylethene molecules shrink in size upon photocyclization reactions and expand upon photocycloreversion reactions. The geometrical structure changes of diarylethene molecules during the photoisomerization reactions are, however, relatively small. Therefore, many diarylethene crystals are able to undergo photoswitching reactions without disintegration of the crystals. If such molecular structure changes of component diarylethenes take place in crystal, the bulk crystal shape is anticipated to follow the shape changes. In other words, the change of shape of individual molecules can be directly linked to the macroscopic shape deformation of the crystal. The most convenient way to know the crystal shape change is to measure the surface morphology by using an atomic force microscope (AFM).

Figure 4.15 shows the morphology changes of (1 0 0) surface of a single crystal of 1,2-bis(2,4-dimethyl-5-phenyl-3-thienyl)perfluorocyclopentene (**5**) upon irradiation with UV (365 nm) and visible (>500 nm) light measured with AFM [38]. Upon irradiation with UV light for more than 10 seconds, new steps appeared on the (1 0 0) surface. No step formation was discerned during the irradiation for the initial 10 seconds, but appeared after the induction period. The step height, about 1.0 nm, corresponds to one molecular layer, as shown in Figure 4.15c. When the irradiation time was prolonged, the number of steps increased and steps with the height of 2.0 and 3.0 nm appeared. The step height was in multiples of the minimum height of 1.0 nm. Any step with a height lower than the unit height (~1.0 nm) was not observed.

The molecular structures of open-ring isomer **5o** and closed-ring isomer **5c** were compared by using X-ray crystallographic analysis. The end-to-end distance between the two phenyl ring edges decreases from 0.141 to 0.139 nm when **5o** converts to

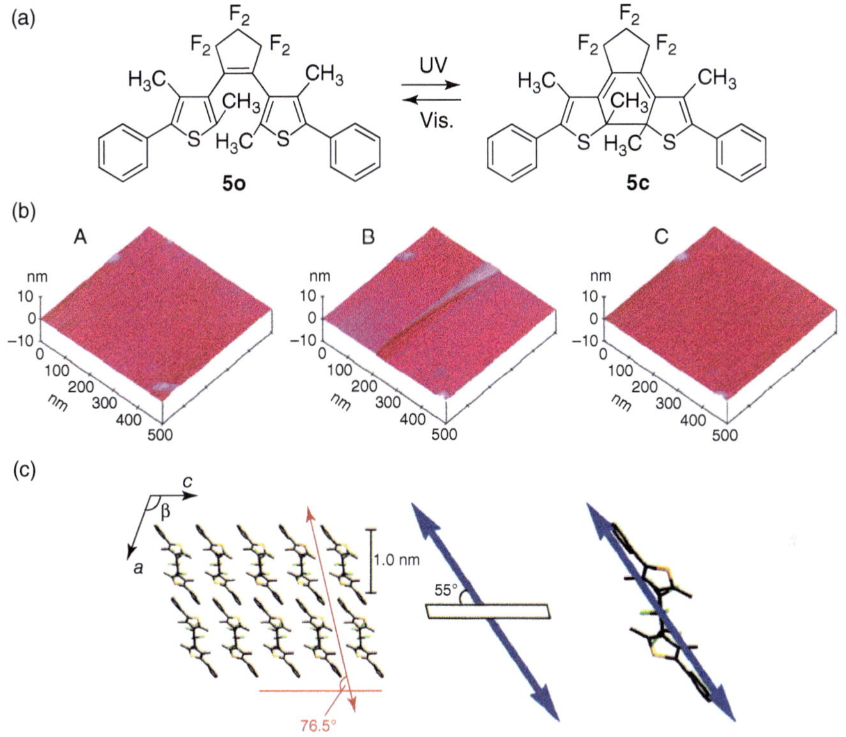

Figure 4.15 (a) Structures of open- and closed-ring isomers of **5**. (b) AFM images of the (100) crystal surface (A) before photoirradiation, (B) after irradiation with 366 nm light for 10 seconds, and (C) after irradiation with visible ($\lambda > 500$ nm) light. (c) Molecular packing of the open-ring isomers in crystal.

5c. The shortening in the distance upon the photocyclization reaction results in the decrease in the thickness of the molecular layers. The shrinkage of the molecular layers produces vacancies in the crystal bulk. Accumulation of the vacancies deep in the crystal allows the surface molecular layer to sink as much as one layer (~1.0 nm). Prolonged irradiation increases the vacancies and causes subsidence of multilayers. The irradiation for 10 seconds induces the photocyclization reactions of 600 molecular layers. The random distribution of the reactions as deep as 5×10^5 layers (=500 μm) produces the one-molecular layer vacancy. This is the explanation of the step formation on the (1 0 0) crystal surface. The structure change of each molecule is linked to the shape deformation of the crystal surface.

4.6.2 Reversible Shape Change

When the crystal size is small in the range of 10–500 μm, not only the surface morphology but also the bulk crystal shape was found to reversibly change upon alternate irradiation with UV and visible light. Figure 4.16 shows the color and shape changes of two small-size crystals, 1,2-bis(2-ethyl-5-phenyl-3-thienyl)-perfluorocyclopentene (**68**) and 1,2-bis(5-ethyl-2-phenyl-4-thiazolyl)perfluorocyclo-

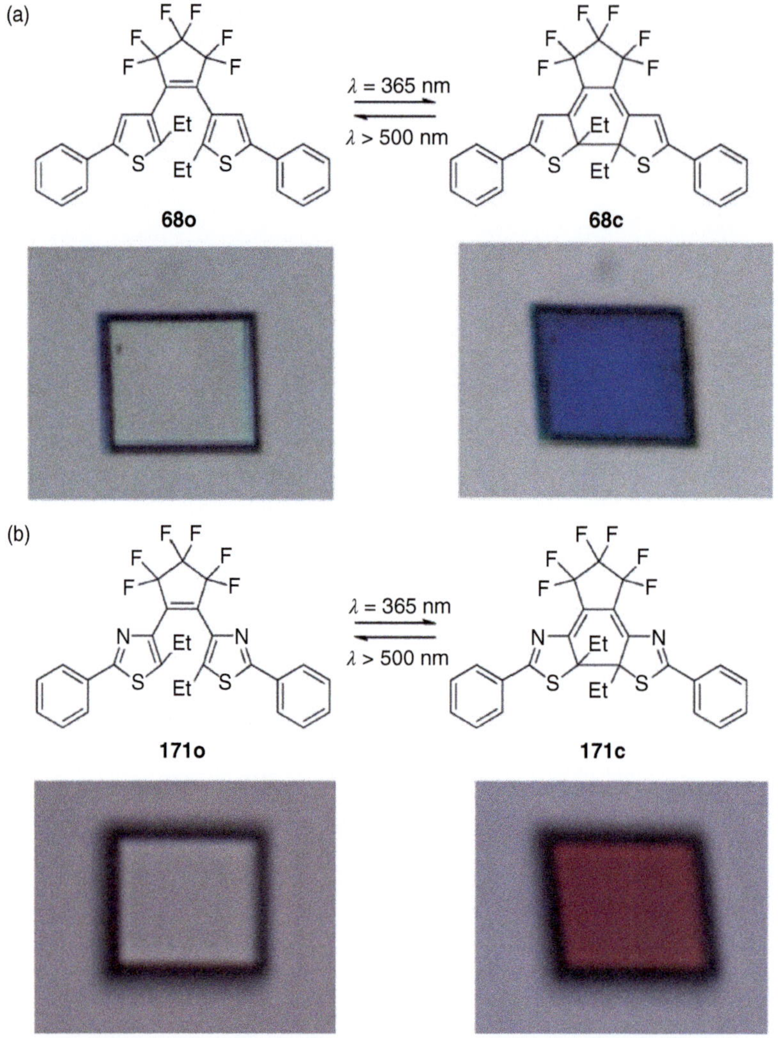

Figure 4.16 Deformation of single crystals upon irradiation with UV (365 nm) and visible ($\lambda > 500$ nm) light. (a) 1,2-Bis(2-ethyl-5-phenyl-3-thienyl)perfluorocyclopentene (**68**) and (b) 1,2-bis(5-ethyl-2-phenyl-4-thiazolyl)perfluorocyclopentene (**171**).

pentene (**171**) [37, 39]. Upon irradiation with 365 nm light, crystal **68** changed the color from colorless to blue, while colorless crystal **171** turned red. Both crystals changed the shape from square to lozenge. For example, the corner angles of crystal **171** changed from 90° and 90° to 86° and 94° [39]. Although the color changes were different, both crystals showed a similar bulk shape deformation. The result indicates that the deformation of crystal bulk shape is ascribed to the geometrical structure change of component molecules instead of the electronic structure change.

To elucidate the relation between the crystal shape deformation and the structure change of individual molecule, X-ray crystallographic study was carried out

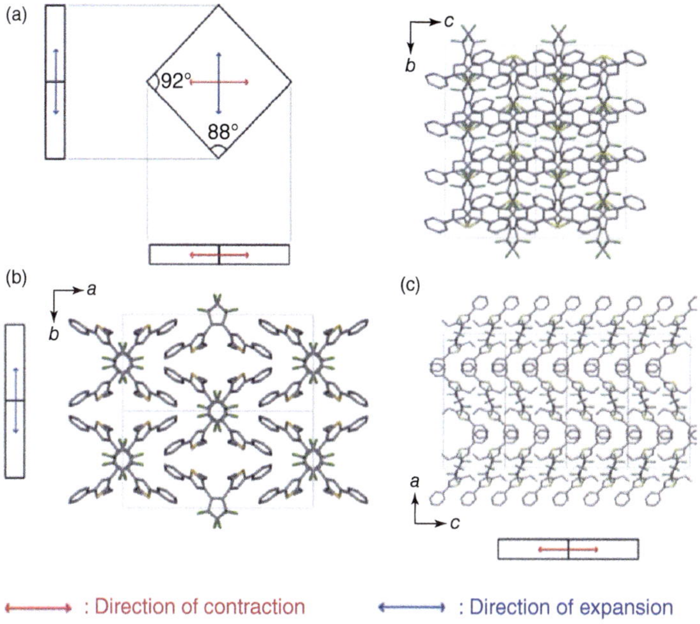

Figure 4.17 Molecular packing of open-ring isomers of 1,2-bis(2-ethyl-5-phenyl-3-thienyl)perfluorocyclopentene (**68o**) in crystal before UV irradiation. The red arrows indicate the direction of contraction and blue arrows the direction of expansion of the crystal upon UV irradiation. *Source*: Reprinted with permission from Ref. [40]. Copyright 2014 American Chemical Society.

[37]. Figure 4.17 shows the molecular packing of crystal **68o** before UV irradiation. The shape deformation from a square to a lozenge indicates that the crystal expands along the *b*-axis and contracts along the *c*-axis. In fact, in situ X-ray crystallographic analysis indicated that the unit cell length of the *b*-axis expands from 10.971 Å to 11.026 Å and that of the *c*-axis contracts from 10.585 Å to 10.542 Å, when the crystal was irradiated with UV light and 13% of **68o** converted to **68c**. Upon irradiation with visible light, the unit cell lengths returned to the original ones.

Figure 4.18 shows the geometrical structures of open-ring isomer **68o** and closed-ring isomer **68c**. These two isomers were isolated by using high-performance liquid chromatography (HPLC) and independently crystallized in *n*-hexane. When the molecule converts from **68o** to **68c**, the height of the triangle shape increases from 0.61 to 0.73 nm, which is in the direction of the *b*-axis. On the other hand, the thickness of molecules is reduced. The cofacial packing of molecules along the *c*-axis indicates that thin layers of closed-ring isomers allow the molecules to be stacked one-by-one, resulting in contraction along the *c*-axis. The structure change of individual molecules directly induced the crystal shape deformation.

Figure 4.19b shows another type of crystal shape deformation. Upon irradiation with UV light, the rectangular crystal **4** contracted along the long axis of the crystal as much as 7%, while it reverted back to the original shape upon irradiation with visible

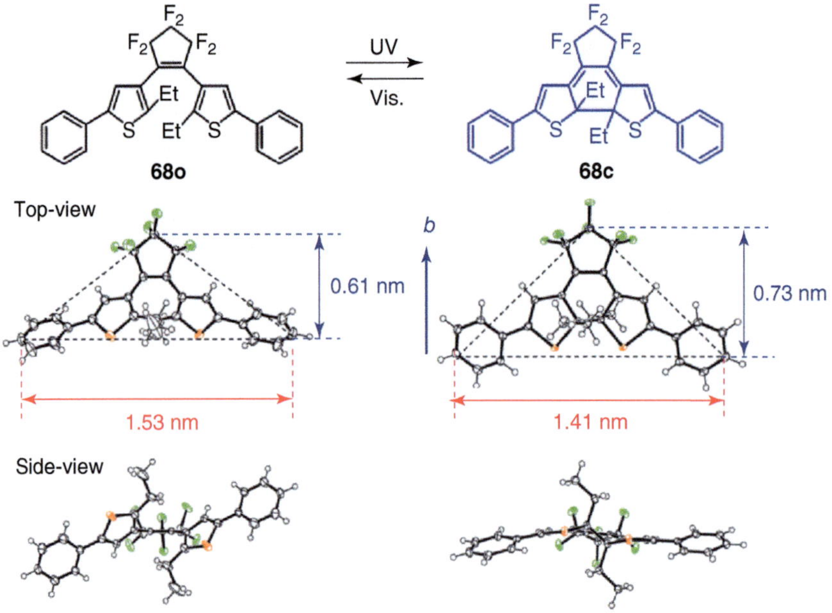

Figure 4.18 Chemical structures of open- and closed-ring isomers of 1,2-bis(2-ethyl-5-phenyl-3-thienyl)perfluorocyclopentene (**68**) and top and side views of the geometrical structures of the two isomers. The two isomers were isolated by using HPLC and independently recrystallized.

light. A rod-like crystal made of **4** bent toward the direction of the incident light, as shown in Figure 4.19c. The bending is due to the gradient in the extent of photoisomerization caused by high absorbance of the crystal. The shrinkage of the irradiated side of the crystal causes the bending as observed in a bimetal. The bent rod-like crystal again became straight upon irradiation with visible light. The rod-like crystal can hit a tiny silica-particle, as shown in Figure 4.20. The relatively large Young's modulus (~10 GPs) of the crystal enabled the crystal-rod to hit the silica-particle as if it were a tennis ball. One of the applications of such a bending crystal is an electric current photoswitch [41]. The surface was coated with gold (thickness: ~6 nm) and the crystal was placed in between two electrodes. The reversible on/off photoswitching of the electric current was demonstrated, as shown in Figure 4.19d.

4.6.3 Bending Response of Mixed Crystals

Above single component crystals lack durability and break in less than 100 bending cycles. To improve the fatigue-resistant property, two-component crystals were prepared [42]. Defects formed in two-component crystals are considered to prevent disintegration of the crystals. Two-component crystals composed of 1-(5-methyl-2-phenyl-4-thiazolyl)-2-(5-methyl-2-*p*-tolyl-4-thiazolyl)perfluorocyclopenetene (**172**) and 1,2-bis(5-methyl-2-*p*-tolyl-4-thiazolyl)perfluorocyclopentene (**173**) were found to exhibit superior fatigue-resistant property. Rod-like crystals composed

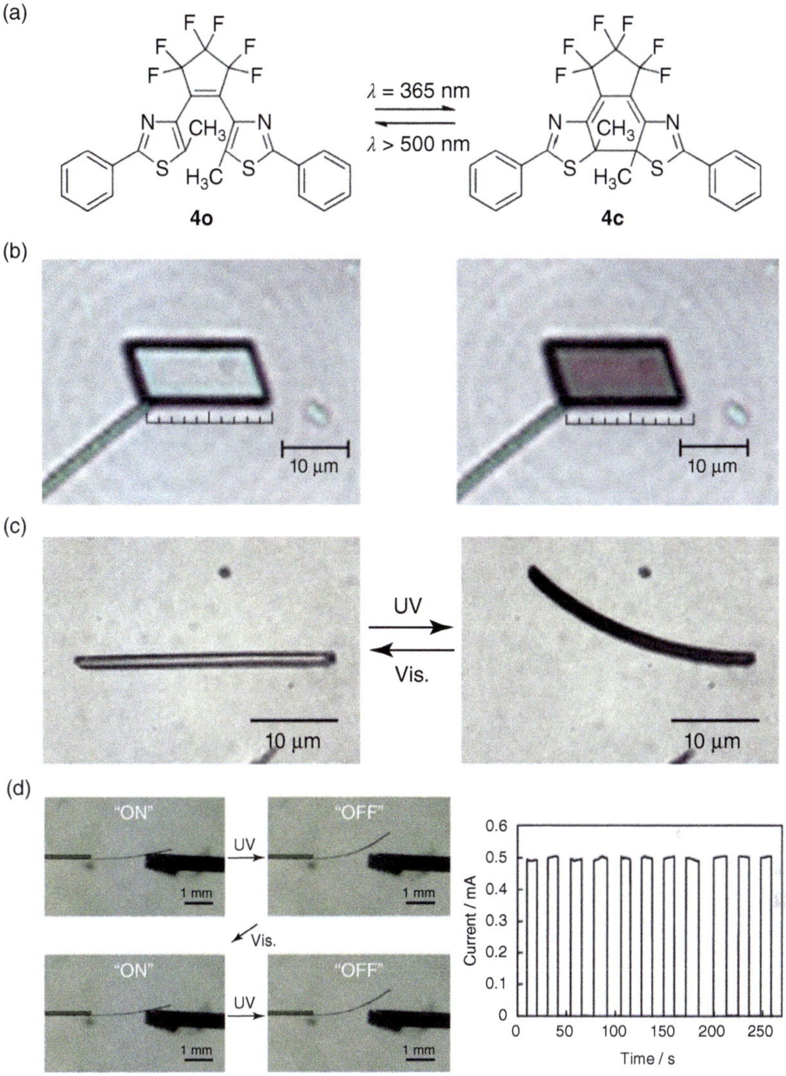

Figure 4.19 (a) Chemical structures of open- and closed-ring isomers of 1,2-bis(5-methyl-2-phenyl-4-thiazolyl)perfluorocyclopentene (**4**). (b) Reversible contraction of single crystal **4** upon alternate irradiation with UV and visible light. (c) Reversible bending of single crystal **4** upon alternate irradiation with UV and visible light. (d) Reversible current switching of single crystal **4** coated with gold (thickness: ~6 μm) upon alternate irradiation with UV and visible light.

of **172** (63%) and **173** (37%) were obtained by recrystallization from an ethanol solution containing equimolar amounts of **172o** and **173o**. Figure 4.21b shows bending cycles of the rod-like crystal (1.3 mm × 25 μm × 13 μm) upon alternate irradiation with UV and visible light. The crystal repeated the bending cycles more than 1000 times without disintegration. The crystal surface remained clear even after 1000

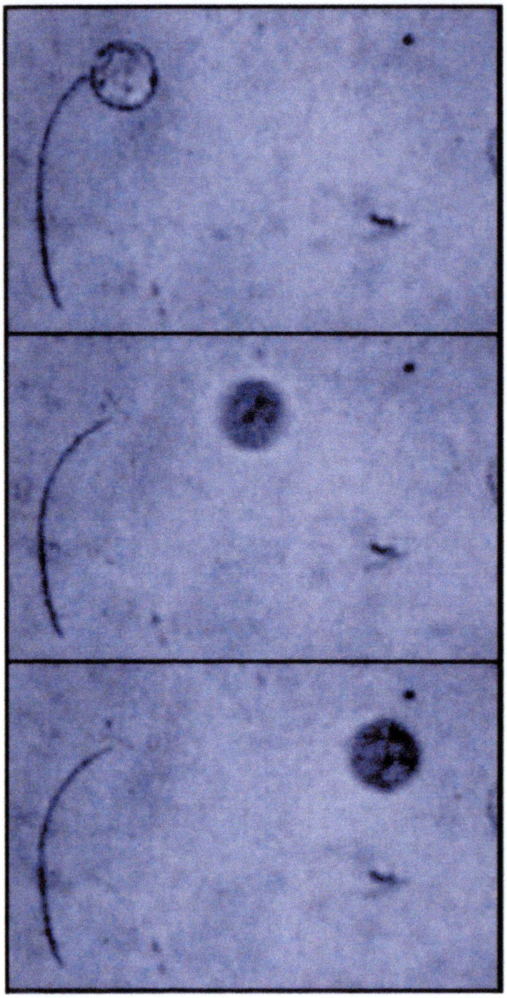

Figure 4.20 A shot of a tiny silica-particle (~100 μm) by crystal-rod of **4** upon irradiation with 365 nm light pulse. *Source*: Reprinted with permission from Ref. [40]. Copyright 2014 American Chemical Society.

cycles and no damage was discerned. A long rod-like crystal (3 mm in length) showed curling into a hairpin shape upon irradiation with UV light, as shown in Figure 4.21c. The crystal retained crystallinity even after the curling and returned to its original straight shape upon irradiation with visible light. This two-component crystal is flexible. One interesting performance of the crystal is photoinduced bending response at low temperature. The rod-like crystal exhibited reversible light-driven bending even at 4.6 K and the bending response time was less than 5 μs.

The direct conversion of light energy to macroscopic mechanical work was achieved using diarylethene cocrystals [43]. Relatively large-sized cocrystals (typically 1–5 mm in length, 0.2–1.5 mm in width, and 10–50 μm in thickness) composed of 1,2-bis(2-methyl-5-(1-naphthyl)-3-thienyl)perfluorocyclopentene (**174**) and perfluorocyclopentene (**FN**) were prepared. The crystal is composed of **174o** and **FN** in the ratio of 1 : 2. The crystal changed the color from colorless to blue and bent moving away from the light source. The bending ceased when the

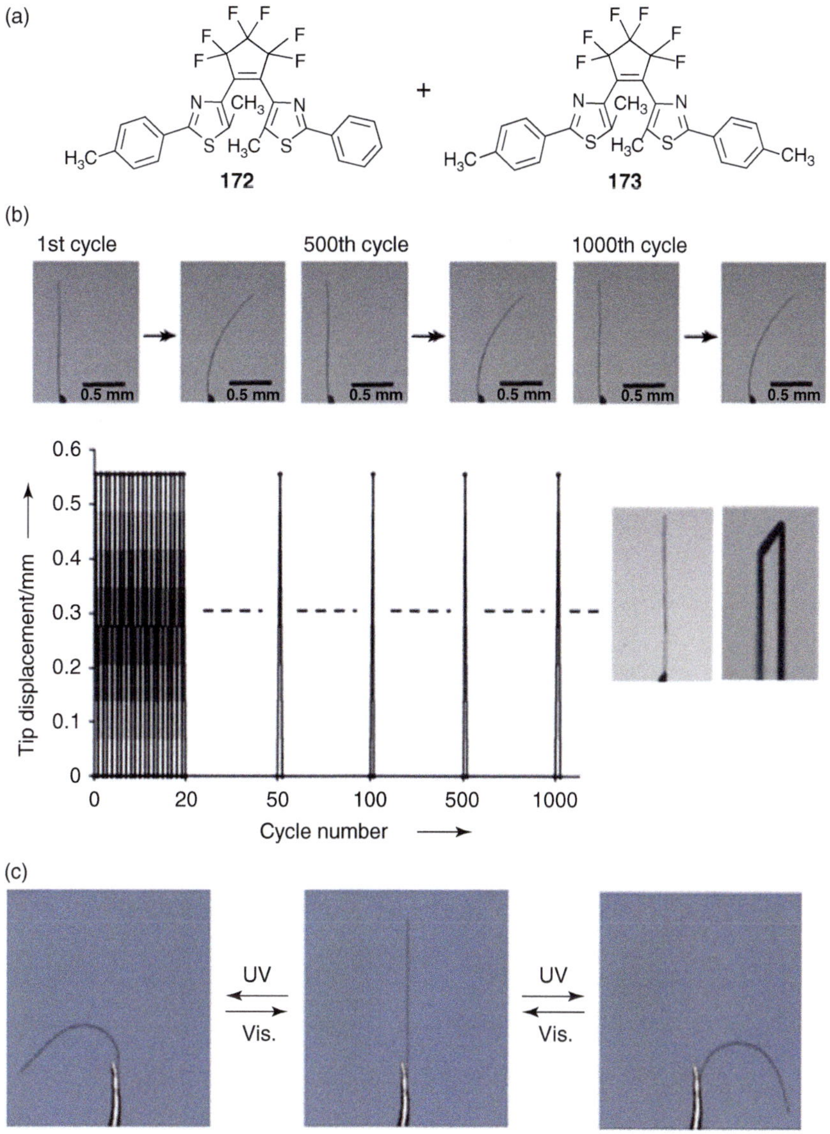

Figure 4.21 (a) Chemical structures of component diarylethenes **172** and **173** in the mixed crystal. (b) Reversible bending of the rod-like mixed crystal of **172** and **173** (**172**/**173** = 63/37) upon alternate irradiation with UV (365 nm) and visible (>500 nm) light. Pairs of photographs of the crystal (1.3 mm × 25 μm × 13 μm) showing 1st, 500th, and 1000th cycles and tip displacement of the crystal during reversible bending cycles. An expanded photograph of the crystal after the 1000th cycle is shown in right side. (c) Reversible curling to a hairpin shape upon irradiation with UV light (crystal length: 3.0 mm). The crystal kept the crystalline state even after the curling and returned to the original straight shape upon irradiation with visible light.

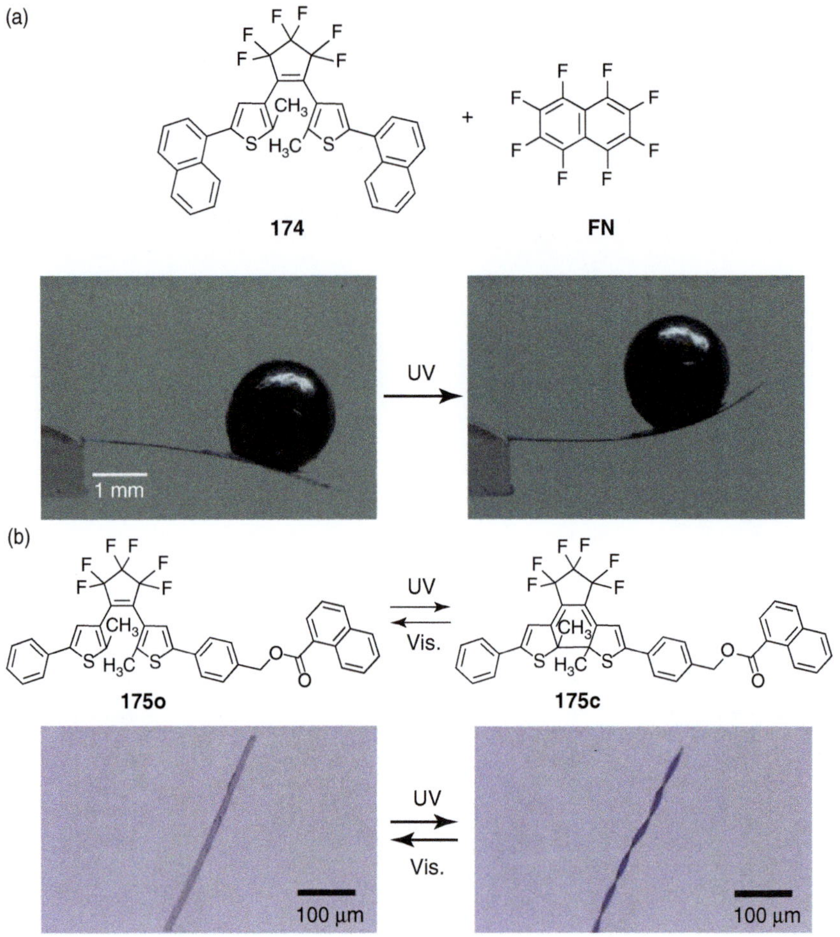

Figure 4.22 (a) Photomechanical work of a molecular crystal cantilever made of a two-component cocrystal of **174** and FN. Illumination of UV (365 nm) light was carried out below the crystal. The lead ball and the crystal cantilever weigh 46.77 and 0.17 mg, respectively. (b) Photoinduced reversible twisting of a needle-like crystal of **175** upon alternate irradiation with UV and visible light.

illumination light was switched off and the crystal kept its bending shape in the dark. Upon visible irradiation, the blue color disappeared and the crystal returned to the original straight shape. A rectangular plate-like cocrystal with the length of 5 mm was fixed at the edge of glass plate as a cantilever arm, and a lead ball was loaded onto the crystal, as shown in Figure 4.22a. The crystal and the ball weigh 0.17 and 46.77 mg, respectively. Upon irradiation with UV light from below, the heavy ball was lifted as high as 0.95 mm. The cantilever arm performed lifting work and the amount of work was 0.43 μJ. The modulus of the crystal was measured to be as large as 11 GPa. This is much larger than those of typical polymer materials (~1 GPa). The large elastic modulus enabled the crystal to generate a strong force and carry out large mechanical work. The maximum stress was estimated to be 44 MPa, which is 100 times larger than that of biological muscles (=0.3 MPa).

One of the fundamental issues to be solved is whether the local shape change of each molecule additively contributes to the strain, or whether cooperative phase transition or domain formation induced by the photoisomerization produces the strain, which causes the bending. To reveal this substantial problem, mixed crystals composed of two kinds of diarylethene derivatives, which have a similar geometrical structure but exhibit different absorption spectra in the closed-ring isomers, were prepared and their bending response by selective photoisomerization of the two-component derivatives with an appropriate wavelength of light was studied [44]. The result clearly indicated that the local shape change of each molecule is additively linked to the macroscopic deformation of the crystals.

Not only bending but also twisting of crystals takes place upon photoirradiation [45]. Figure 4.22b shows one such example. The colorless needle-like crystal of 1-(2-methyl-5-(4-(1-naphthoyloxymethyl)phenyl)-3-thienyl)-2-(2-methyl-5-phenyl-3-thienyl)perfluorocyclopentene (**175**) was found to undergo the reversible twisting response upon irradiation with UV light. The twisted crystal relaxed back to its original straight shape in a few seconds under visible light irradiation. The photoinduced crystal twisting was caused by contraction of the crystal in a diagonal direction, as can be seen from a relationship between the face irradiated with UV light and the direction of the crystal twisting. The heterogeneity of the photoisomerization reactions in the thickness direction induced the twisting shape change. The photoinduced twisting was also found to depend on the illumination direction. Changing the UV illumination direction with respect to the crystal resulted in the different twisting modes, ranging from helicoid to cylindrical [46].

Fabrication of single crystalline assemblies is a promising approach to integrate photoswitchable molecules into macroscopic materials that perform significant photomechanical works. The photomechanical crystals are stable in water but dissolved in organic solvents. Therefore it is hard to apply crystal to micro-actuators working in nonagueous liquids. This drawback was overcome by coating the crystal surface with ultrathin polymer films, such as poly(vinyl alcohol), using a stepwise assembly method [47]. The crystals are now ready for practical use.

References

1 Golden, J.H. (1961). Bi(anthracene-9,10-dimethylene)(tetrabenzo[2,2] paracyclo phane). *J. Chem. Soc.*: 3741–3748.

2 Hadjoudis, E., Vittorakis, M., and Moustakali-Mavridis, I. (1987). Photochromism. *Tetrahedron* 43 (7): 1345–1360.

3 Trozzolo, A.M., Leslie, T.M., Sarpotdar, A.S. et al. (1979). Photochemistry of some three-membered heterocycles. *Pure Appl. Chem.* 51 (2): 261–270.

4 Maeda, K. and Hayashi, T. (1970). The mechanism of photochromism, thermochromism and piezochromism of dimers of triarylimidazolyl. *Bull. Chem. Soc. Jpn.* 43 (2): 429–438.

5 Ichimura, K. and Watanabe, S. (1976). Photocyclization of diphenylmalenonitrile in the crystalline state. *Bull. Chem. Soc. Jpn.* 49 (8): 2220–2223.

6 Mori, Y., Ohashi, Y., and Maeda, K. (1989). Crystal structure and photochemical behavior of 2,2,4,6-tetraphenyldihydro-1,3,5-triazine and its inclusion compounds. *Bull. Chem. Soc. Jpn.* 62 (10): 3171–3176.

7 Sixl, H. and Warta, R. (1985). Reaction mechanism of photochromic 2-(2′,4′-dinitrobenzyl)pyridine (DNBP) single crystals. *Chem. Phys.* 94 (1–2): 147–155.

8 Irie, M., Uchida, K., Eriguchi, T., and Tsuzuki, H. (1995). Photochromism of single crystalline diarylethene. *Chem. Lett.* 899-890.

9 Kobatake, S., Yamada, T., Uchida, K. et al. (1999). Photochromism of 1,2-bis(2,5-dimethyl-3-thienyl)perfluorocyclopentene in a single crystalline phase. *J. Am. Chem. Soc.* 121 (11): 2380–2386.

10 Kobatake, S., Yamada, M., Yamada, T., and Irie, M. (1999). Photochromism of 1,2-bis(2-methyl-6-nitro-1-benzothiophen-3-yl)perfluorocyclopentene in a single-crystalline phase: dichroism of the closed-ring form isomer. *J. Am. Chem. Soc.* 121 (37): 8450–8456.

11 Yamada, T., Kobatake, S., and Irie, M. (2000). X-ray crystallographic study on single-crystalline photochromism of 1,2-bis(2,5-dimethyl-3-thienyl)perfluorocyclopentene. *Bull. Chem. Soc. Jpn.* 73 (10): 2179–2184.

12 Kobatake, S., Morimoto, M., Asano, Y. et al. (2002). Absorption spectra of colored isomer of diarylethene in single crystals. *Chem. Lett.* 1224-1225.

13 Yamada, T., Kobatake, S., Muto, K., and Irie, M. (2000). X-ray crystallographic study on single-crystalline photochromism of bis(2,5-dimethyl-3-thienyl)perfluorocyclopentene. *J. Am. Chem. Soc.* 122 (8): 1589–1592.

14 Shibata, K., Muto, K., Kobatake, S., and Irie, M. (2002). Photocyclization/cycloreversion quantum yields of diarylethenes in single crystals. *J. Phys. Chem. A* 106 (1): 209–214.

15 Higashiguchi, K., Matsuda, K., Kobatake, S. et al. (2000). Fatigue mechanism of photochromic 1,2-bis(2,5-dimethyl-3-thienyl)perfluorocyclopentene. *Bull. Chem. Soc. Jpn.* 73 (10): 2389–2394.

16 Irie, M., Lifka, T., Kobatake, S., and Kato, N. (2000). Photochromism of 1,2-bis(2-methyl-5-phenyl-3-thienyl)perfluorocyclopentene in a single-crystalline phase. *J. Am. Chem. Soc.* 122 (20): 4871–4876.

17 Yamada, T., Muto, K., Kobatake, S., and Irie, M. (2001). Crystal structure-reactivity correlation in single-crystalline photochromism of 1,2-bis(2-methyl-5-phenyl-3-thienyl)perfluorocyclopentene. *J. Org. Chem.* 66 (189): 6164–6168.

18 Kobatake, S., Shibata, K., Uchida, K., and Irie, M. (2000). Photochromism of 1,2-bis(2-ethyl-5-phenyl-3-thienyl)perfluorocyclopentene in a single-crystalline phase. Thermal cycloreversion of the closed-ring isomer. *J. Am. Chem. Soc.* 122 (20): 12135–12141.

19 Morimoto, M., Kobatake, S., and Irie, M. (2003). Polymorphism of 1,2-bis(2-methyl-5-*p*-methoxyphenyl-3-thienyl)perfluorocyclopentene and photochromic reactivity of the single crystals. *Chem. Eur. J.* 9 (3): 621–627.

20 Morimoto, M., Kobatake, S., and Irie, M. (2003). Aryl-perfluoroaryl interaction in photochromic diarylethene crystals. *Crys. Growth Des.* 3 (5): 847–854.

21 Sumi, T., Takagi, Y., Yagi, A. et al. (2014). Photoirradiation wavelength dependence of cycloreversion quantum yields of diarylethenes. *Chem. Commun.* 50: 3928–3930.

22 Irie, M., Sakemura, K., Okinaka, M., and Uchida, K. (1995). Photochromism of dithienylethenes with electron-donating substituents. *J. Org. Chem.* 60 (25): 8305–8309.

23 Uchida, K., Guillaumont, D., Tsuchida, E. et al. (2002). Theoretical study of an intermediate, a factor determining the quantum yield in photochromism of diarylethene derivatives. *J. Mol. Struc.* 579: 115–120.

24 Kobatake, S., Uchida, K., Tsuchida, E., and Irie, M. (2002). Single-crystalline photochromism of diarylethenes: reactivity-structure relationship. *Chem. Commun.*: 2804–2805.

25 Ohara, H., Morimoto, M., and Irie, M. (2010). Photochromism of dithienylethene single crystals having anthracene substituents. *Photochem. Photobiol. Sci.* 9: 1079–1081.

26 Takami, S., Kuroki, L., and Irie, M. (2007). Photochromism of mixed crystals containing bisthienyl-, bisthiazolyl-, and bisoxazolylethene derivatives. *J. Am. Chem. Soc.* 129 (23): 7319–7326.

27 Kobatake, S., Matsumoto, Y., and Irie, M. (2005). Conformational control of photochromic reactivity in a diarylethene single crystal. *Angew. Chem. Int. Ed.* 44: 2148–2151.

28 Fukaminato, T., Kawai, T., Kobatake, S., and Irie, M. (2003). Fluorescence of photochromic 1,2-bis(3-methyl-2-thienyl)ethene. *J. Phys. Chem. B* 107 (33): 8372–8377.

29 Fukaminato, T., Kobatake, S., Kawai, T., and Irie, M. (2001). Three-dimensional erasable optical memory using a photochromic diarylethene single crystal as the recording medium. *Proc. Jpn. Acad. Ser. B* 77: 30–35.

30 Morimoto, M., Kashihara, R., Mutoh, K. et al. (2016). Turn-on mode fluorescence photoswitching of diarylethene single crystals. *CrystEngComm.* 18: 7241–7248.

31 Merian, E. (1966). Steric factors influencing the dyeing of hydrophobic fibers. *Text. Res. J.* 36: 612–618.

32 Kay, E.R., Leigh, D.A., and Zerbetto, F. (2007). Synthesis of molecular motors and mechanical machines. *Angew. Chem. Int. Ed.* 46: 72–191.

33 Anelli, P.L., Spencer, N., and Stoddart, J.F. (1991). A molecular shuttle. *J. Am. Chem. Soc.* 113 (13): 5131.

34 Liu, Y., Flood, A.H., Bonvallet, P.A. et al. (2005). Linear artificial molecular muscles. *J. Am. Chem. Soc.* 127 (27): 9745–9759.

35 Jiménez, M.C., Dietrich-Buchecker, C., and Sauvage, J.-P. (2000). Towards synthetic molecular muscles: contraction and stretching of a linear rotaxane dimer. *Angew. Chem. Int. Ed.* 39: 3284–3287.

36 Koumura, N., Zijlstra, R.W.J., van Delden, R.A. et al. (1999). Light-driven monodirectional molecular motor. *Nature* 401: 152–155.

37 Kobatake, S., Takami, S., Muto, H. et al. (2007). Rapid and reversible shape changes of molecular crystals on photoirradiation. *Nature* 446: 778–781.

38 Irie, M., Kobatake, S., and Horichi, M. (2001). Reversible surface morphology changes of a photochromic diarylethene single crystal by photoirradiation. *Science* 291: 1769–1771.

39 Kuroki, L., Takami, S., Yoza, K. et al. (2010). Photoinduced shape changes of diarylethene single crystals: correlation between shape changes and molecular packing. *Photochem. Photobiol. Sci.* 9: 221–225.

40 Irie, M., Fukaminato, T., Matsuda, K., and Kobatake, S. (2014). Photochromism of diarylethene molecules and crystals: memories, switches and actuators. *Chem. Rev.* 114: 12174–12277.

41 Kitagawa, D. and Kobatake, S. (2015). Photoreversible current ON/OFF switching by the photoinduced bending of gold-coated diarylethene crystals. *Chem. Commun.* 51: 4421–4424.

42 Terao, F., Morimoto, M., and Irie, M. (2012). Light-driven molecular-crystal actuators: rapid and reversible bending of rodlike mixed crystals of diarylethene derivatives. *Angew. Chem. Int. Ed.* 51: 901–904.

43 Morimoto, M. and Irie, M. (2010). A diarylethene cocrystal that converts light into mechanical work. *J. Am. Chem. Soc.* 132 (40): 14172–14178.

44 Ohshima, S., Morimoto, M., and Irie, M. (2015). Light-driven bending of diarylethene mixed crystals. *Chem. Sci.* 6: 5746–5752.

45 Kitagawa, D., Nishi, H., and Kobatake, S. (2013). Photoinduced twisting of a photochromic diarylethene crystal. *Angew. Chem. Int. Ed.* 52: 9320–9322.

46 Kitagawa, D., Tsujioka, H., Tong, F. et al. (2018). Control of photomechanical crystal twisting by illumination direction. *J. Am. Chem. Soc.* 140: 4208–4212.

47 Chiba, H. Morimoto, M., and Irie, M. (2021). Stepwise assembly of ultrathin poly(vinyl alcohol) films on photoresponsive diarylethene crystals. *Chem. Lett.* 50: 84–86.

5
Memory

Molecular photoswitches are potentially applicable to various types of optical recording. The molecules undergo reversible isomerization reactions between two isomers upon photoirradiation. The photoisomerization reactions can be used to transfer optical information into chemical information. *On/off* optical information is fixed to the chemical structure change of the molecules. The photochemical optical memory is based on a photon-mode recording method, which differs from the heat-mode recording employed in current optical media. The photon-mode recording has various advantages over heat-mode recording in terms of resolution, speed of writing, and multiplex recording, such as wavelength, polarization, and phase. Diarylethenes are the most promising molecules for the photon-mode recording media because of thermal stability of both isomers and high resistance to photofatigue.

5.1 Single-Molecule Memory

One of the characteristic features of molecular photoswitches is that even a single molecule can work as an information storage unit. Owing to the tremendous progress of optical detection techniques, it is now possible to observe photoswitching performance of a single molecule. An appealing feature of single molecules is their small size (~1 nm) and confinement of electrons in the isolated small domain. If a single molecule would work as one bit memory, the ultimate high-density (1 P bit/in.^{2}) optical memory could be realized.

The first step to achieve the single-molecule optical memory is to prepare photoswitchable fluorescent molecules, which undergo thermally irreversible photoisomerization reactions between fluorescent and nonfluorescent isomer states. Thermal irreversibility is an indispensable issue for the memory media. A convenient and frequently used approach is the integration of a thermally irreversible P-type photoswitchable diarylethene and a fluorescent unit in a molecule. Förster-type resonance energy transfer (FRET) mechanism can be adopted for the molecule to reversibly switch the fluorescence, as shown in Figure 3.13 [1, 2]. Dyad **126** was dispersed in polymer matrices and photoswitching of its fluorescence was observed at the single-molecule level. Figure 5.1 shows the fluorescence

Diarylethene Molecular Photoswitches: Concepts and Functionalities, First Edition. Masahiro Irie.
© 2021 WILEY-VCH GmbH. Published 2021 by WILEY-VCH GmbH.

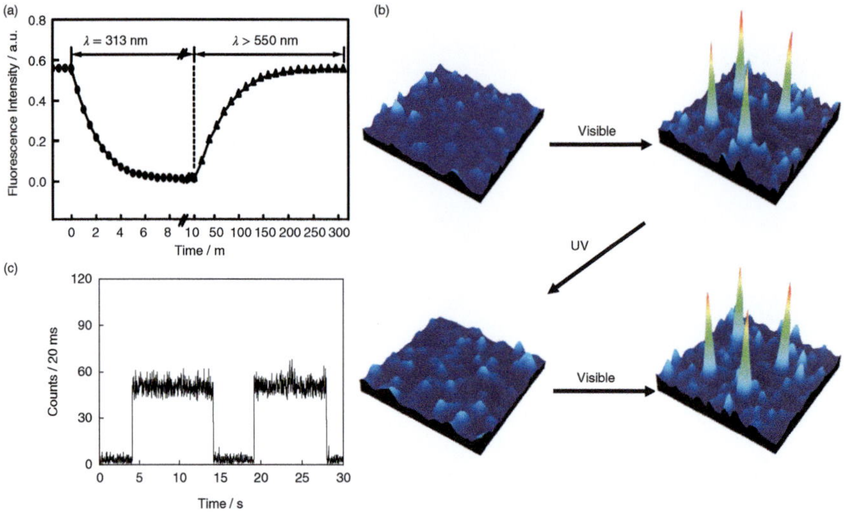

Figure 5.1 (a) Fluorescence photoswitching of dyad **126** upon alternate irradiation with UV (313 nm) and visible (>550 nm) light in the ensemble system in toluene. (b) The images of fluorescence photoswitching of four single molecules in a Zeonex polymer film upon irradiation with visible (488 nm) and UV (325 nm) light. (c) The time trace of the fluorescence at the single-molecule level in a Zeonex polymer film under irradiation with UV (325 nm) and visible (488 nm) light.

photoswitching in the ensemble system in toluene and at the single-molecule level. In the ensemble system composed of vast numbers of molecules, the intensity of fluorescence gradually decreased upon irradiation with visible (313 nm) light and then returned to the original intensity upon irradiation with visible (>550 nm) light. On the other hand, at the single-molecule level the fluorescence showed digital *on/off* photoswitching between two discrete states, as shown in Figure 5.1b,c. The images of the photoswitching of four molecules indicate that the single molecules work as one-bit optical memory. The time trace definitely indicates that the molecular photoswitch has photoswitchable bistable states.

Durable fluorescent diarylethene dyad **176** was prepared and a detailed analysis of single-molecule photoswitching was carried out in polymer matrices, which have different *Tg*, such as poly (*n*-butyl methacrylate) (PnBMA), poly(methyl methacrylate) (PMMA), and amorphous polyolefin (Arton) [3]. The time trace of the fluorescence was measured under irradiation with both UV (325 nm) and visible (488 nm) light. The histograms of *on* and *off* response times are shown in Figure 5.2. When the polymer matrix had a low *Tg*, such as PnBMA ($Tg = 21\,°C$), the histograms were normal exponential shapes. The exponential distribution indicates that the quantum yield of the photoswitching reaction is constant. On the other hand, when *Tg* of the polymer matrix was high, such as PMMA ($Tg = 105\,°C$) and Arton ($Tg \sim 300\,°C$), the histograms showed a peak. The abnormal distribution indicates that the quantum yield of the photoreactions is not constant. The reactions take place after several excitation processes. The appearance of the peak in the histograms suggests that the quantum yield increases with an increasing number of absorbed photons.

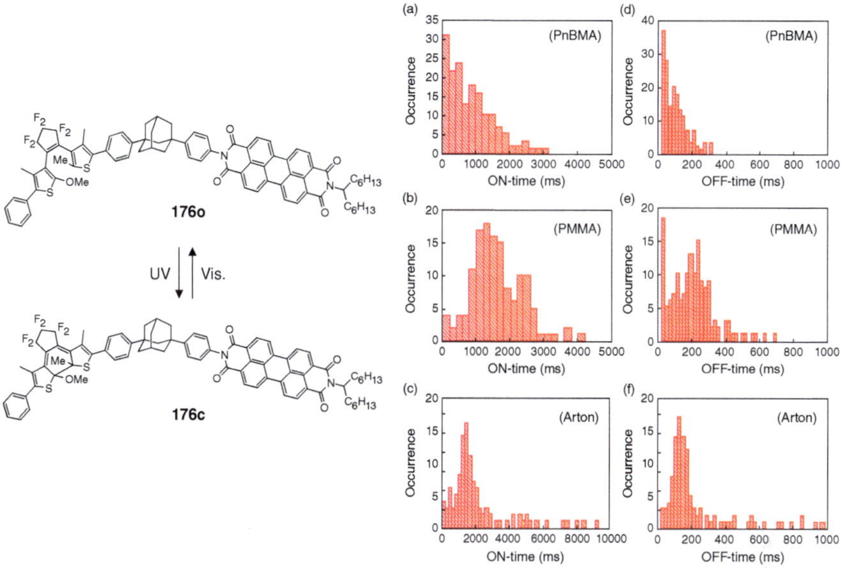

Figure 5.2 Histograms of (a–c) on time and (d–f) off time in PnBMA (a, d), PMMA (b, e), and Arton (c, f). The histograms were constructed from the time trace of fluorescence intensity of **176** in the polymer matrices irradiated with both 488 nm and 325 nm light.

A possible interpretation of the abnormal histograms is given by a multilocal minima model, as shown in Figure 5.3. The polymer chains surrounding the molecular photoswitches provide steric hindrance. The steric hindrance modifies the potential energy surfaces of both ground and excited states. In the absence of such steric hindrance in low-*Tg* polymer matrices, the molecule undergoes one-step photoisomerization, resulting in a constant quantum yield, as shown in Figure 5.3a. In high-*Tg* polymer matrices, on the other hand, multilocal potential surfaces prevent the one-step photoreaction and require multistep photoreactions to reach the final reaction process, as shown in Figure 5.3b. It requires many excitation steps for the molecule to undergo the isomerization reaction. Therefore, the quantum yield depends on the number of excitation processes or the absorbed photons. The results indicate the photoreactions in polymer matrices strongly depend on *Tg* of the polymers.

A drawback of the above molecular systems is that the optical information (the fluorescence *off* state) is destroyed during reading the information. The fluorescence quenching by the FRET mechanism induces an undesirable photoreaction to convert the *off* state to the *on* state with reading light. To avoid the destructive readout, an intramolecular electron transfer (IET) mechanism was employed as the fluorescence quenching mechanism [4]. A photoswitchable fluorescent molecule **177** composed of a sulfone derivative of diarylethene as the electron acceptor and a fluorescent perylenebisimide derivative as the electron donor was prepared, as shown in Figure 5.4. The fluorescence of the closed-ring isomer remarkably decreased when the dielectric constant of the solvent increased above 10, while that of the open-ring isomer remained constant. In polar solvents or in polar polymer matrices, therefore,

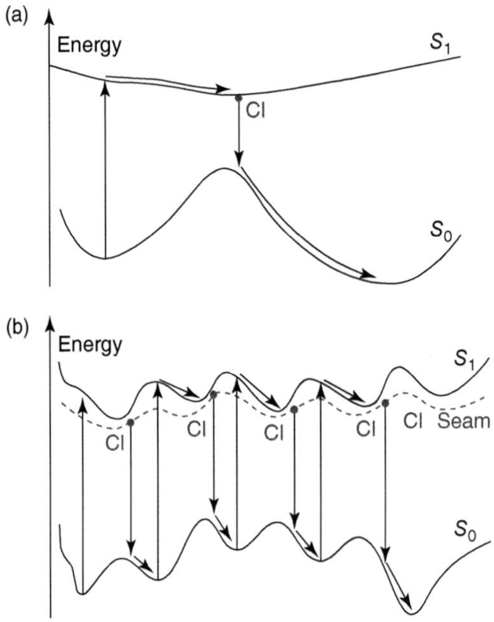

Figure 5.3 A schematic diagram of the potential energy surfaces of a diarylethene (a) in low-*Tg* polymer matrices and (b) in high-*Tg* polymer matrices. CI indicates conical intersection. *Source*: Reprinted with permission from Ref. [3]. Copyright 2007 American Chemical Society.

the fluorescence was reversibly switched by alternate irradiation with visible and UV light. The time trace in a polar poly(methyl acrylate) film is shown in Figure 5.4. The fluorescence *off* state was initially observed under irradiation with 532 nm light. The fluorescence *off* state changed to the fluorescence *on* state in one step upon UV irradiation. After stopping UV light irradiation, the fluorescence *on* state remained stable during irradiation with 532 nm light. The fluorescence *on* state turned back to the *off* state when the sample was irradiated with visible (440–490 nm) light. After stopping visible light irradiation, the dark state remained stable under irradiation with 532 nm light until the next irradiation with UV light. It is worth noting that both fluorescence *off* and *on* states remain stable during excitation with 532 nm light. The result clearly indicates that by using the IET mechanism, the nondestructive readout of fluorescence *on/off* states is accomplished at the single-molecule level.

5.2 Near-Field Optical Memory

One of the problems to be solved in the single-molecule optical memory is how to address optical information into single molecules. It is required to convey optical information into single molecules, which are placed separately at a distance of nanometers. The spot size of focused laser light in conventional optical recording is restricted by the diffraction limit, wavelength of light, and numerical aperture of the lens. The size cannot be reduced to less than one-half of the wavelength of light. The minimum spot size is in the order of a few hundred nanometers. One of the approaches to reduce the spot size is to employ near-field optics. If the light source is brought into close proximity with the surface of the recording medium (less than

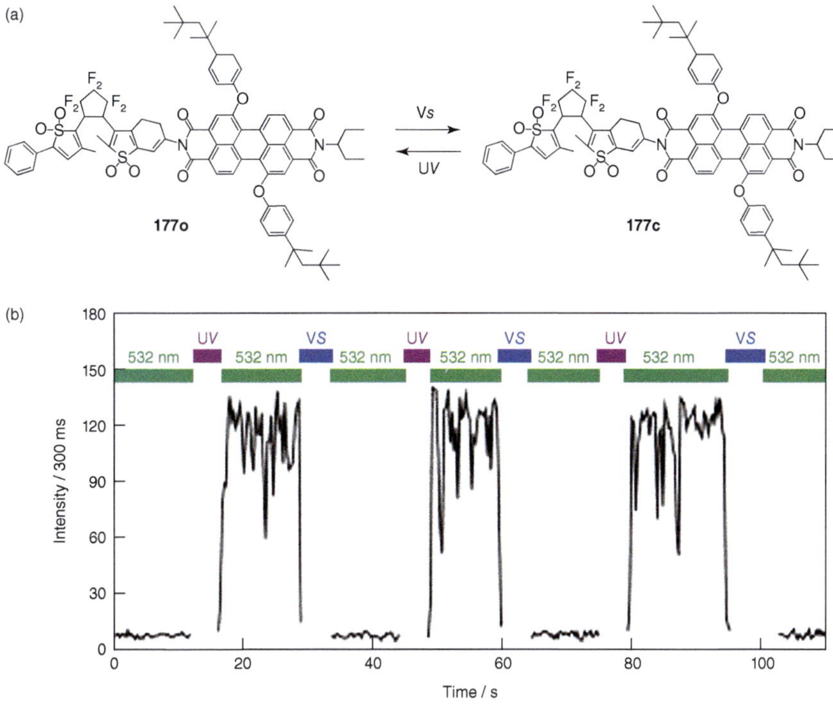

Figure 5.4 Single-molecule fluorescence photoswitching in a poly(methyl acrylate) thin film containing **177**. (a) Chemical structures of open- and closed-ring isomers of **177**. (b) Time trace of the fluorescence intensity of a single **177** molecule under excitation with 532 nm light. Reversible photoswitching was observed upon alternate irradiation with UV (330–380 nm) light and visible (440–490 nm) light.

the wavelength of light, that is near-field), the mark size depends on the diameter of the light source and the distance between the light source and the surface.

The near-field recording using a transmittance-detection method was attempted by using a polystyrene film containing diarylethene **13** (20 wt%) with 1.5 μm thickness [5]. For the optical recording, the film was initially uniformly colored by irradiation with UV light and then bleached using a small micropipette light source. The mark size (~1 μm wide and ~4 μm long) was, however, much larger than the diameter of the micropipette (~100 nm) due to light scattering from the thick film.

In order to obtain smaller marks, the medium thickness should be reduced. One possible approach to reduce the thickness is to use bulk amorphous or crystalline diarylethenes. A clear amorphous film of diarylethene **178** was prepared by spin coating the toluene solution on quartz substrates [6]. The film thickness was ~300 nm. The thickness is enough to obtain a transmittance change of 80%. The amorphous film was initially colored to blue by irradiation with UV light and then bleached with 633 nm laser light from a small micropipette light source. Figure 5.5 shows the images of the small marks. The marks were able to be bleached by irradiation with UV light. The mark size diameter was less than 100 nm, which corresponds to the recording density of >70 Gbit/in.^{2}.

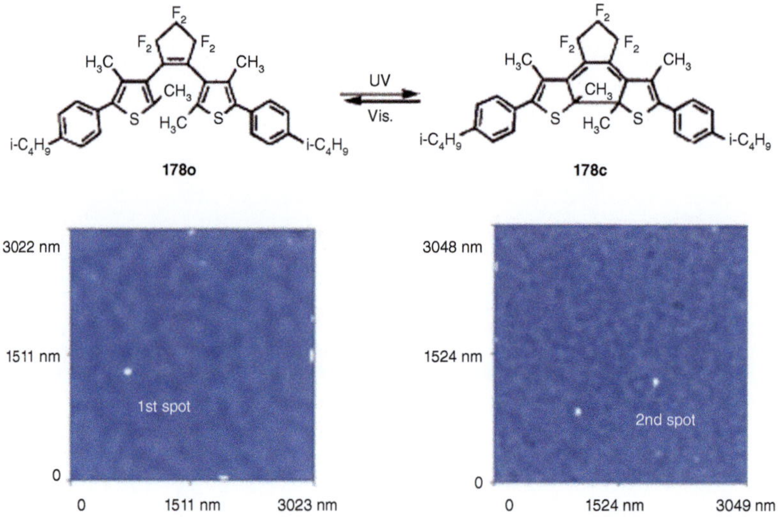

Figure 5.5 Near-field optical recording on an amorphous film of **178**. Thickness of the film was 300 nm.

Nanometer-scale photoswitching reactions on the surface of a diarylethene crystal were also demonstrated using an optical near-field assisted AFM [7]. Upon irradiation with a small light source from the probe tip, a nanometer-scale concavity (~30 nm) was reversibly formed on the surface. The results suggest the fundamental ability of diarylethene crystals in dynamic memorization by using nanometer-scale light–matter interactions.

5.3 Three-Dimensional Optical Memory

Current optical disk memory is based on two-dimensional (2D) (x–y space) recording. Optical information is stored in 2D arrays of marks on the surface. If the 2D memory layer is stacked one by one into multi-layers, the memory density increases with an increase in the number of layers (longitudinal z-axis). The three-dimensional (3D) optical recording systems are highly promising for future high-capacity optical memory. 3D optical memory using diarylethene derivatives was demonstrated on diarylethene-doped films, amorphous films, or crystals.

Toriumi and Kawata [8] recorded 3D bit marks in a diarylethene-doped PMMA film by using a mode-locked Ti–sapphire laser (760 nm) with a short pulse width (=100 fs). The high-intensity laser pulse can induce photochemical reactions only at the focused position of the laser beam, where two-photon absorption (TPA) takes place. The diarylethene molecules excited by the TPA undergo the reaction, just as when excited with 380 nm light. When such near-infrared (IR) laser light is used, the focus position can be controlled at will deep in the recording medium.

It would be more desirable to use a laser diode (LD) as the light source. The TPA recording was carried out using the LD light source with a pulse width of 2 ns on a

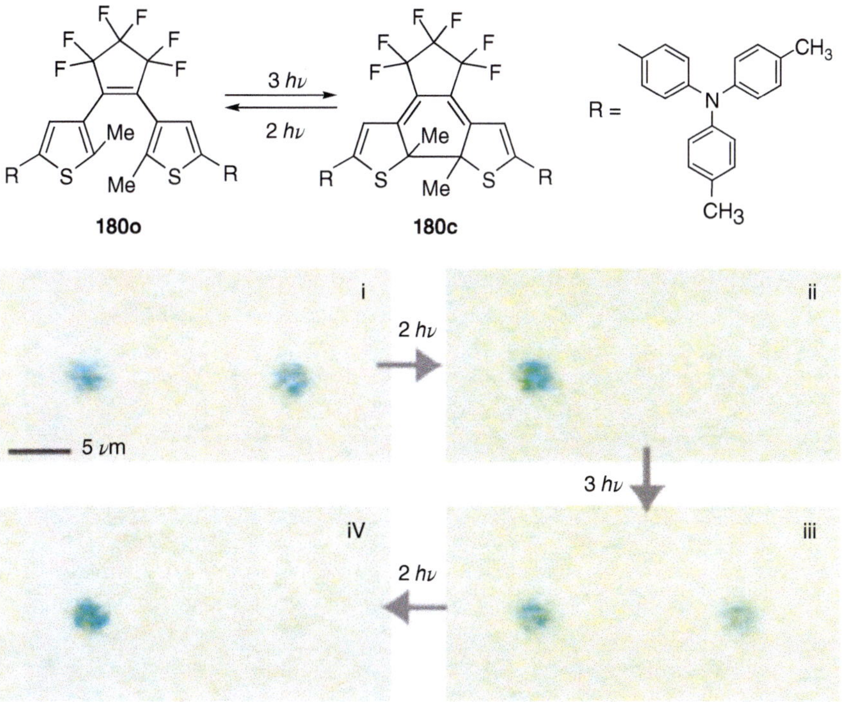

Figure 5.6 The images of an amorphous film of diarylethene **180** irradiated with Cr:F laser (1.28 μm). Irradiation of the NIR laser with weaker intensity of 48 pJ/pulse (100 kHz) for 180 seconds led to decoloration ((i) → (ii), (iii) → (iv)), while the laser with high intensity of 0.7 nJ/pulse (100 kHz) for 50 ms induced coloration ((ii) → (iii)). *Source*: Reprinted with permission from Ref. [10]. Copyright 2014 American Chemical Society.

bulk amorphous film composed of *cis*-1,2-dicyano-1,2-bis(2,4,5-trimethyl-3-thienyl)-ethene (**179**) with thickness of ~100 μm [9]. IR LD light (785 nm) was used as the recording light source and green LD light (520 nm) was used to erase the memory. The bit marks were readout using the refractive index changes in the IR region. Reversible recording of pit arrays by using 785 and 520 nm light was successfully demonstrated.

Multiphoton absorption was applied to induce both cyclization and cycloreversion reactions of diarylethene **180** using a single near-IR femtoseconds laser pulse at 1.28 μm with 35 fs full width at half-maximum, as shown in Figure 5.6 [11]. Simultaneous three-photon absorption by irradiation with high-intensity laser pulse (~1 nJ/pulse) caused the cyclization reaction (coloration), while TPA by irradiation with weaker laser pulse induced the cycloreversion reaction (decoloration). Single-wavelength laser (1.28 μm) light can induce both cyclization and cycloreversion reactions by controlling the laser light intensity.

Holographic recording is another approach to increase the memory capacity. The holographic recording utilizes characteristic features of light. The interference patterns between object and reference beams are recorded on the memory media and the patterns are readout using the reference beam. The sensitivity of the

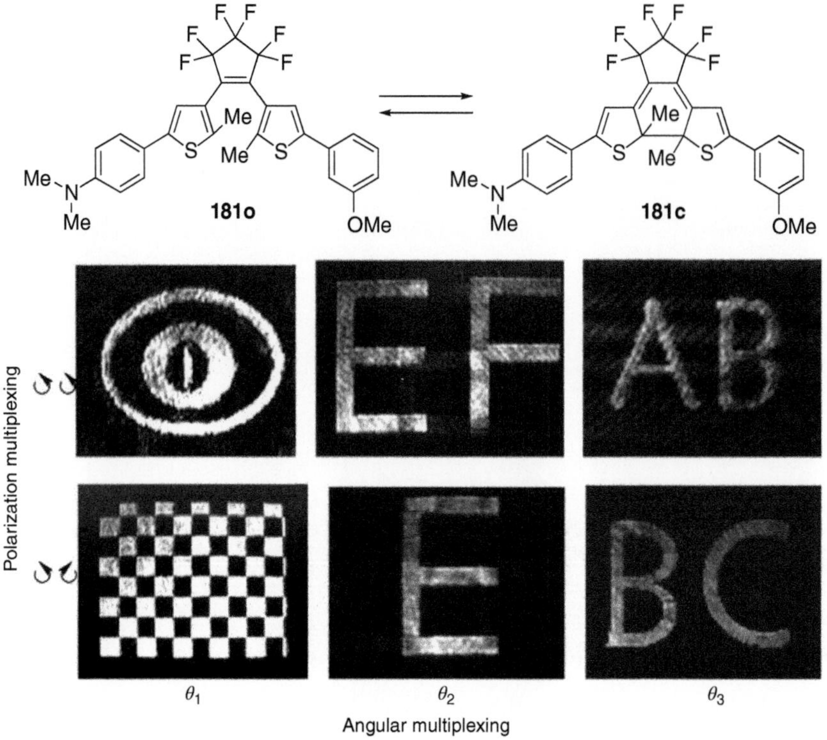

Figure 5.7 Holographic recording on a PMMA film containing **181** based on a polarization modulation technique. Angular multiplex and polarization multiplexing were combined. *Source*: Reprinted with permission from Ref. [10]. Copyright 2014 America Chemical Society.

holographic recording is dependent on the extent of refractive index modulation by photoirradiation. Some of diarylethene derivatives were used as the memory media and multiplex recording was demonstrated. Figure 5.7 shows an example [12]. Polarization holographic recording on a PMMA film containing **181** was carried out. By combining angular and polarization multiplexing, six holograms were recorded in the same place of the film with good contrast and no crosstalk. The recorded data were stable at room temperature in the dark for more than 2 months, and no remarkable degradation was found in the resolution of the reconstructed patters after 12 write/erase cycles.

5.4 Readout Using Infrared Absorption, Raman Scattering, and Refractive Index Changes

In molecular photoswitch memory, it is essential to develop readout methods, which can avoid the destruction of signals during the reading process. When absorption spectral changes in UV and visible regions are used as the recording signals, the signals are destroyed by the reading light. A possible way to avoid

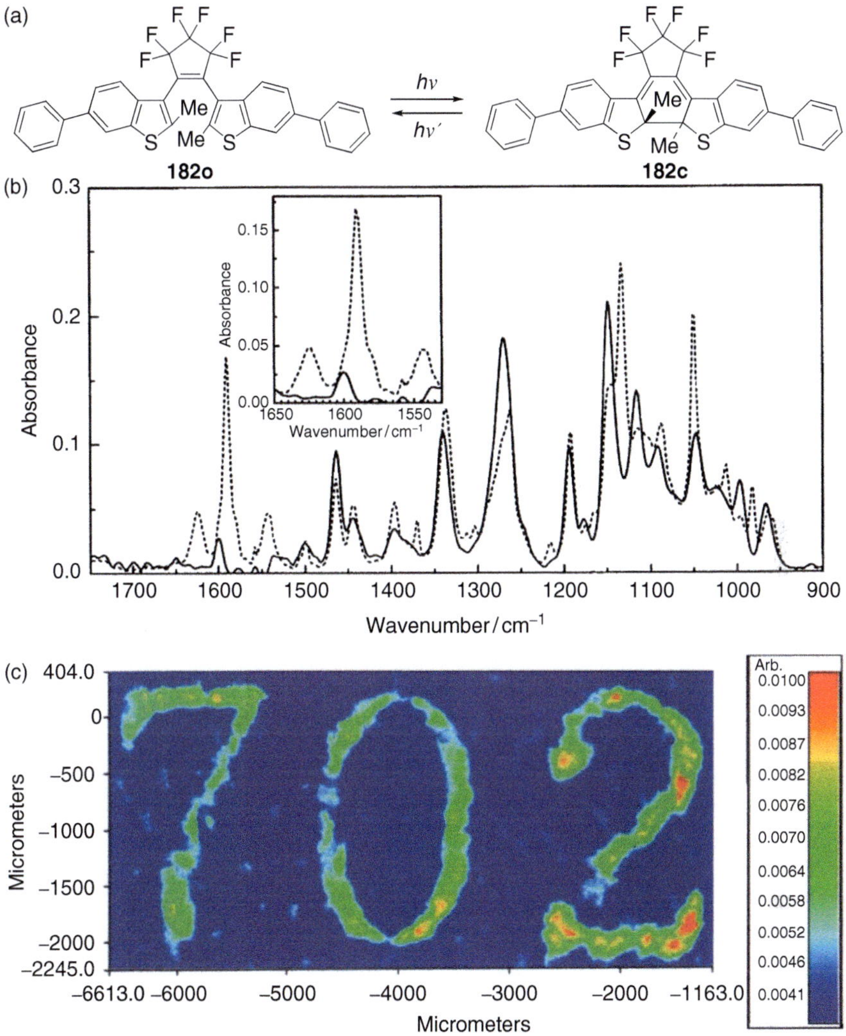

Figure 5.8 (a) Chemical structures of open- and closed-ring isomers of **182**. (b) IR spectra of **182o** (solid line) and **182c** (broken line). (c) IR image recorded on a Zeonex polyolefin film containing 4 wt% **182**. The IR image was monitored by IR light of 1590 cm^{-1}, where only the closed-ring isomer shows strong absorption.

the destructive readout is to use IR absorption or Raman spectral changes as the readout signals. If the difference between the open- and closed-ring isomers could be detected by using IR light, the recording information would be read many times without destruction. Uchida and coworkers [13] found that diarylethene **182** has a strong absorption peak at 1590 cm^{-1} only in the closed-ring isomer, as shown in Figure 5.8. Diarylethene **182** was doped into a Zeonex polyolefin film, which has no absorption between 1500 and 1700 cm^{-1}. Initially the film was uniformly colored with UV light. Then, a mask patter, in which 702 is written, was placed on the film

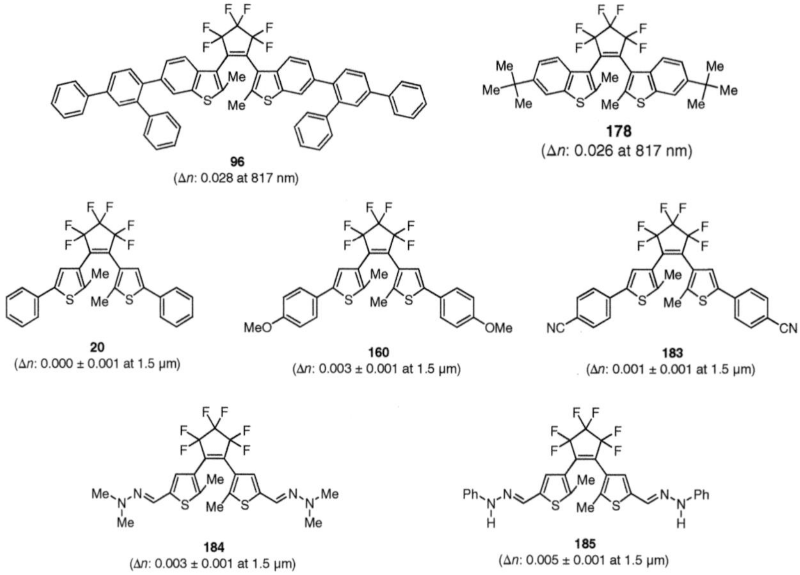

Scheme 5.1 Refractive index changes (Δ*n*) of diarylethene derivatives at 817 nm and 1.5 μm.

and irradiated with visible light. After visible light irradiation, purple number 702 pattern remained. The image was monitored by IR light at 1590 cm^{-1}, as shown in Figure 5.8c. The green areas have stronger absorption at the wavenumber. Even after prolonged readout of the image using the IR light, no destruction of the image was observed. Upon irradiation with visible light, the image disappeared and the recorded information was erased. Raman spectral changes can also be used for nondestructive readout (see Section 8.1).

Diarylethene molecules change not only absorption and fluorescence spectra but also refractive index, as shown in Scheme 5.1 [14–16]. The refractive index changes can be used for nondestructive readout, because the changes can be detected by light, the wavelength of which is longer than the absorption spectra of both open- and closed-ring isomers. A Diarylethene derivative having 2,4-diphenylphenyl substituents **96** forms a clear amorphous film by spin coating the toluene solution. Upon alternate irradiation with UV and visible light, the refractive indices of the amorphous film of **96** was modulated by as large as 0.028 between 1.617 and 1.645 at 817 nm, where both open- and closed-ring isomers have no absorption band [15]. The refractive change at 817 nm can be used as the readout signal.

References

1 Irie, M., Fukaminato, T., Sasaki, T. et al. (2001). A digital fluorescent molecular photoswitch. *Nature* 420: 759–760.

2 Fukaminato, T., Sasaki, T., Kawai, T. et al. (2004). Digital photoswitching of fluorescence based on the photochromism of diarylethene derivatives at a single-molecule level. *J. Am. Chem. Soc.* 126 (45): 14843–14849.
3 Fukaminato, T., Umemoto, T., Iwata, Y. et al. (2007). Photochromism of diarylethene single molecules in polymer matrices. *J. Am. Chem. Soc.* 129 (18): 5932–5938.
4 Fukaminato, T., Doi, T., Tamaoki, N. et al. (2011). Single-molecule fluorescence photoswitching of a diarylethene-perylenebisimide dyad: non-destructive fluorescence readout. *J. Am. Chem. Soc.* 133: 4984–4990.
5 Hamano, M. and Irie, M. (1996). Rewritable near-field optical recording on photochromic thin films. *Jpn. J. Appl. Phys.* 35 (3): 1764–1767.
6 Kim, M.-S., Sakata, T., Kawai, T., and Irie, M. (2003). Amorphous photochromic films for near-field optical recording. *Jpn. J. Appl. Phys.* 42 (6A): 3676–3681.
7 Nakagomi, R., Uchiyama, K., Suzuki, H. et al. (2018). Nanometre-scale pattern formation on the surface of a photochromic crystal by optical near-field induced photoisomerization. *Sci. Rep.* 8: 14468.
8 Toriumi, A. and Kawata, S. (1998). Reflection confocal microscope readout system for three-dimensional photochromic optical data storage. *Opt. Lett.* 23 (24): 1924–1926.
9 Shiono, T., Itoh, T., and Nishino, S. (2005). Two-photon absorption recording in photochromic diarylethenes using laser diode for three-dimensional optical memory. *Jpn. J. Appl. Phys.* 44 (5B): 3559–3563.
10 Irie, M., Fukaminato, T., Matsuda, K., and Kobatake, S. (2014). Photochromism of diarylethene molecules and crystals: memories, switches and actuators. *Chem. Rev.* 114: 12174–12277.
11 Mori, K., Ishibashi, Y., Matsuda, H. et al. (2011). One-color reversible control of photochromic reactions in a diarylethene derivative: three-photon cyclization and two-photon cycloreversion by a near-infrared femtosecond laser pulse at 1.28 μm. *J. Am. Chem. Soc.* 133: 2621–2625.
12 Li, H., Pu, S., Liu, G. et al. (2010). Polarization holographic optical recording based on a new photochromic diarylethene compound. *Front. Chem. China* 5 (2): 234–240.
13 Uchida, K., Saito, M., Murakami, A. et al. (2003). Non-destructive readout of the photochromic reactions of diarylethene derivatives using infrared light. *Adv. Mater.* 15 (2): 121–125.
14 Chauvin, J., Kawai, T., and Irie, M. (2001). Refractive index change of an amorphous bisbenzothienylethene. *Jpn. J. Appl. Phys.* 40 (4A): 2518–2522.
15 Kim, M.-S., Maruyama, H., Kawai, T., and Irie, M. (2003). Refractive index changes of amorphous diarylethenes containing 2,4-diphenylphenyl substituents. *Chem. Mater.* 15 (24): 4539–4543.
16 Bertarelli, C., Bianco, A., D'Amore, F. et al. (2004). Effect of substitution on the change of refractive index in dithienylethenes: an ellipsometric study. *Adv. Funct. Mater.* 14 (4): 357–363.

6
Switches

The photoswitching reaction of diarylethenes is one of electrocyclic reactions between two isomers, hexatriene and cyclohexadiene ones. Diarylethene molecular photoswitches inherently possess bistability. A striking feature of the bistability is that while π-systems of the two aryl rings are disconnected in the open-ring hexatriene isomer, the closed-ring cyclohexadiene isomer has a planar structure and π-electrons are delocalized through the molecule. The difference in the two electronic and geometrical structures brings about physical and chemical property changes of materials when the bistable molecules are incorporated into the materials.

6.1 Single-Molecule Conductance Photoswitch

Photoswitching of conductance of a single molecule has been a long-standing dream in the field of molecular electronics. Conductance switches are basic components of almost all electronic devices. Manufacturing reliable molecular size switches is the first step toward molecular electronics. Various approaches have been proposed to accomplish single-molecule photoswitching. When a single photoswitchable bistable molecule is inserted between two nanogap electrodes, the single-molecule junction works as a photoswitch.

One-way conductance photoswitching using diarylethene derivative **186** was reported for the first time by van Wees and co-workers [1]. The closed-ring isomer of diarylethene **186** with thiol substituents was inserted in a gold nanogap prepared by a mechanically controllable break junction and the conductance photoswitching was measured along with the photocycloreversion reaction, as shown in Figure 6.1. By irradiation of the closed-ring isomer with visible light, the resistance increased about 3 orders of magnitude. Although in solution reversible photoisomerization reactions between the closed- and open-ring isomers upon irradiation with UV and visible light were observed, the cyclization reaction with UV light was prohibited when the molecule was connected to the gold electrodes. Only one-way conductance decrease was observed upon irradiation with visible light.

Nuckolls and co-workers [2] prepared diarylethene-bridged single-walled carbon nanotubes (SWNTs) having **187** or **188** as the bridge, as shown in Figure 6.2. SWNT

Diarylethene Molecular Photoswitches: Concepts and Functionalities, First Edition. Masahiro Irie.
© 2021 WILEY-VCH GmbH. Published 2021 by WILEY-VCH GmbH.

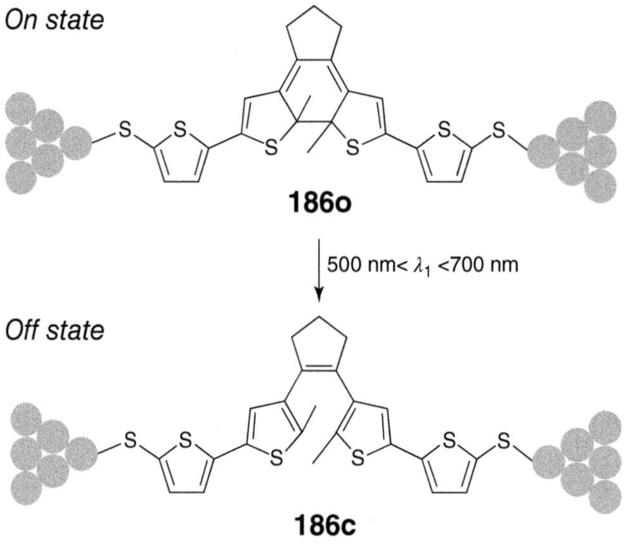

Figure 6.1 One-way switching of conductance from the on state to the off state of a single-molecule photoswitch. By exposing the switch to light of wavelength in the range of 500 nm < λ < 700 nm, the molecule switched from **186o** to **186c**. The reverse switching from the off state to the on state was not observed.

was oxidatively cut through a lithographic mask defined with an electron beam. The amide bonds were formed between the diarylethene diamines and carboxylic acids that terminate the SWNT. By irradiation with UV light, the conductance of the switch bridged with **188** increased by more than 2 orders of magnitude. The pyrrole based compound **188** is known to undergo a T-type thermally reversible photocycloreversion reaction. Therefore, the high conductance state returned to the low conductance state in the dark. In this case, light can switch on the conductance, while the reverse switching off process takes place thermally.

Guo and co-workers [3] prepared a graphene–diarylethene–graphene junction, which photoswitches the conductance reversibly at room temperature. A single diarylethene molecule **189** was inserted into graphene sheets, as shown in Figure 6.2a. Although several studies have demonstrated the single-molecule photoswitching using diarylethene derivatives, they failed to produce actual bidirectional conductance photoswitching. To achieve the reversible bidirectional photoswitching, three methylene (CH_2) groups were incorporated into each side of the diarylethene molecular switch. Theoretical analysis suggested that this structural modification reduces the electrode–molecule coupling to a level that may avoid the quenching of the excited state by the electrodes. The current–voltage characteristics measurement demonstrated that onset of the current through the closed-ring isomer occurred at ~0.3 V, whereas no appreciable conduction through the open-ring isomer was observed up to 1 V, as shown in Figure 6.2b. The diarylethene-inserted junction was found to undergo reversible photoswitching at a single-molecule level in a robust and reproducible way, as shown in Figure 6.2c. The *on/off* ratio was 107.1 ± 56.3 when light with different wavelengths (UV and

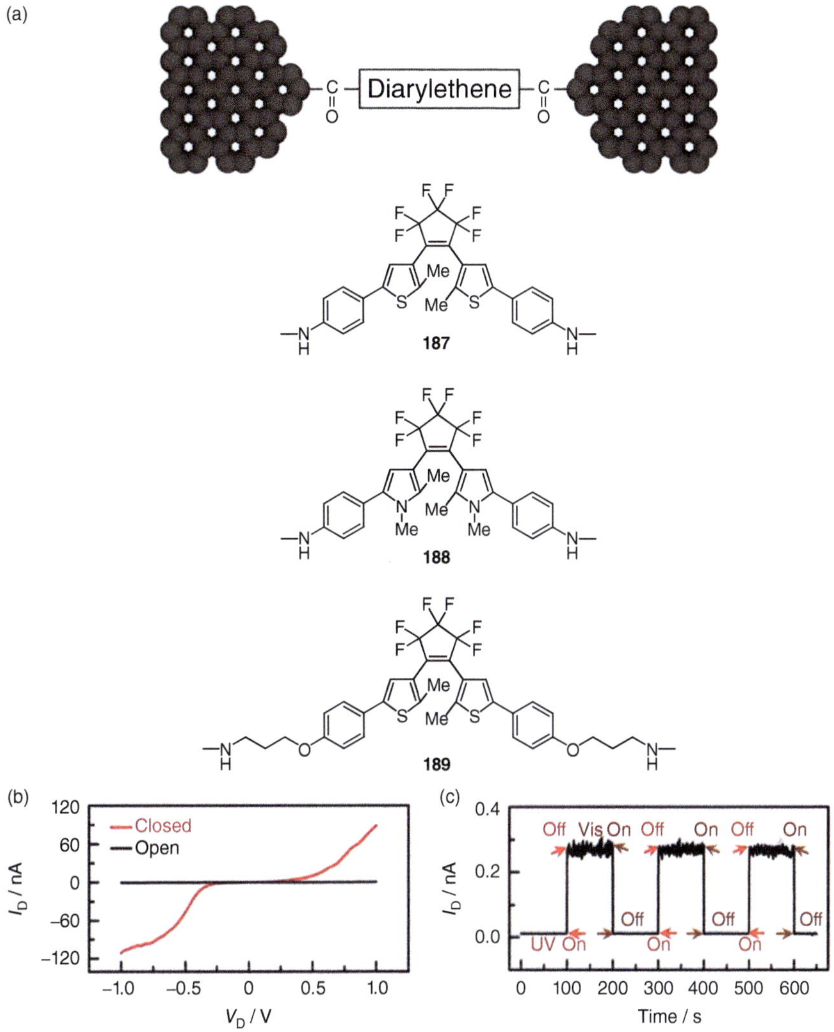

Figure 6.2 Photoswitching of conductance of diarylethene derivatives inserted between carbon electrodes (single-walled carbon nanotubes or graphene sheets). (a) Schematic representation of the junction. (b) *I*–*V* characteristics of individual diarylethene **189** in open-ring (black line) and closed-ring (red line) forms at gate voltage $V_G = 0$ V. V_D, drain voltage; I_D, drain current. (c) Real-time measurement of the current through a diarylethene **189** that reversibly switched between the closed-ring and open-ring forms, upon exposure to UV and visible light, respectively. $V_D = 100$ mV and $V_G = 0$ V.

visible) was toggled back and forth. The results indicate the crucial importance of the molecule–electrode coupling strength in determining the device performance.

Theoretical study on the conductance photoswitching of a single bistable molecule has been extensively carried out [4–8]. Yoshizawa and co-workers [4, 5] and Tsuji and Hoffmann [8] calculated the conductance of individual molecules based on molecular orbital calculations combined with the Green's function theory of conductance.

HOMO-LUMO gaps, π-electron conjugations, and phase and amplitude of the frontier molecular orbitals at the sites connected to electrodes are important parameters in determining transmission through the π-electron systems. The theoretical calculations predicted the *on*/*off* ratio of conductance for the open- and closed-ring isomers of diarylethenes is of the order magnitude 10^2, which is qualitatively consistent with the above experimental results.

Another technique used to detect conductance photoswitching of single molecules is scanning tunneling microscopy (STM). STM is developed for imaging surfaces at a single-molecule level. The apparent height of the STM image is a convolution of electronic and topographic characteristics. Therefore, the change of the apparent height along with photoisomerization can provide information on the electronic structure change of a single molecule. Feringa, van Wees, and co-workers [9] prepared a mixed self-assembled monolayer, which incorporates dodecanethiols and asymmetric diarylethene **190** in the closed-ring form on Au (111), as shown in Figure 6.3, and observed that the STM image changes upon alternate irradiation with visible (>420 nm) and UV (313 nm) light. Reversible bidirectional photoconversion between two isomer states was successfully observed by using the asymmetric diarylethene. After 1 minute of irradiation with visible light, the bright protrusion disappeared, suggesting that the single molecule in the closed-ring "*on*" form switched to open-ring "*off*" form. Within 4 minutes of UV irradiation, the transition from the "*off*"- to the "*on*" state took place and bright protrusion was restored. Reversible photoswitching of the STM apparent height of

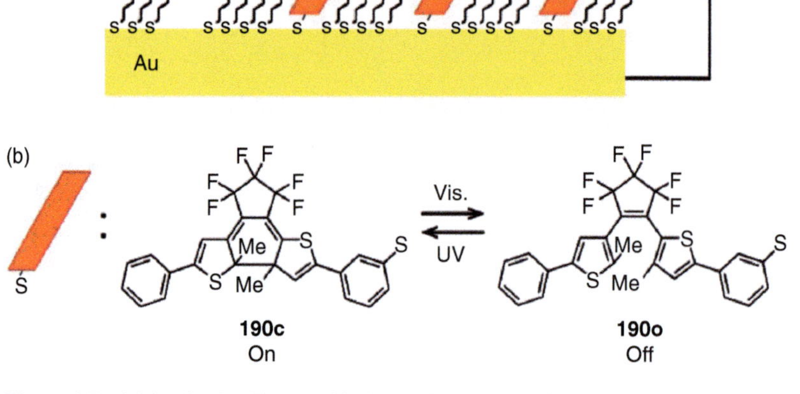

Figure 6.3 (a) A mixed self-assembled monolayer, which incorporates dodecanethiol and asymmetric diarylethene **192c** on Au (111) surface. (b) The chemical structures of the asymmetric diarylethene in the closed-ring form **192c** and the open-ring form **192o**.

individual diarylethene molecules was reproducible and could be repeated for a number of irradiation cycles. The single molecules that photoswitch between two stable conductance states can be used as ultrahigh-density optical memory.

6.2 Optical Switch Based on Refractive Index Change

The closed-ring isomer has higher polarizability and refractive index in comparison with the open-ring isomer. Therefore, it is possible to modulate the refractive index upon alternate irradiation with UV and visible light. For example, the refractive indices of **96** and **178** change by as large as 0.02–0.03 at 817 nm. Scheme 5.1 shows the refractive index changes (Δn) between the open- and closed-ring isomers of **96**, **178**, **20**, **160**, **183**, **184**, and **185**. Although Δn of compound **20** at 1.5 μm is lower than 10^{-3}, Δn increases to $(3 \pm 1) \times 10^{-3}$ for compound **160**. Compound **185** showed Δn as large as 10^{-2} between 800 and 1000 nm. Donor substituents on diarylethenes give rise to an increase in Δn.

Waveguide-type optical switches were prepared using the difference in the refractive indexes between the open- and closed-ring isomers [10]. Figure 6.4 shows one of such switches. The Mach–Zehnder modulator is composed of polycarbonate doped with diarylethene **13** in the core and a thick light blocking metal layer on it, as shown in Figure 6.4. The metal layer was opened on one arm of the modulator, so that only one arm of the modulator can be irradiated with light, resulting in a differential phase shift. The optical modulator exhibited an extinction ratio of about −12 dB at a wavelength of 1.55 μm.

6.3 Magnetism

The electronic structure difference between the two isomers of diarylethenes influences intramolecular magnetic interactions. Figure 6.5 shows open-ring isomer **191o** and closed-ring isomer **191c** of a radical-substituted diarylethene with simplified structures. Although there is no resonant closed-shell structure for **191o**, there exists a resonant quinoid-type closed-shell structure for **191c**, as in the structures shown below. In the open-ring isomer **191o**, the bond-alternation is disconnected

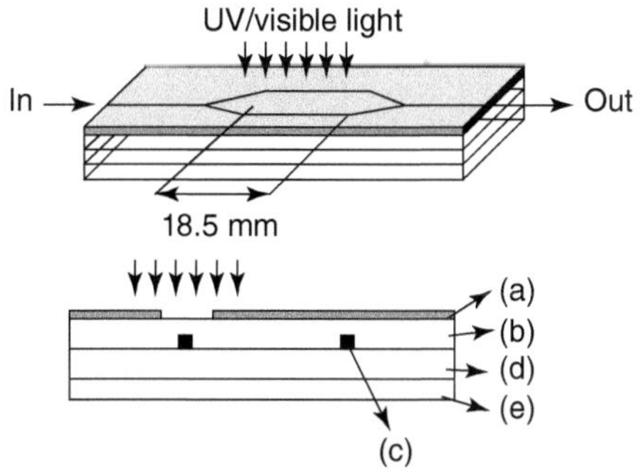

Figure 6.4 Configuration of the Mach–Zehnder modulator using diarylethene **13**. (a) Au layer (200 nm), (b) UV-curable VTC-2 (25 μm), (c) polycarbonate doped with diarylethene **13** (2 μm, 37 wt%), (d) Cyclotene 3022 (benzocyclobutene) (18 μm), (e) silicon wafer. *Source*: Reprinted with permission from Ref. [10]. Copyright 2002 AIP Publishing.

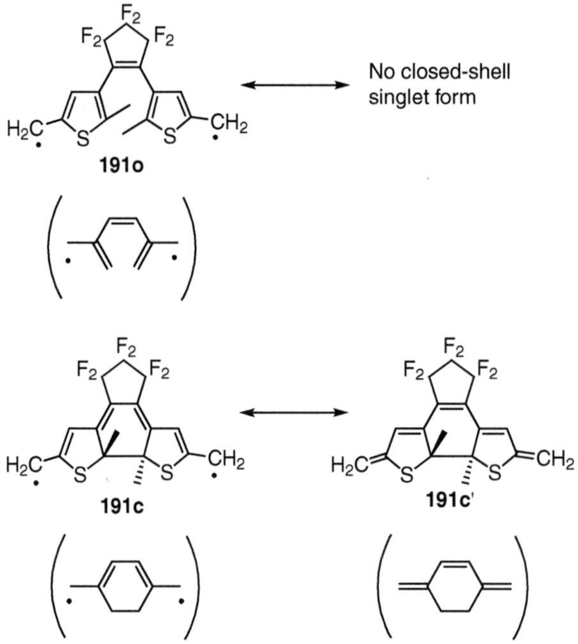

Figure 6.5 The open-ring (**191o**) and closed-ring (**191c**) isomers of radical-substituted diarylethene **191**.

at the 3-position of the thiophene rings, but in the closed-ring isomer **191c** the ground state electronic structure has no unpaired electrons. The conjugation of the open-ring isomer is a typical disjoint biradical, in which the intramolecular radical–radical interaction is weak. On the other hand, in the closed-ring isomer strong antiferromagnetic interaction is expected to operate. Therefore, switching of the intramolecular interaction is expected to take place upon photoisomerization between the open- and closed-ring isomers. The open-ring isomer is in an "*off*" state, while the closed-ring isomer is in an "*on*" state.

Two radicals were introduced to diarylethene **192**, as shown in Scheme 6.1, in which 1,2-bis(2-methyl-1-benzothiophen-3-yl)perfluorocyclopentene **13** was employed as a photoswitching spin coupler and nitronyl nitroxides as spin sources [11]. The measurement of magnetic susceptibilities revealed that spins of **192c** have a remarkable antiferromagnetic interaction ($2J/\mathbf{k}_B = -11.6$ K), although the interaction between the two spins is small in the open-ring isomer **192o** ($2J/\mathbf{k}_B = -2.2$ K). The result agrees with the theoretical prediction that the open-ring isomer is in "*off*" state and the closed-ring isomer is in "*on*" state.

Although the switching of the exchange interaction was detected by susceptibility measurement, both the open- and closed-ring isomers of **192** had 9-line ESR spectra, which means that the exchange interaction between the two radicals was much stronger than the hyperfine coupling constant in both isomers. The nitronyl nitroxide radical itself has two nitrogen atoms to give a 5-line ESR spectrum with relative intensities 1:2:3:2:1. When two nitronyl nitroxides are magnetically coupled

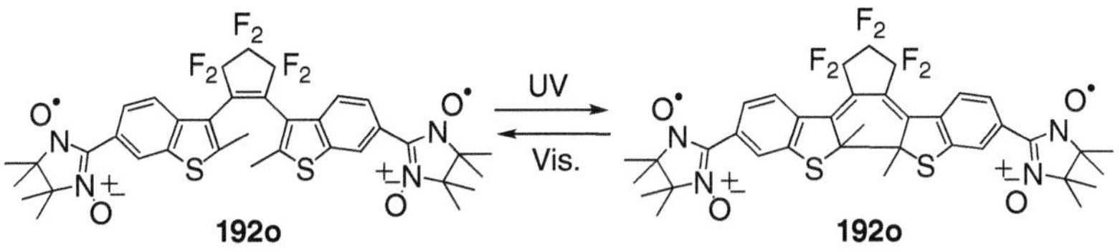

Scheme 6.1 Open-ring (**192o**) and closed-ring (**192c**) isomers of radical-substituted diarylethene **192**.

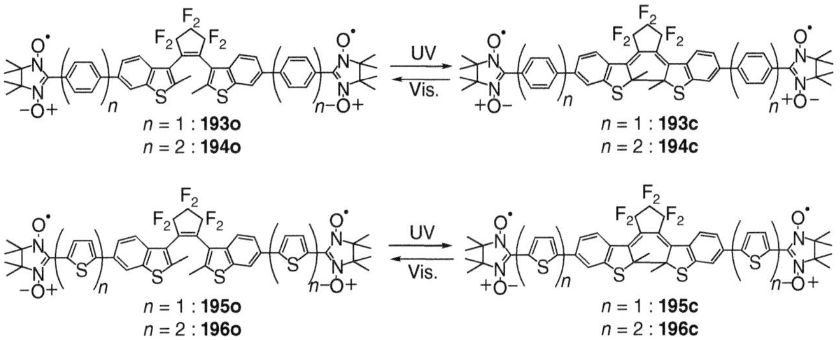

Scheme 6.2 Open-ring (**193o, 194o, 195o, 196o**) and closed-ring (**193c, 194c, 195c, 196c**) isomers of radical-substituted diarylethenes **193, 194, 195,** and **196**.

via exchange interaction, the biradical gives a 9-line ESR spectrum with relative intensities 1:4:10:16:19:16:10:4:1. The ESR spectrum gives information concerning the magnitude of exchange interaction.

To decrease the exchange interaction, biradicals **193** ($n = 1$) and **194** ($n = 2$), in which *p*-phenylene spacers are introduced, were prepared (Scheme 6.2) [12, 13]. The changes in the ESR spectra accompanying the photoisomerization reaction were examined for **193**, as shown in Figure 6.6. Before UV irradiation when the diarylethene is in the open-ring isomer the ESR spectrum showed a complex of 15 lines. This indicates that the two spins of the nitronyl nitroxide radicals are coupled by an exchange interaction that is comparable to the hyperfine coupling constant. Upon irradiation with 366 nm light, the spectrum converted completely to a 9-line spectrum. The 9-line spectrum indicates that the exchange interaction between the two spins in **193c** is much larger than the hyperfine coupling constant. The open-ring isomer **194o** showed a 5-line spectrum, while the closed-ring isomer **194c** had a distorted 9-line spectrum. Based on the simulation of the ESR spectra, the dependence of the exchange interaction on the π-conjugation chain length was estimated and is shown in Table 6.1.

Diarylethenes **195** ($n = 1$) and **196** ($n = 2$), which have oligothiophene spacers, were also prepared and their ESR spectra of the open- and closed-ring isomers were measured [14]. **196o** gave a 5-line spectrum, while a 9-line spectrum was observed when converted to closed-ring isomer **196c**. Table 6.1 lists the exchange interactions between the two diarylethene-bridged nitronyl nitroxide radicals in **193–196**.

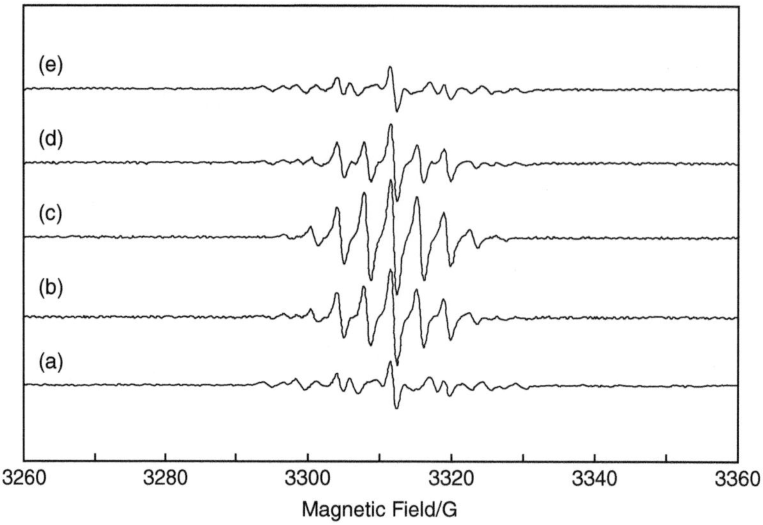

Figure 6.6 ESR spectral changes for **193** ($n = 1$) along with photoisomerization. (a) Initial, (b) irradiation with 366 nm light for 1 minute, (c) 4 minutes, (d) irradiation with >520 nm light for 20 minutes, (e) 50 minutes. *Source*: Reprinted with permission from Ref. [12]. Copyright 2000 American Chemical Society.

Table 6.1 Magnetic interactions between two nitronyl nitroxide radicals in diarylethene photoswitches, which have various types of spacers.

	Open-ring form isomer		Closed-ring form isomer	
	ESR line shape	$\|2J/k_B\ K\|$	ESR line shape	$\|2J/k_B\ K\|$
192	9 lines	2.2	9 lines	>11.6
193	15 lines [a)]	1.2×10^{-3} $< 3 \times 10^{-4}$	9 lines	>0.04
194	5 lines	$< 3 \times 10^{-4}$	Distorted 9 lines	0.010
195	13 lines [a)]	5.6×10^{-3} $< 3 \times 10^{-4}$	9 lines	>0.04
196	5 lines	$< 3 \times 10^{-4}$	9 lines	>0.04

a) Antiparallel and parallel conformers show different exchange interactions.

The closed-ring isomers have stronger interactions than the open-ring isomers. The exchange interactions through the oligothiophene spacers were larger than those of the corresponding biradicals with oligophenylene spacers. The exchange interaction difference between the open- and closed-ring isomers was estimated to be more than 150-fold for diarylethene **196** having oligothiophene spacers.

The intramolecular interaction between the radicals was found to depend on the substitution position of the thiophene rings to the ethene moiety. The open-ring

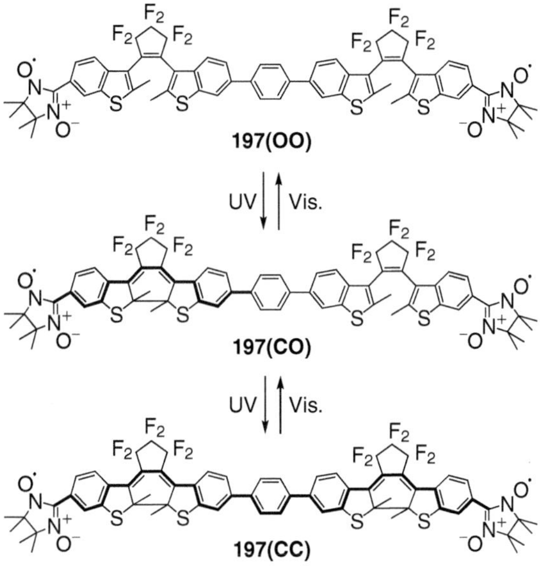

Figure 6.7 Photoisomerization of diarylethene dyad **197** upon photoirradiation.

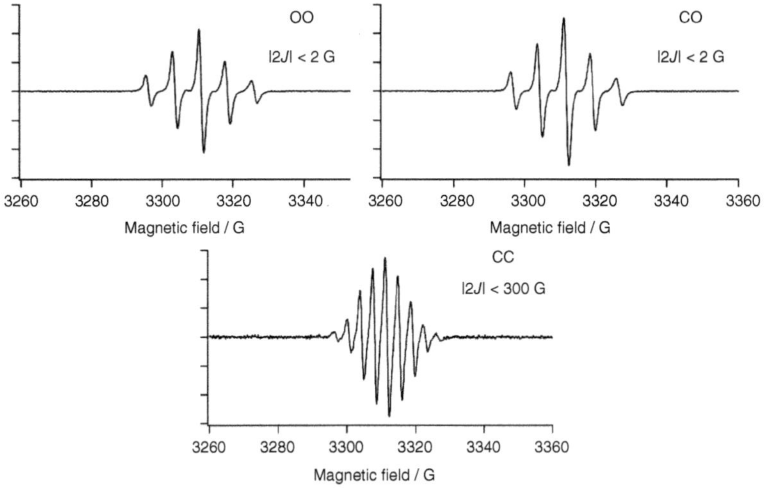

Figure 6.8 ESR spectra of (a) **197 (OO)**, (b) **197 (CO)**, and (c) **197 (CC)**. *Source*: Adapted with permission from Ref. [16]. Copyright 2001 American Chemical Society.

isomer showed larger interaction than the closed-ring isomer, when the thiophene rings were attached to the ethene bridge at 2-position [15]. This result is explained by the switching of the connectivity of π-bonds.

Diarylethene dimer was also used as a switching unit, as shown in Figure 6.7 [16]. In molecular switch **197,** there exist three kinds of isomer states: open–open (**OO**), closed–open (**CO**), and closed–closed (**CC**). Bond alternation is disconnected at the open-ring form moieties. Therefore, the two spins at both ends of **197 (OO)** and **197**

(CO) cannot interact with each other. On the other hand, π-conjugation is delocalized throughout the molecule in **197 (CC)** and the exchange interaction between the two terminal radicals is expected to take place.

ESR spectra of isolated **197 (OO)**, **197 (CO)**, and **197 (CC)** were measured in benzene at room temperature and are shown in Figure 6.8. The spectra of **197 (OO)** and **197 (CO)** showed 5-line type, indicating that the exchange interaction between the two terminal radicals is much smaller than the hyperfine coupling constant ($2J/\mathbf{k}_B < 3 \times 10^{-4}$ K). On the other hand, the spectrum of **197 (CC)** showed a clear 9-line type, indicating that the exchange interaction between the two radicals is much larger than the hyperfine constant ($2J/\mathbf{k}_B > 0.04$ K). These results proved that each diarylethene chromophore plays the role of a switching unit to control the magnetic interaction. Not only monomeric diarylethenes but also arrays of diarylethenes work as molecular photoswitches.

References

1 Dulić, D., van der Molen, S.J., Kudernac, T. et al. (2003). One-way optoelectronic switching of photochromic molecules on gold. *Phys. Rev. Lett.* 91 (20): 207402.

2 Whalley, A.C., Steigerwald, M.L., Guo, X., and Nuckolls, C. (2007). Reversible switching in molecular electronic devices. *J. Am. Chem. Soc.* 129 (42): 12590–12591.

3 Jia, C., Migliore, A., Xin, N. et al. (2016). Covalently bonded single-molecule junctions with stable and reversible photoswitched conductivity. *Science* 352: 1443–1445.

4 Kondo, M., Tada, T., and Yoshizawa, K. (2005). A theoretical measurement of the quantum transport through an optical molecular switch. *Chem. Phys. Lett.* 412: 55–59.

5 Staykov, A., Nozaki, D., and Yoshizawa, K. (2007). Photoswitching of conductivity through a diarylperfluorocyclopentene nanowire. *J. Phys. Chem. C* 111 (8): 3517–3521.

6 Zhuang, M. and Ernzerhof, M. (2009). Reversibility and transport properties of dithienylethenes photoswitches. *J. Chem. Phys.* 130: 114704.

7 Ashraf, M.K., Bruque, N.A., Tan, J.L. et al. (2011). Conductance switching in diarylethenes bridging carbon nanotubes. *J. Chem. Phys.* 134: 024524.

8 Tsuji, H. and Hoffmann, R. (2014). Frontier orbital control of molecular conductance and its switching. *Angew. Chem. Int. Ed.* 53: 4093–4097.

9 Arramel, Pijper, T.C., Kudernac, T. et al. (2013). Reversible light induced conductance switching of asymmetric diarylethenes on gold: Surface and electronic studies. *Nanoscale* 5: 9277–9282.

10 Kang, J.-W., Kim, J.-J., and Kim, E. (2002). All-optical Mach-Zehnder modulator using a photochromic dye-doped polymer. *Appl. Phys. Lett.* 80 (10): 1710–1712.

11 Matsuda, K. and Irie, M. (2000). A diarylethene with two nitronyl nitroxides: photoswitching of intramolecular magnetic interaction. *J. Am. Chem. Soc.* 122 (30): 7195–7201.

12 Matsuda, K. and Irie, M. (2000). Photoswitching of intramolecular magnetic interaction using a photochromic spin coupler: an ESR study. *J. Am. Chem. Soc.* 122 (34): 8309–8310.

13 Matsuda, K. and Irie, M. (2001). Photochromism of diarylethenes with two nitronyl nitroxides: photoswitching of an intramolecular magnetic interaction. *Chem. Eur. J.* 7 (16): 3466–3473.

14 Matsuda, K., Matsuo, M., and Irie, M. (2001). Photoswitching of intramolecular magnetic interaction using diarylethene with oligothiophene π-conjugated chain. *J. Org. Chem.* 66 (26): 8799–8803.

15 Matsuda, K., Matsuo, M., Mizoguchi, S. et al. (2002). Reversed photoswitching of intramolecular magnetic interaction using a photochromic bis(2-thienyl)ethene spin coupler. *J. Phys. Chem. B* 106: 11218–11225.

16 Matsuda, K. and Irie, M. (2001). Photoswitching of intramolecular magnetic interaction using a diarylethene dimer. *J. Am. Chem. Soc.* 123 (40): 9896–9897.

7
Surface Properties

7.1 Surface Wettability

Surface wettability plays an important role in various applications, such as adhesion, adsorption, mechanical friction, dewetting, lithography, and biomedical engineering. There are two factors that control wettability on solid surfaces, chemical and topographical factors.

When photoisomerizable chromophores that reversibly change their polarity are placed on the surface, the surface wettability is chemically photomodulated. Surface modification with photoswitchable azobenzene or spirobenzopyran derivatives was already studied in the 1980s and applied to photocontrol the adsorption/desorption behavior of proteins or organic substances onto the polymer surfaces [1]. The surface wettability of poly(2-hydroxyethyl methacrylate) having azobenzene pendant groups was measured by placing a water droplet on the film surface [2]. Azobenzene is known to increase the dipole moment from 0.5 D to 3.1 D when isomerized from the *trans*- to the *cis*-form. The water contact angle (CA) was found to decrease from 77° to 64° upon UV irradiation, while it returned to 77° upon irradiation with visible light. Similar reversible wettability change was observed when spirobenzopyran residues were incorporated into polymers [3].

The wettability change based on the chemical approach is moderate. To enlarge the photoinduced CA change, photocontrol of surface topography was explored using diarylethene crystals. Superhydrophobic surfaces of lotus or taro leaves are governed by the topographic factor. The surface of the lotus leaf is covered with projections having a diameter of around 10 μm that exhibit a super water-repelling property. If such surface roughness is changed by photoirradiation, the change is expected to affect wettability.

Uchida and co-workers [4–7] demonstrated reversible changes of surface topology by the growth of microfibrils on a diarylethene crystal upon photoirradiation, as shown in Figure 7.1. The single crystal of **198o** was irradiated with UV light and then stored in the dark at 30 °C for 24 hours. Upon irradiation with UV light, the crystal turned deep blue and microfibrils grew during the storage in the dark. The crystal surface was fully covered with the fibrils. The fibrils were melted by irradiation with visible light, and the rough surface reverted to the initial flat surface. The formation of needle-shaped microfibrils can be explained by using a

Diarylethene Molecular Photoswitches: Concepts and Functionalities, First Edition. Masahiro Irie.
© 2021 WILEY-VCH GmbH. Published 2021 by WILEY-VCH GmbH.

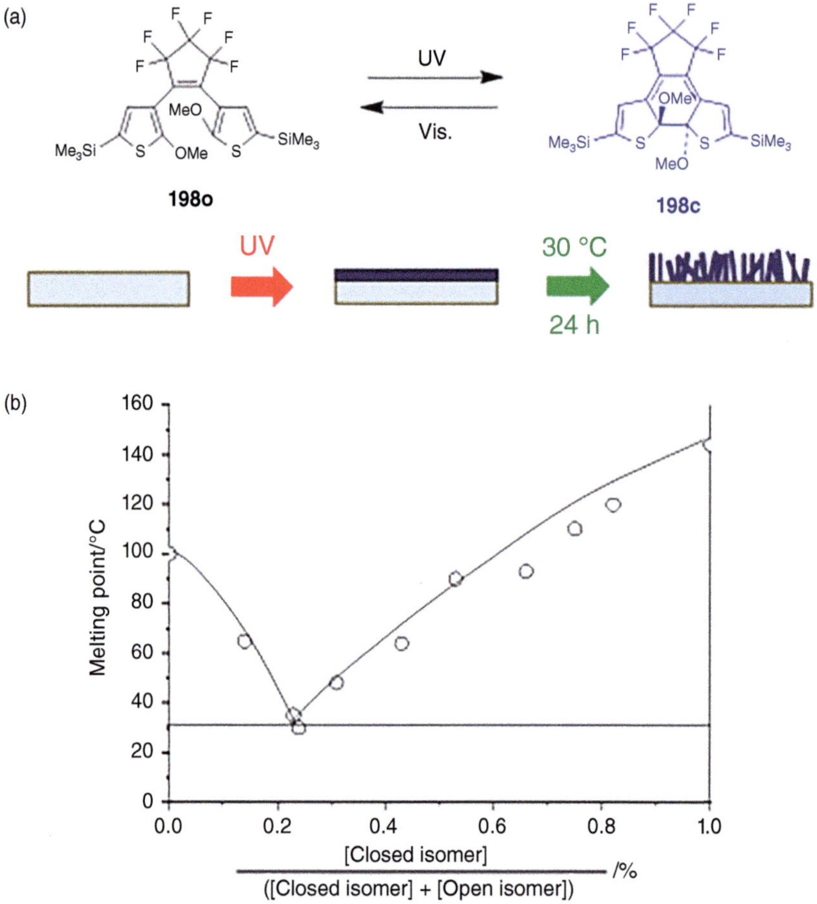

Figure 7.1 (a) Schematic illustration of the fabrication procedure of the crystal surface, which shows "lotus leaf" effect. Diarylethene crystal **198** was used for the fabrication of needle-shaped microfibrils. (b) The phase diagram of **198** upon photoisomerization between open- and closed-ring isomers.

phase diagram shown in Figure 7.1b. The melting point of the crystal decreases from 100 to 30 °C upon irradiation with UV light due to the formation of closed-ring isomers as impurities in the crystal. When the conversion from open-ring isomer **198o** to closed-ring isomer **198c** exceeded 24%, the excess closed-ring isomers started to form microfibrils on the surface. Upon irradiation with visible light, the closed-ring isomers returned to the open-ring isomers and the fibrils disappeared.

When the surface was covered with the needle-shaped microfibrils, the surface showed superhydrophobicity. The CA and sliding angle (SA) of a water droplet on the surface were 163° and less than 2°, respectively. These values resemble those of a water droplet on a natural lotus leaf. The surface wettability was reversibly changed by alternate irradiation with UV and visible light.

Another surface modification mimicking "rose petals" was also demonstrated [6]. Rose petals show super water-repellency with a CA of 152°. At the same time, these

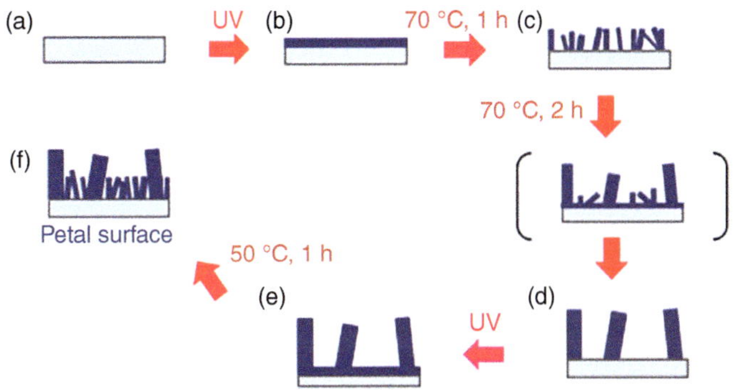

Figure 7.2 Schematic illustration of the fabrication procedure of the crystal surface, which shows "rose petal" effect. Diarylethene crystal **198** was used for the fabrication of the composite surface consisting of both needle-shaped fibrils and rod-shaped crystals.

droplets are pinned to the surface and cannot fall off, even when the petal is turned upside down. To mimic the rose petal surface, a composite surface having both larger cavities, which can contain water, and smaller cavities, which do not allow penetration of water, was prepared as shown in Figure 7.2. Initially, UV-irradiated surface was stored at 70 °C for 3 hours in the dark. The needle-shaped fibrils were converted to relatively large rod-shaped crystals, as shown in Figure 7.2d. Then, the surface was irradiated with UV light and stored at 50 °C for 1 hour in the dark. The surface between the large rod-shaped crystals was covered with newly formed fibrils. Figure 7.2f shows the composite surface. The surface consisted of both needle-shaped fibrils and rod-shaped crystals showing superhydrophobicity (CA > 150°), and the water droplet was fixed on the surface even when the surface was turned upside down. The rose petal effect was realized. Elaborated fabrication procedures further succeeded in the preparation of a double roughness structure on the crystal surface [7]. The surface showed water droplet bounding just like a natural lotus leaf.

The phase diagram shown in Figure 7.1 indicates that the melting point of diarylethene crystals changes upon photoisomerization. The melting point of the open-ring isomer **198o** crystal is 100 °C. The melting point decreases with increase in the content of closed-ring isomer **198c** in the crystal and reaches as low as 30 °C at the eutectic crystal ratio [open-ring isomer]/[closed-ring isomer] of 76/24. When the content of the closed-ring isomer exceeds 24%, the melting point of the crystal again starts to increase and the closed-ring isomer **198c** crystal shows the melting point of 144 °C. Not only the melting point but also the glass transition temperature (*Tg*) of amorphous diarylethenes was found to change upon photoisomerization.

7.2 Selective Metal Deposition

Tg of the amorphous diarylethene film of open-ring isomer **199o**, shown in Figure 7.3a, is 32°, while it increases to 95° in the closed-ring isomer **199c** film [9, 10]. The dramatic *Tg* change caused selective deposition of metals on the surface.

The selective deposition means that vapor metal atoms are deposited selectively on a hard amorphous diarylethene surface but not on a soft surface. Tsujioka and co-workers [8–12] developed the selective metal deposition technique on amorphous diarylethenes as well as on polymers containing diarylethenes. Figure 7.3 shows examples of the metal deposition. An amorphous diarylethene **199o** film was prepared on a glass substrate using a conventional vacuum evaporation method. The partially colored film was exposed to Mg vapor in a vacuum. Mg was deposited on the colored areas, which are shown as black stripes, but not on the colorless areas. The colored areas are considered to become hard, because *Tg* of the photogenerated closed-ring isomer is as high as 95°. The atoms on the colorless surface desorbed from the surface due to active molecular motion caused by the low *Tg*, while Mg atoms remained on the colored hard surface areas.

Patterning was also carried out by applying a laser beam on the amorphous diarylethene **199o** bulk film and also on a polymer film doped with **199**, as shown in Figure 7.3b, c, respectively. The amorphous film was prepared by a vacuum evaporation method. A focused laser beam of visible light (630 nm) was applied to the blue surface generated by UV exposure, and uncolored lines were prepared. After that, Mg was evaporated. Nondeposited lines, whose line widths are less than 10 μm, were formed, as shown in Figure 7.3b, left side. Conversely, lines were formed by Mg deposition on the colored surface generated by a focused violet laser (410 nm) applied over the uncolored surface, as shown in Figure 7.3b, right side.

The mobility of polymer chains can also be controlled by doping diarylethene molecules into polymer matrices. A polystyrene film doped with 5 wt% diarylethene **199o** was prepared on a glass substrate by a casting method and dried at 60 °C for 1 hour. The film thickness was 3 μm. The laser beam with a wavelength of 410 nm was scanned on the film. Then, Mg was evaporated to the surface without a shadow mask. Parallel line Mg patterns with a width of 15 μm were obtained, as shown in Figure 7.3c [8]. Selective metal deposition was also carried out on photoswitchable crystal surfaces.

Selective metal deposition can be applied to various electronic devices. Figure 7.4 shows one such example, that is fabrication of μ-fuses [12]. An amorphous film of diarylethene **199o** on the glass substrate was irradiated with UV light (365 nm) to prepare a whole colored surface. Then, uncolored narrow lines were prepared on it using 635 nm laser scanning. Between the uncolored narrow lines, colored lines were formed by scanning a violet laser (410 nm). Pb was evaporated onto the whole area without a shadow mask. Figure 7.4 shows one of the μ-fuses. The black parts show the uncolored lines, where Pb is not deposited. A micro-size Pb bridge was formed between left-side and right-side Pb deposited areas. When voltage was applied to the device, the bridge part was broken as shown in the right-side figure. The current–voltage characteristics of the device indicate that the current is interrupted at around 300 μA at 2.4 V. The result indicates that the μ-fuse has a very low interrupted current level (I_{int}). I_{int} can be controlled by adjusting the cross section of the Pb-bridge part.

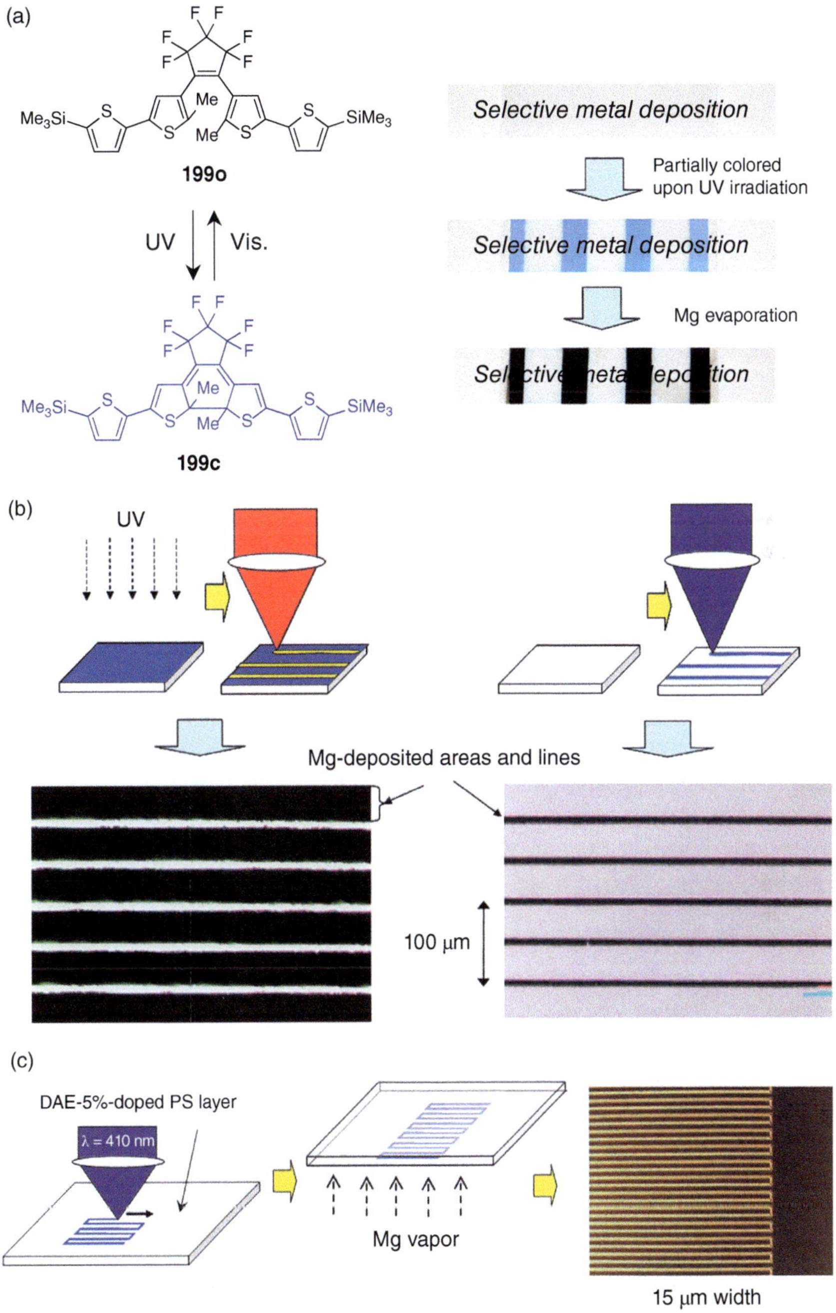

Figure 7.3 (a) Selective metal deposition on the UV-irradiated colored surface of an amorphous diarylethene **199** film. (b) Mg patterning by scanning a laser beam (left-side 630 nm, right-side 410 nm) on the amorphous films followed by evaporation of Mg. (c) Mg patterning by scanning a 410 nm laser beam on a polystyrene film doped with 5 wt% **199o** followed by evaporation without a shadow mask.

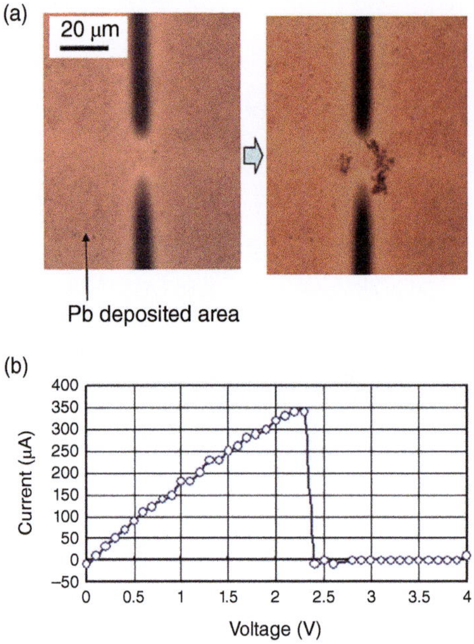

Figure 7.4 (a) μ-Fuse and its operation. (b) Current–voltage characteristics of the μ-fuse.

7.3 Subwavelength Nanopatterning

Optical lithography is a key technology for fabricating nanostructures in electronic devices. Various approaches have been developed to increase the resolution of nanopatterning. As described in Section **5.2,** the resolution of microscale optical patterning is limited to one-half of the wavelength of light. Although high resolution is achieved using near-field optics, the technique requires inflexible and costly photomasks or scanning of nanoscale tips in close proximity to the sample. To overcome these limitations, the absorption spectral change of photoswitchable molecules was used to restrict the size of the writing beam for the nanopatterning.

Figure 7.5 shows the principle of the nanopatterning [13]. On top of the recording photoresist layer, a thin PMMA film containing diarylethene **200** is placed. The diarylethene undergoes photoswitching between the open- (**200o**) and closed-ring (**200c**) isomers upon irradiation with UV (λ_1) and visible (λ_2) light. Both wavelengths of light are applied to the sample in an interference pattern that overlaps peaks at λ_1 with nodes at λ_2, Absorption at λ_1 makes the film transparent at that wavelength, but regions exposed to λ_2 revert the closed-ring isomers to the initial open-ring isomers and continue to absorb at λ_1, protecting the photoresist. Only at the λ_2 nodes, a stable transparent aperture exists. Photons at λ_1 penetrate this aperture, forming a nanoscale writing beam that can pattern the underlying photoresist. The size of the aperture decreases as the ratio of the intensity at λ_2 with respect to that at λ_1 increases. This absorbance modulation technique can confine light to spatial dimensions far smaller than the wavelength.

Figure 7.5b shows the results. The sample consisted of a silicon substrate spin-coated with 200 nm of antireflection coating, 200 nm of photoresist, 8 nm of

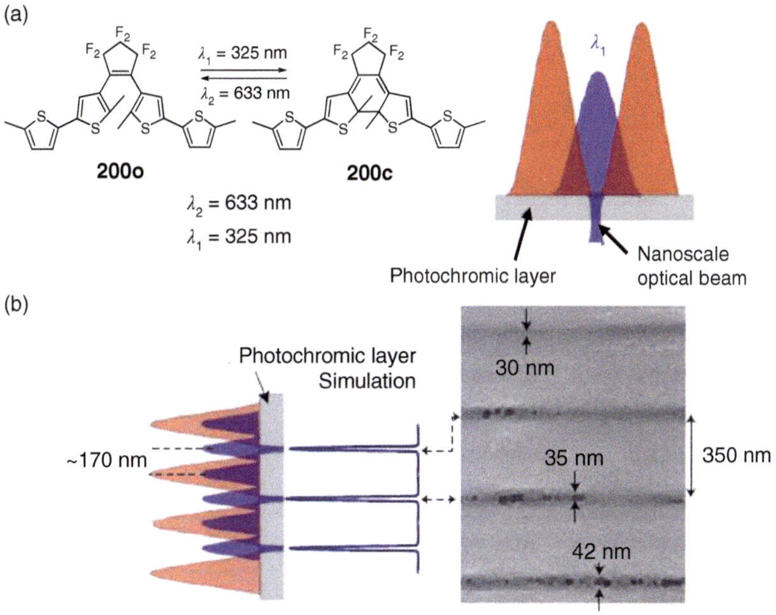

Figure 7.5 (a) Scheme of absorption modulation. Absorption at λ_1 makes the film transparent at that wavelength, while regions exposed to λ_2 revert to the initial isomer and continue to absorb at λ_1. A subwavelength transparent region is formed through which photons at λ_1 penetrate, forming a nanoscale writing beam. (b) Scanning electron micrograph images of subwavelength lines produced using the absorption modulation.

PVA, and 410 nm of a PMMA film containing **200**. The sample was illuminated by two overlapping standing waves with periods of 350 nm (λ_2 = 633 nm) and 170 nm (λ_1 = 325 nm), respectively. The nodes of the λ_2 standing wave approximately coincided with every other peak of the λ_1 standing wave. The simulation indicates that the transmitted light at λ_1 is substantially narrower than the diffraction limit. The scanning electron microscope image of the lines indicated that the average width of the lines recorded in the photoresist is 36 nm, which is close to 1/10 of λ_1. Furthermore, the narrow lines were spaced by 350 nm, which corresponds to the period of the λ_2 standing wave. Although this demonstration utilizes one-dimensional standing waves, it is anticipated to extend to two-dimensional peaks and nodes, which can be generated with diffractive micro-optics. These nanoscale optical beams may be useful for optical nanoscopy.

References

1 Irie, M. (1990). Photoresponsive polymers. *Adv. Polym. Sci.* 94: 27–67.

2 Ishihara, K., Hamada, N., Kato, S., and Shinohara, I. (1983). Photo-induced change in surface free energy of azoaromatic polymers. *J. Polym. Sci. Polym. Chem. Ed.* 21: 1551–1555.

3 Hayashida, S., Sato, H., and Sugawara, S. (1986). Photo-induced wettability change in films containing spiropyran microcrystals. *Polym. J.* 18 (3): 227–235.

4 Uchida, K., Nishimura, R., Hatano, E. et al. (2018). Photochromic crystalline systems mimicking bio-functions. *Chem. Eur. J.* 24 (34): 8491–8506.

5 Uchida, K., Izumi, N., Sukata, S. et al. (2006). Photoinduced reversible formation of microfibrils on a photochromic diarylethene microcrystalline surface. *Angew. Chem. Int. Ed.* 45: 6470–6473.

6 Uchida, K., Nishikawa, N., Izumi, N. et al. (2010). Phototunable microcrystalline surfaces: lotus and petal effects upon wetting. *Angew. Chem. Int. Ed.* 49: 5942–5944.

7 Nishimura, R., Hyodo, K., Sawaguchi, H. et al. (2016). Fractal surfaces of molecular crystals mimicking lotus leaf with phototunable double roughness structures. *J. Am. Chem. Soc.* 138: 10299–10303.

8 Tsujioka, T., Takagi, R., and Shiozawa, T. (2010). Light-controlled metal deposition on photochromic polymer films. *J. Mater. Chem.* 20: 9623–9627.

9 Tsujioka, T., Sesumi, Y., Takagi, R. et al. (2008). Selective metal deposition on photoswitchable molecular surfaces. *J. Am. Chem. Soc.* 130 (32): 10740–10747.

10 Masui, K., Takagi, R., Sesumi, Y. et al. (2009). Selective metal deposition for a structure with a thin intermediate layer on a photochromic diarylethene film. *J. Mater. Chem.* 19: 3176–3180.

11 Tsujioka, T. (2014). Selective metal deposition on organic surfaces for device applications. *J. Mater. Chem. C* 2: 221–227.

12 Tsujioka, T. and Dohi, M. (2012). Light-controlled selective Pb deposition on photochromic surfaces. *Appl. Phys. Exp.* 5: 041603.

13 Andrew, T.L., Tsai, H.-Y., and Menon, R. (2009). Confining light to deep subwavelength dimensions to enable optical nanopatterning. *Science* 324: 917–921.

8

Polymers and Liquid Crystals

8.1 Polymers

For practical applications of molecular photoswitches to photonic devises, it is indispensable to incorporate molecular photoswitches into solid media, such as polymers, or to make use of bulk amorphous films. Although doping the photoswitchable molecules into polymer matrices is the most convenient way to prepare such photoresponsive materials, increasing the content of the doped molecules worsens the material transparency due to the scattering coming from aggregation. In addition, in many cases segregation takes place during long-term storage, when the content is high. The development of proper polymeric versions of diarylethene molecular photoswitches, which have following properties, is a key step for the applications.

1. The polymer contains a high content of diarylethene units and keeps transparency.
2. The polymer should be soluble in a variety of solvents.
3. The molecular architecture does not limit the ability to fine-tune the optoelectronic properties.
4. The polymer should give photoswitching performance under a wide range of conditions.

Considering these requirements, various polymers having diarylethene units in the main chains or pendant groups have been designed and synthesized. Tables 8.1 and 8.2 show examples of polymers having diarylethene units in the main chain [1–19] and in the side groups [6, 20–39], respectively.

π-Conjugation polymers having diarylethene units in the main chain have been prepared by Wittig condensation, Suzuki–Miyaura coupling, Stille coupling, Knoevenagel polycondensation, Horner reaction, and Friedländer condensation, as shown in Scheme 8.1. Such π-conjugation polymers can modulate fluorescence and electronic properties upon photoirradiation. Kim and co-workers [3, 4] combined the diarylethene units with fluorescent *p*-phenylenevinylene (PPV) derivatives to prepare oligomers **p2** and **p3** (*Mw*: 3800 (**p2**), 9000 (**p3**)) using Wittig condensation (Scheme 8.1c). Interestingly, in **p2** the content of antiparallel conformers was 5 times higher than that of parallel ones. This is ascribed to the rigid polymer structure, which suppresses the rotation of benzothiophene rings. The cyclization

Diarylethene Molecular Photoswitches: Concepts and Functionalities, First Edition. Masahiro Irie.
© 2021 WILEY-VCH GmbH. Published 2021 by WILEY-VCH GmbH.

Table 8.1 Examples of polymers having diarylethene units in the main chain.

Structure	Polymerization method	Research category	References
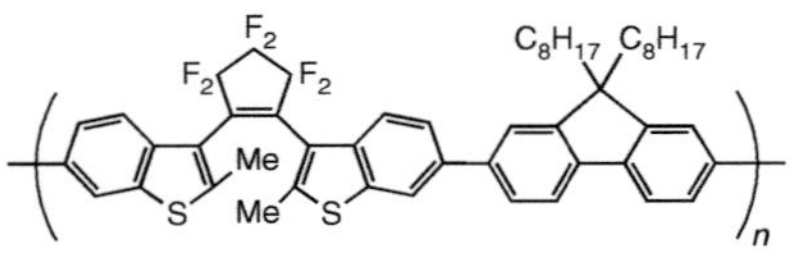 **p1**	Suzuki–Miyaura coupling	Conductivity, fluorescence	[1, 2]
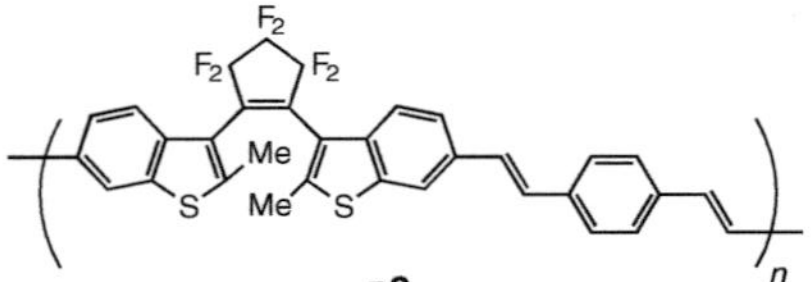 **p2**	Wittig condensation	Fluorescence, conductivity	[3, 4]
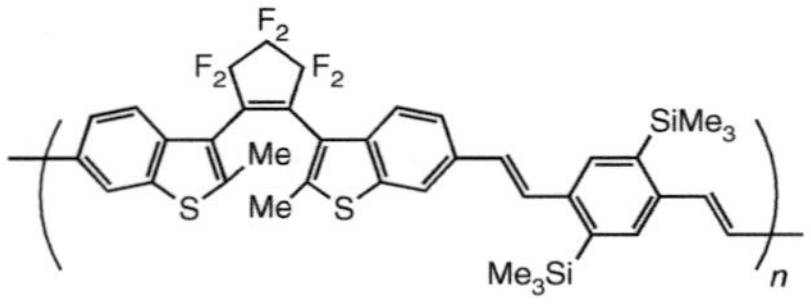 **p3**	Wittig condensation	Conductivity	[4]
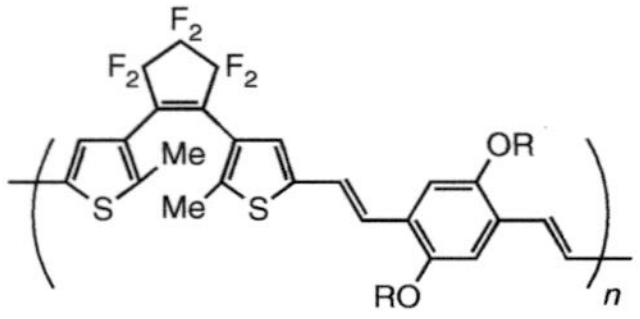 **p4**: R = C_6H_{13} **p5**: R = $C_{12}H_{25}$	Horner reaction	Polymer synthesis	[5]

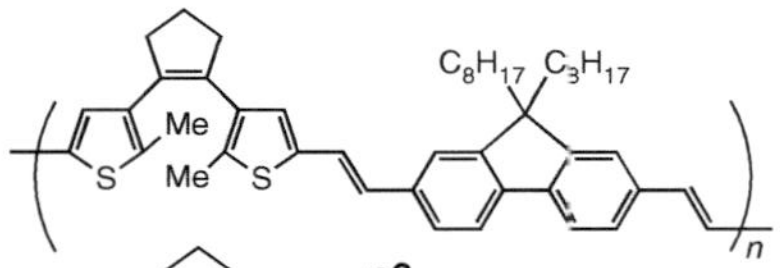 **p6** 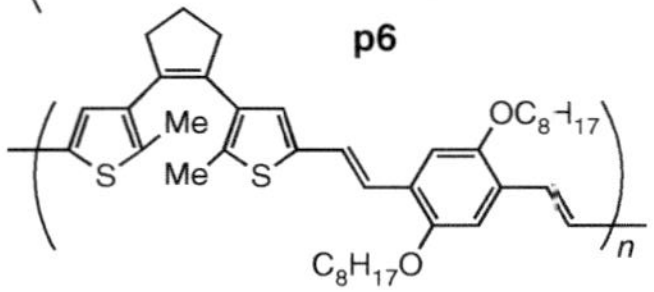 **p7**	Wittig condensation	Fluorescence	[6]
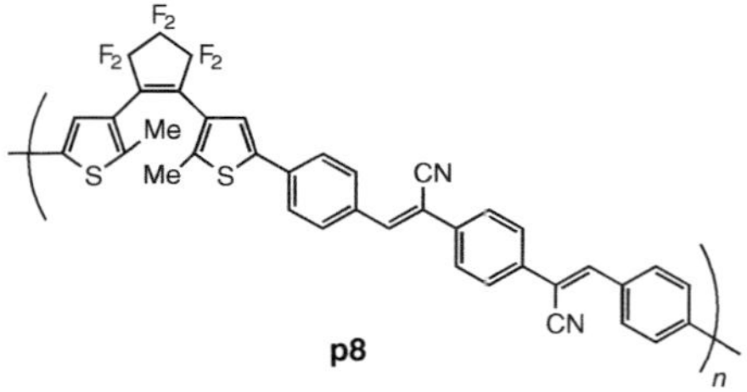 **p8**	Knoevenagel condensation	Fluorescence	[7]
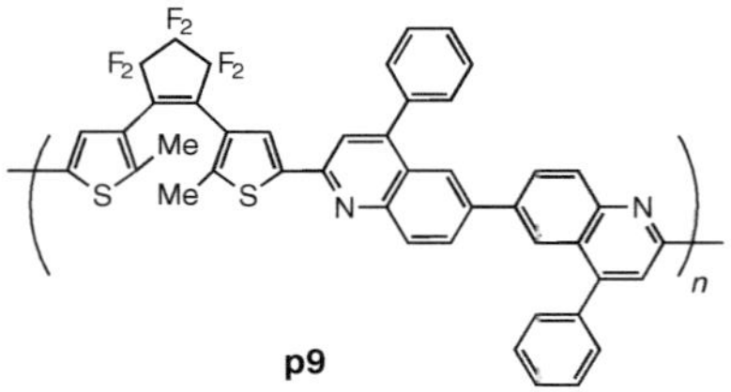 **p9**	Friedländer condensation	Conductivity	[8]

(*Continued*)

Table 8.1 (Continued)

Structure	Polymerization method	Research category	References
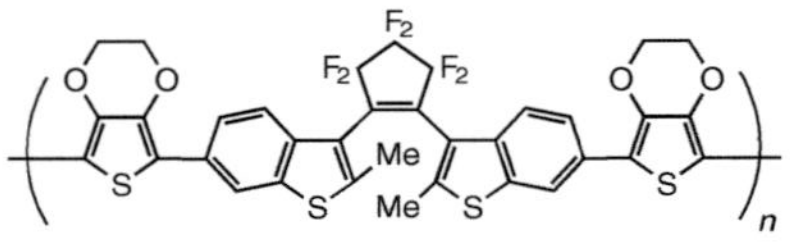 p10	Electrochemical polymerization	New polymerization method	[9]
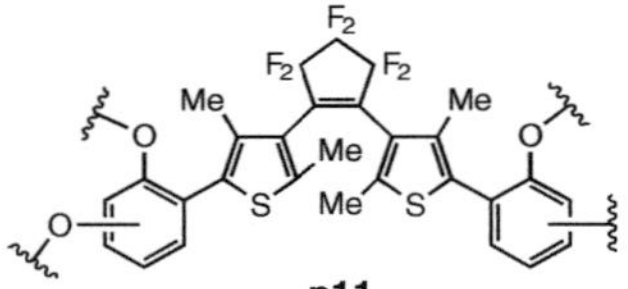 p11	Oxidation polymerization	New polymerization method	[10]
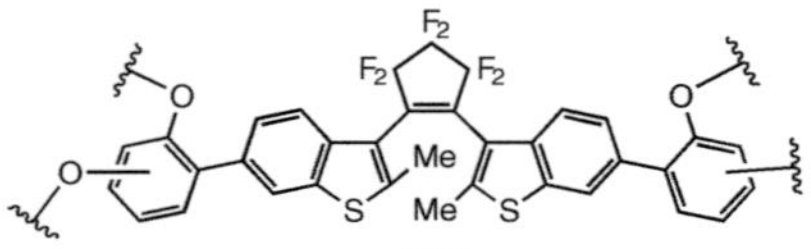 p12	Oxidation polymerization	New polymerization method	[11, 12]
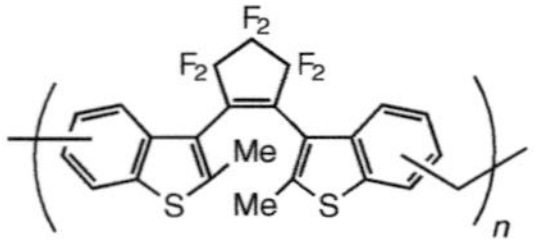 p13	Friedel–Crafts alkylation	New polymerization method	[13]

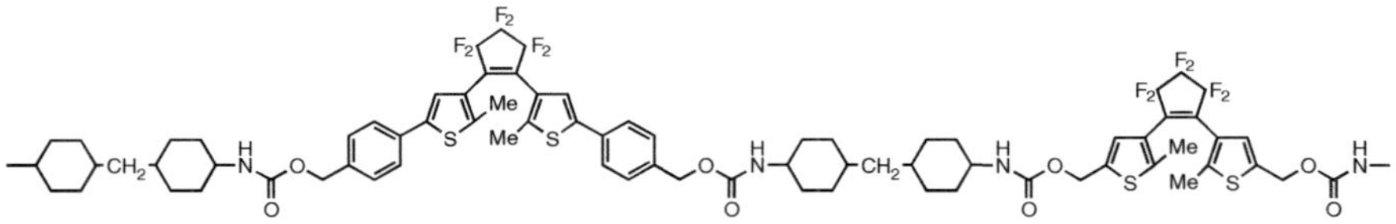 **p20**	Polycondensation	Optical memory, Raman spectroscopy	[14]
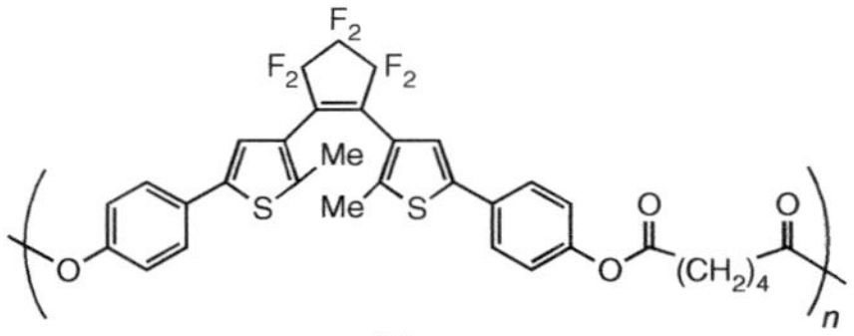 **p15**	Polycondensation	Conductivity	[15]
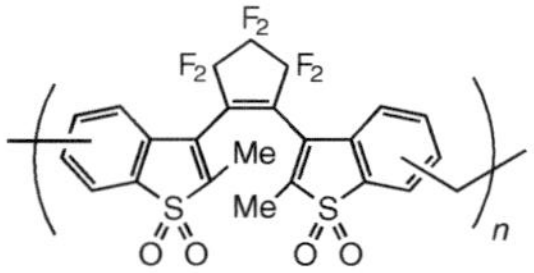 **p16**	Friedel–Crafts alkylation	Fluorescence	[16]
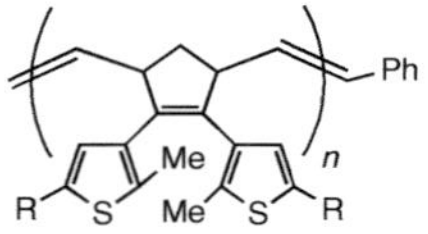 **p17**: R = Cl **p18**: R = CO_2Me **p19**: R = CO_2H	Ring-opening metathesis polymerization	Polymer synthesis	[17]

(*Continued*)

Table 8.1 (Continued)

Structure	Polymerization method	Research category	References
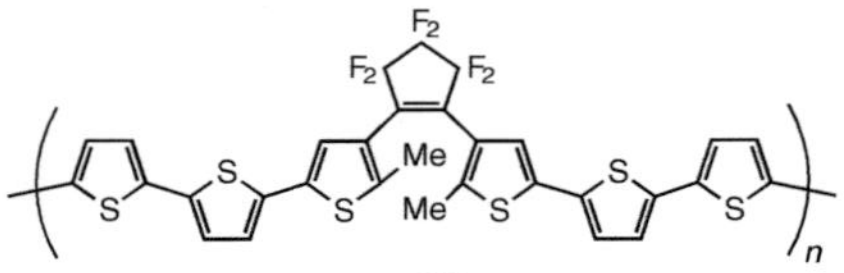 **p20**	Electropolymerization	Electro-polymerizability	[18]
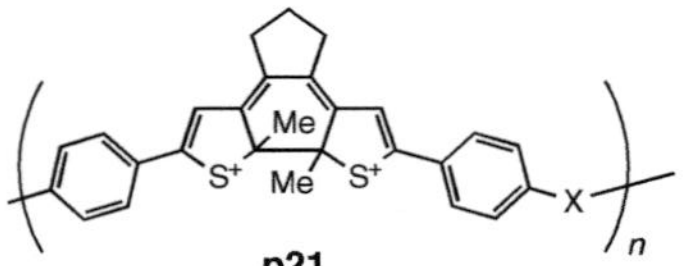 **p21** X = the polymer backbone	Electropolymerization	Electrochemistry	[19]

Table 8.2 Examples of polymers having diarylethene units in the side groups.

Structure	Polymerization method	Research category	References
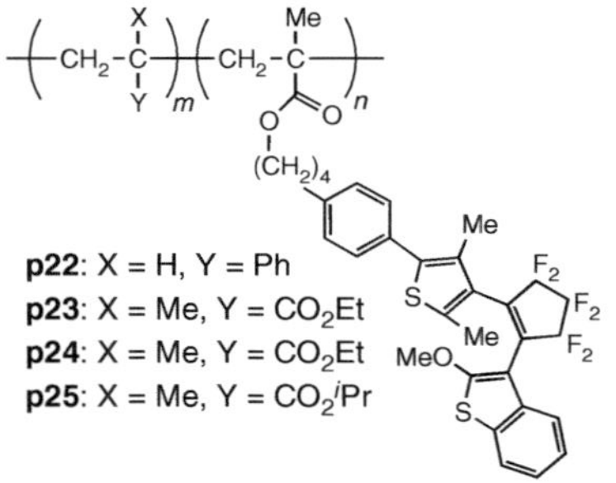 p22: X = H, Y = Ph; p23: X = Me, Y = CO_2Et; p24: X = Me, Y = CO_2Et; p25: X = Me, Y = $CO_2{}^iPr$	Radical polymerization	Polymer synthesis	[20]
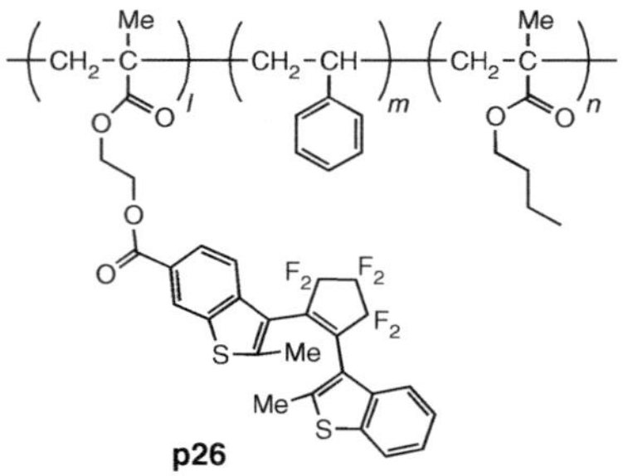 p26	Radical polymerization	Refractive index	[21]
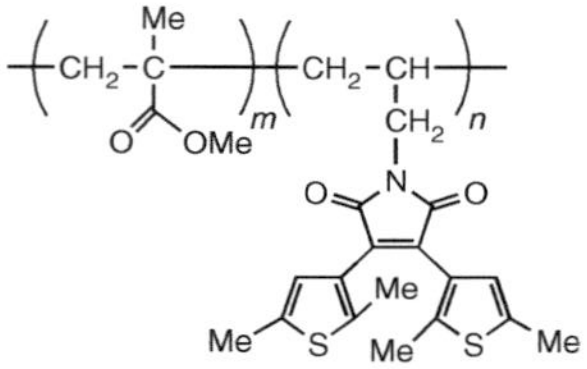 p27	Radical polymerization	Polymer synthesis	[22]

(*Continued*)

Table 8.2 (Continued)

Structure	Polymerization method	Research category	References
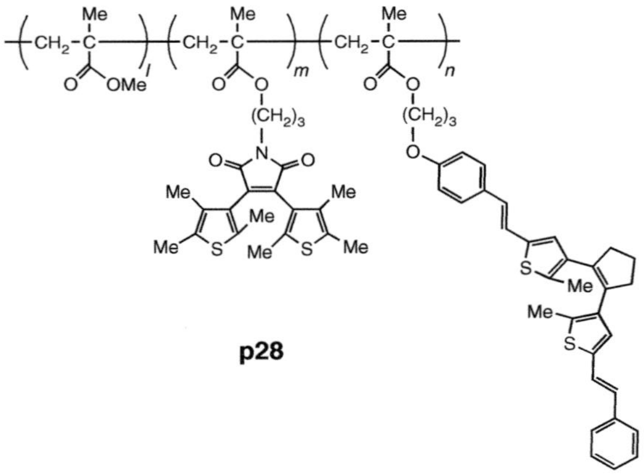 **p28**	Radical polymerization	Polymer synthesis	[6]
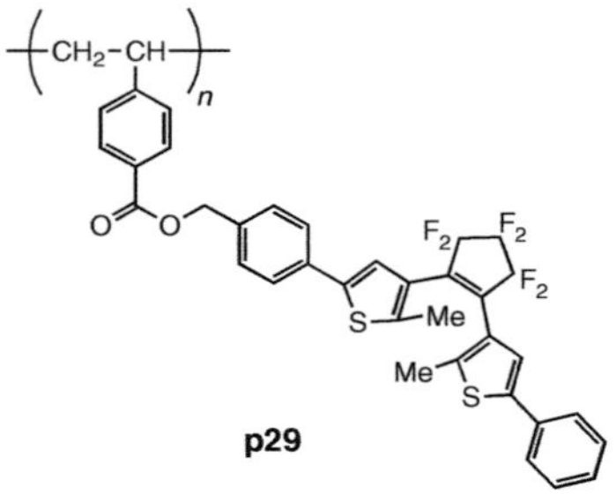 **p29**	Radical polymerization	Polymer synthesis	[23]

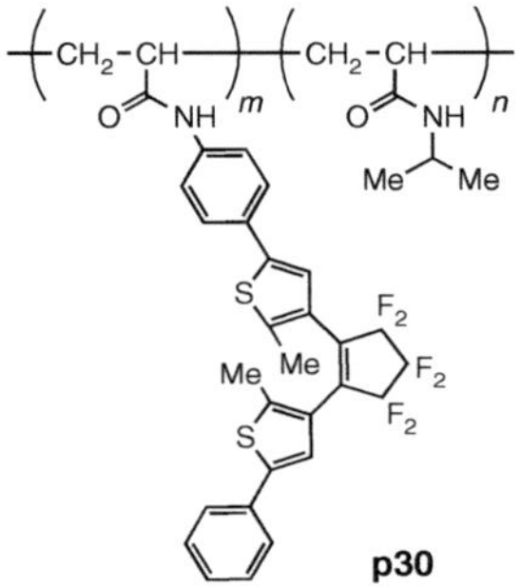 **p30**	Radical polymerization	Nanoparticles	[24]
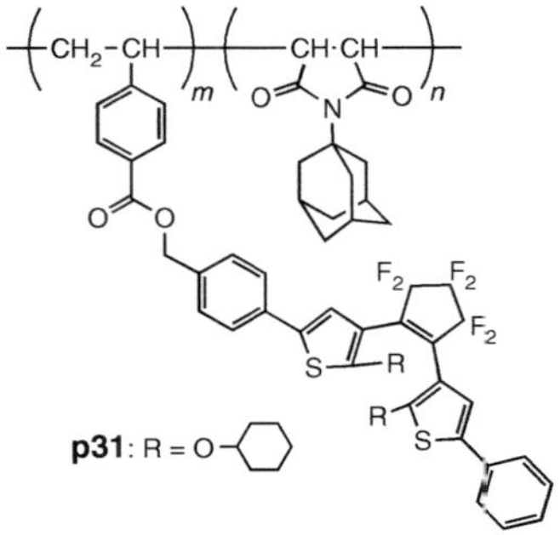 **p31**: R = O–	Radical polymerization	Write-by-light/ erase-by-heat recording	[25]
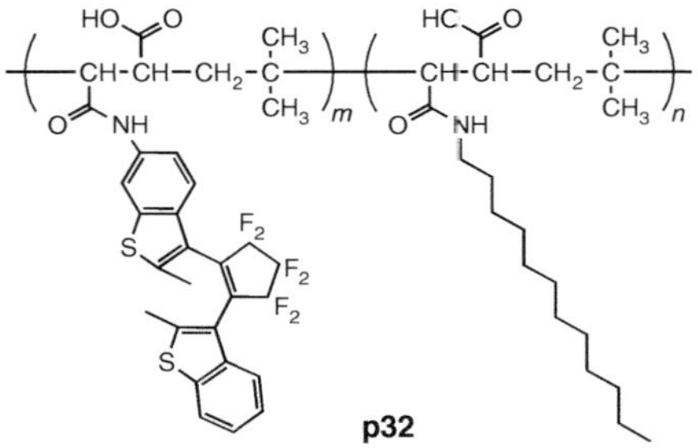 **p32**	Polymer reaction	Fluorescent water-soluble quantum dots	[26]

(Continued)

Table 8.2 (Continued)

Structure	Polymerization method	Research category	References
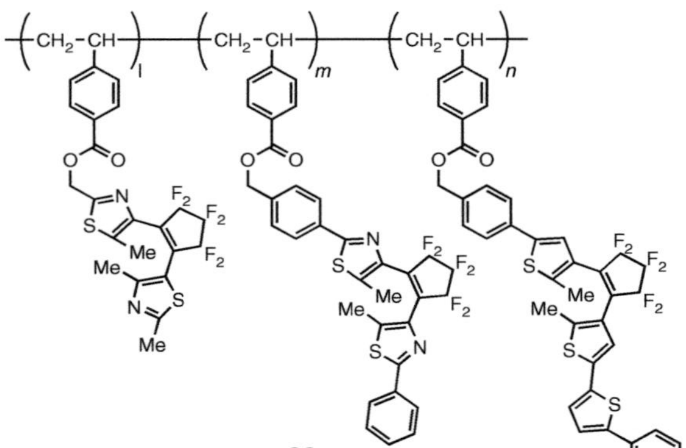 **p33**	Radical polymerization	Multicolor photoswitching	[27]
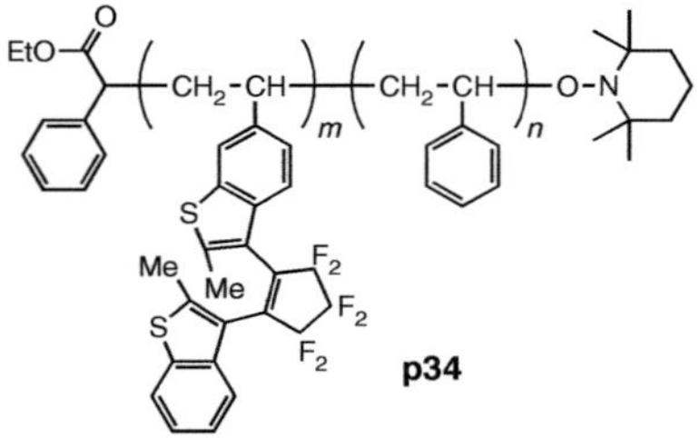 **p34** 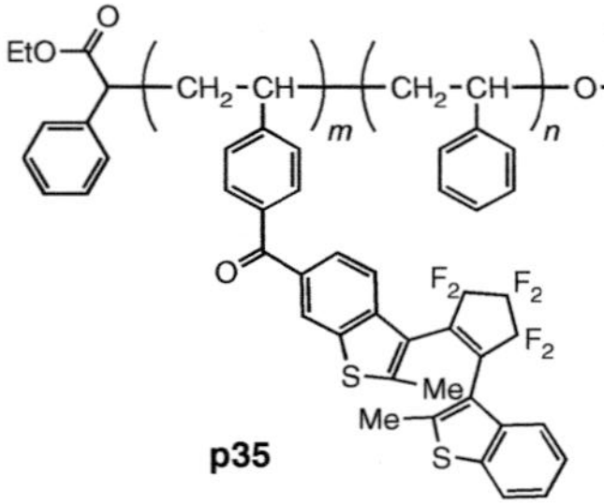 **p35**	Living radical polymerization	Polymer synthesis	[28]

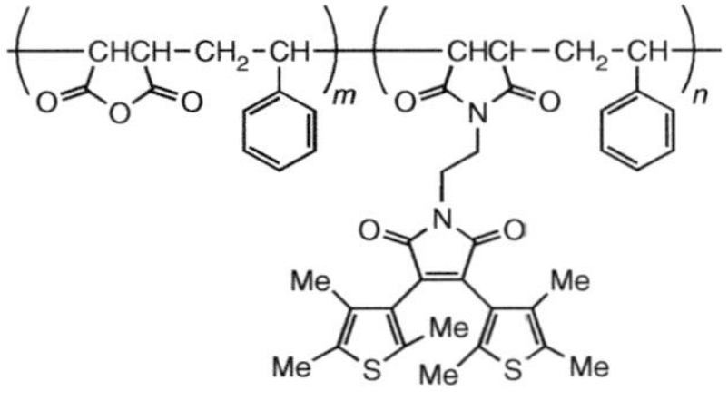 **p36**	Polymer reaction	Polymer synthesis	[29]
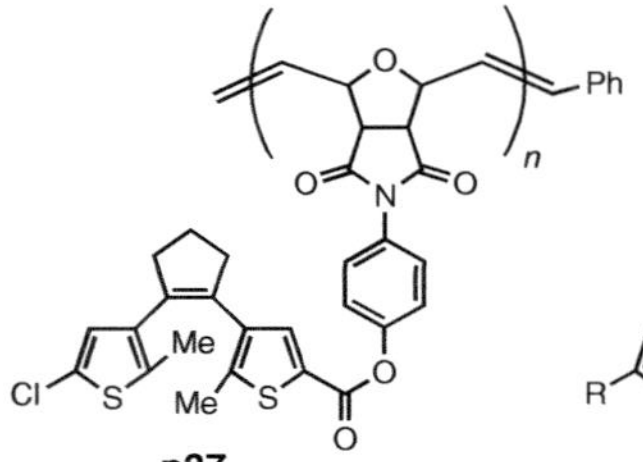 **p37** 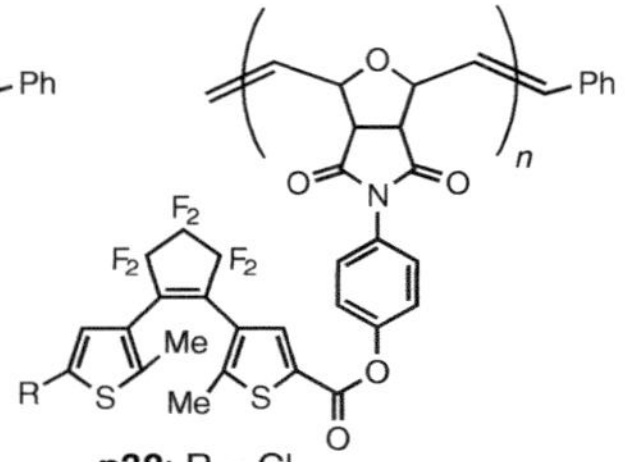 **p38**: R = Cl **p39**: R = Ph 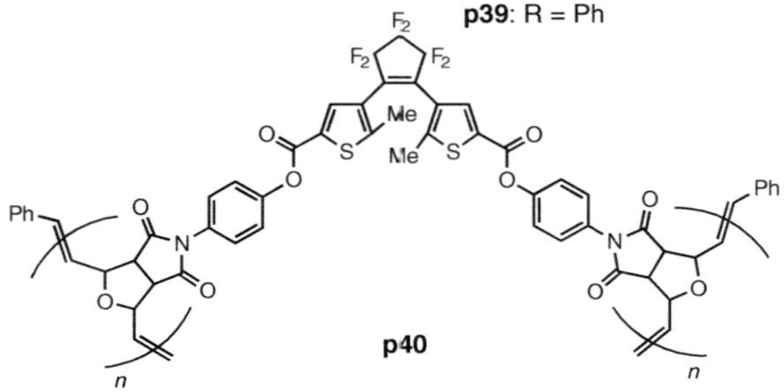 **p40**	Ring-opening metathesis polymerization	Polymer synthesis	[30, 31]

(*Continued*)

Table 8.2 (Continued)

Structure	Polymerization method	Research category	References
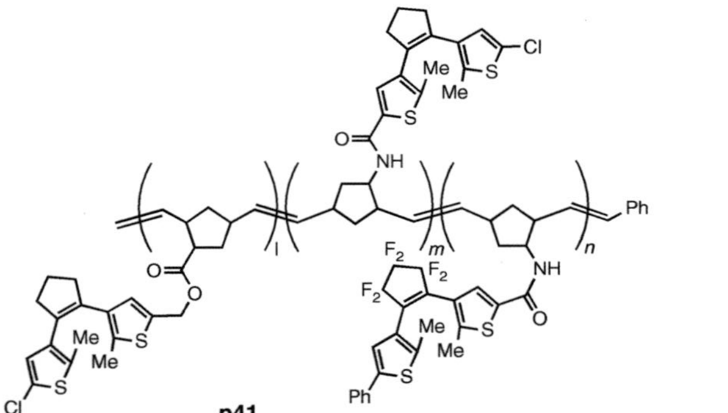 **p41**	Ring-opening metathesis polymerization	Multicolor photoswitching	[32]
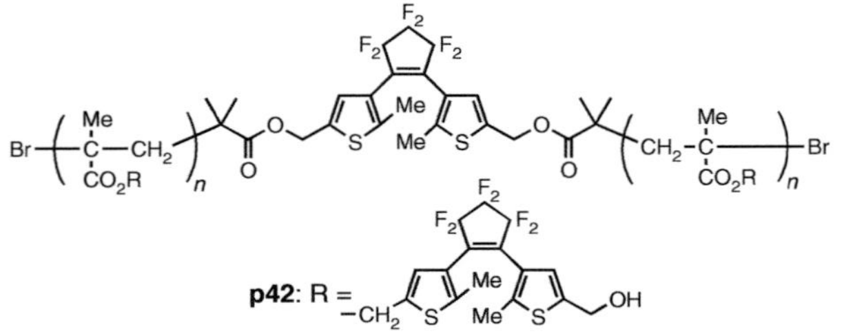 **p42**: R =	Atom transfer radical polymerization	Polymer synthesis	[33]
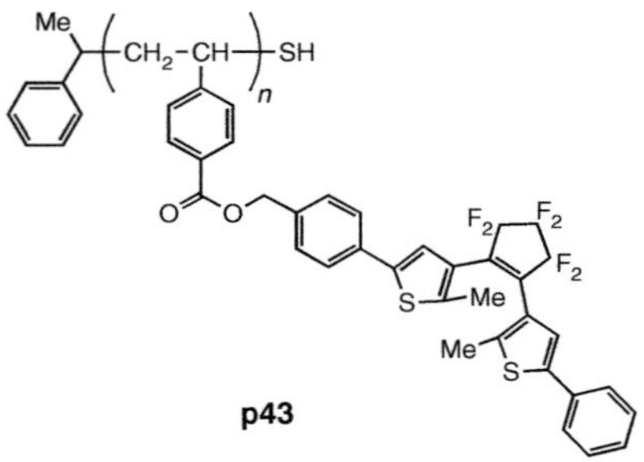 **p43**	Reversible addition-fragmentation chain transfer polymerization	Surface plasmon resonance of gold nanoparticles	[34, 35]

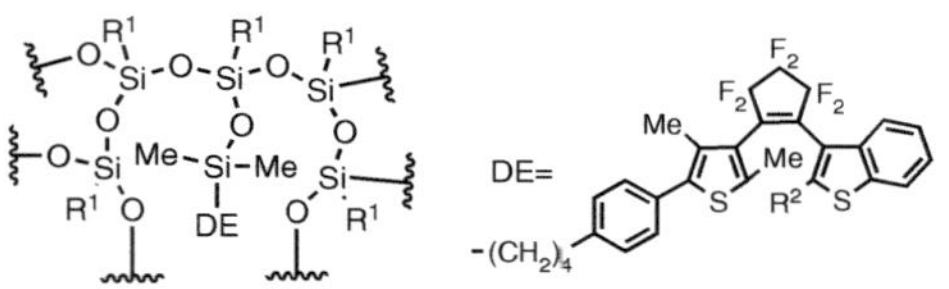 **p44**: R^1 = Me, R^2 = Me **p45**: R^1 = Ph, R^2 = Me **p46**: R^1 = Me, R^2 = MeO **p47**: R^1 = Ph, R^2 = MeO 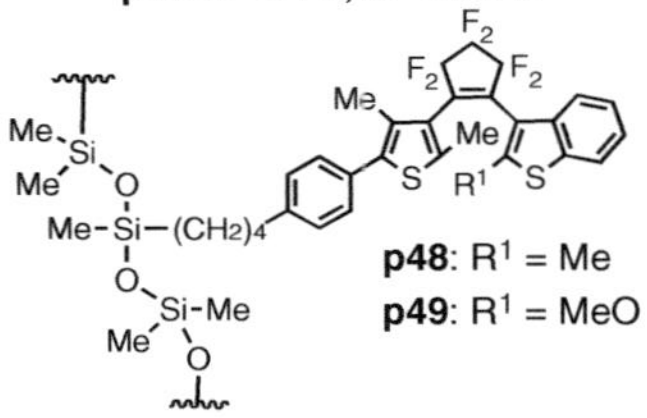 **p48**: R^1 = Me **p49**: R^1 = MeO	Pt-catalyzed hydrosilylation	Polymer synthesis	[36]
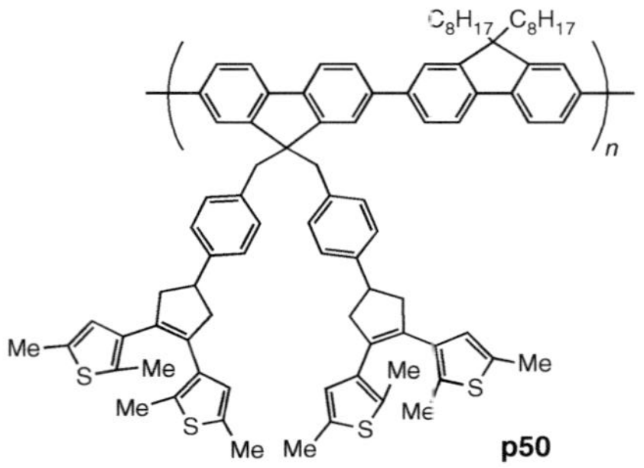 **p50**	Pd-catalyzed cross-coupling	Polymer synthesis Fluorescence	[37]

(*Continued*)

Table 8.2 (Continued)

Structure	Polymerization method	Research category	References
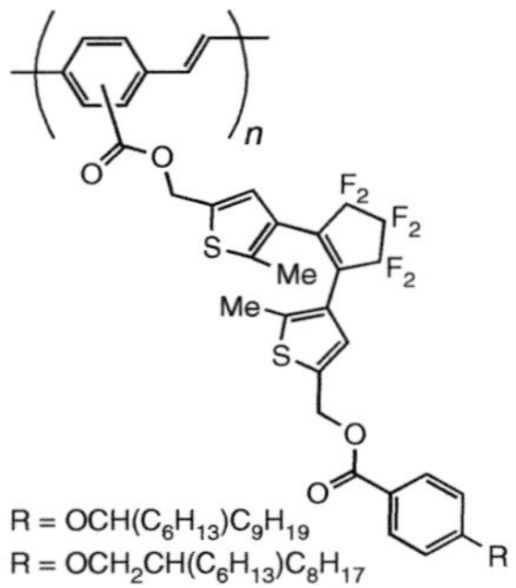 **p51**: R = $OCH(C_6H_{13})C_9H_{19}$ **p52**: R = $OCH_2CH(C_6H_{13})C_8H_{17}$	Stille coupling	Fluorescence	[38]
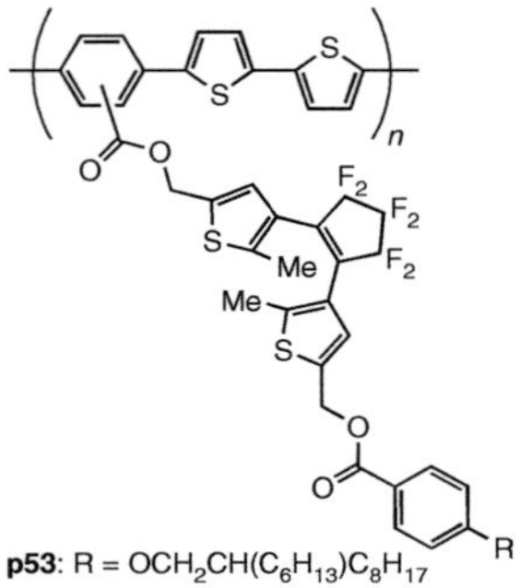 **p53**: R = $OCH_2CH(C_6H_{13})C_8H_{17}$	Stille coupling	Fluorescence	[38]
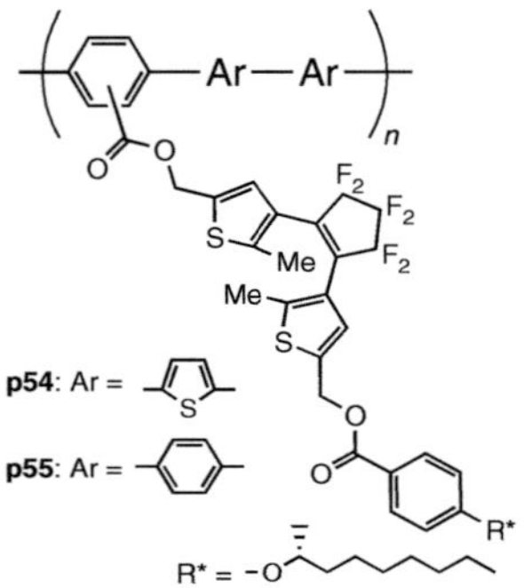 **p54**: Ar = **p55**: Ar = R* =	Stille coupling, Pd-catalyzed coupling	Photoswitching of circularly polarized fluorescence	[39]

reaction and the fluorescence quantum yields in $CHCl_3$ were measured to be 0.38 and 0.53, respectively. Thus, photoswitching efficiency and the fluorescence yield were increased by incorporating both diarylethene and PPV units in the main chain of the polymer, as compared to those of monomeric diarylethenes. The fluorescence of the polymer was quenched to less than half when the open-ring isomers of the diarylethene units converted to the closed-ring isomers upon UV irradiation. The fluorescence quantum yield difference between the open- and closed-ring isomers was much larger than that of monomeric diarylethenes. The polymer is potentially applicable to fluorescence readout memory media.

Not only fluorescence but also electrical conductivity was modulated when trimethyl silyl groups were introduced to PPV [4]. Neat polymer **p3** was placed between ITO and Au electrodes (cell structure: Au/polymer/ITO, polymer thickness: 200 nm) and its photoinduced electric property change was examined. The slope of the current-voltage (*I-V*) curve was much larger when the cell was irradiated with UV light. The current at 2 V of the colored cell, consisting of the closed-ring isomers of **p3**, was ~10 times larger than that observed for the bleached cell, consisting of the open-ring isomers. This indicates that the extension of π-conjugation between the benzothiophene rings and the phenyl ring through the ethene bridge in the closed-ring isomer leads to a higher current response than that in the open-ring isomer, where such π-conjugation is limited.

Polymer **p8** is another example, which modulates its fluorescence upon photoirradiation. Park and co-workers [7] prepared **p8** by the Knoevenagel condensation of 1,4-phenylenediacetonitrile and end-formyl substituted diarylethene (Scheme 8.1f). The weight average molecular weight was 13 000. As described in **3.4.1**, cyanostilbene unit exhibits AIEE effect. AIEE molecules are virtually nonfluorescent in the isolated molecular state but strongly fluorescent in the bulk solid phase due to specific intermolecular interactions. A cyanostilbene-based AIEE fluorophore with covalently bonded diarylethene switch formed strongly fluorescent nanoparticles as well as molecular aggregates in PMMA film, which showed reversible and bistable switching upon alternate irradiation with UV and visible light [40]. Polymer **p8** is the polymeric version of the photoswitchable AIEE fluorophore. Upon irradiation with 290 nm light, the open-ring isomer of **p8** film underwent photoisomerization to the closed-ring isomer. Along with the photoisomerization the fluorescence of **p8** film was significantly diminished. The integrated fluorescence intensity was reduced to 10% of the open-ring isomer state. The outstanding quenching is ascribed to an efficient energy transfer in the polymer film. Upon irradiation with visible light, the absorption as well as fluorescence spectra were reversibly and completely restored to the initial open-ring isomer state. One of the drawbacks of the polymer is the low cycloreversion quantum yield (7.5×10^{-3}). Long π-conjugation of the cyanostilbene group suppresses the cycloreversion reaction.

Polymer **p9** was prepared in an attempt to obtain photoswitchable electron-transport materials, which can be used in organic semiconductor memory [8]. Diacetyl-substituted dithienylethene was copolymerized with 3,3-dibenzoylbenzidine to yield **p9**, as shown in Scheme 8.1h. The polymer was placed between

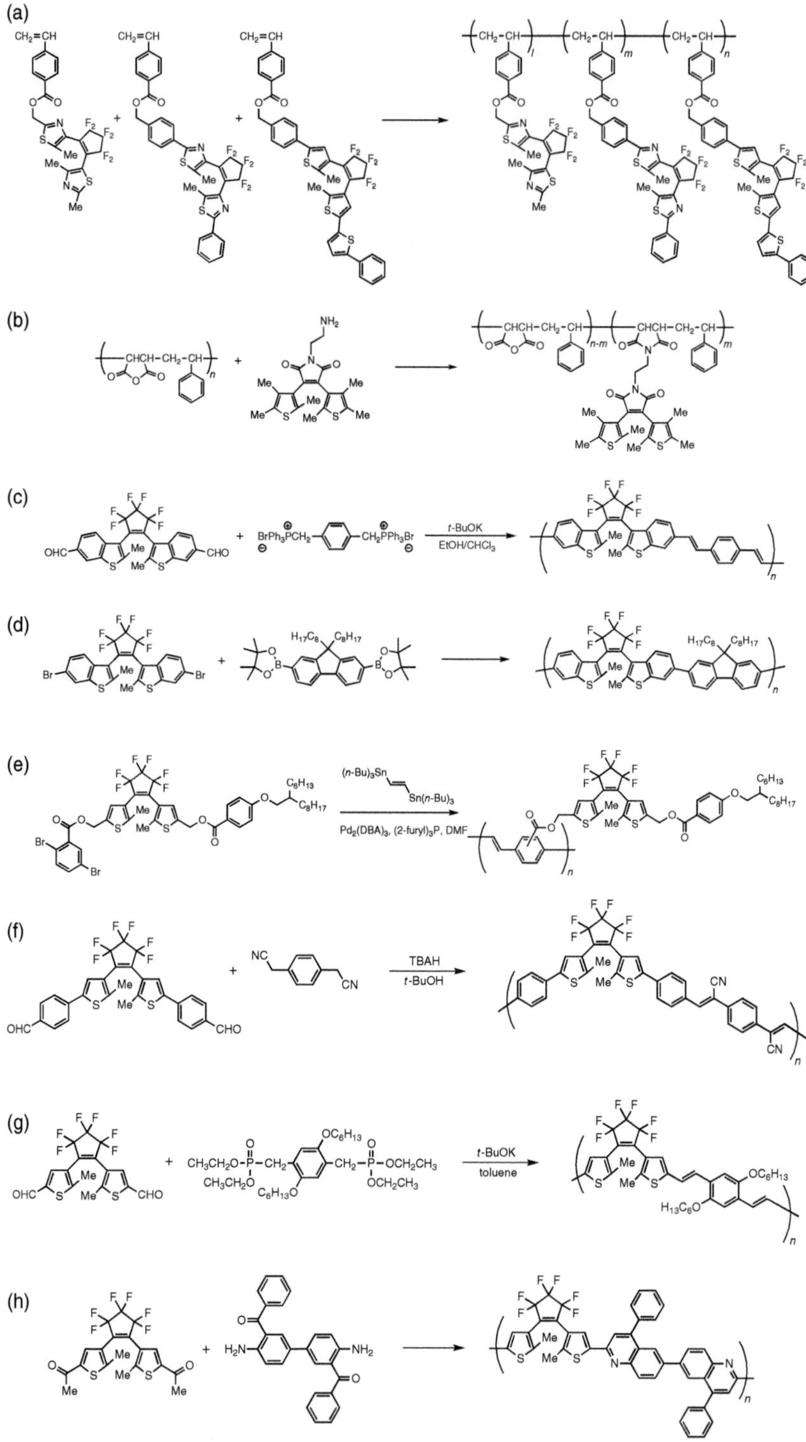

Scheme 8.1 Examples of synthetic methods of polymers having diarylethene units. (a) Radical polymerization [27]. (b) Polymer reaction [29]. (c) Wittig condensation [3, 4]. (d) Suzuki–Miyaura coupling [1, 2]. (e) Stille coupling [38]. (f) Knoevenagel polycondensation [7]. (g) Horner reaction [5]. (h) Friedländer condensation [8].

ITO and Au electrodes (cell structure: Au/polymer/ITO, polymer thickness: 200 nm) and its current–voltage characteristics were measured. The current of the cell consisting of the closed-ring isomer state at the applied voltage of 2 V was 2 times larger than that of the cell consisting of the open-ring isomer state. The high current decreased upon irradiation with visible light. Reversible photoswitching upon alternate irradiation with UV and visible light was observed.

As the photoswitching performance is highly dependent on the content of molecular photswitches in materials, diarylethene polymers containing a high effective concentration of diarylethene units are desirable. To increase the content, Ahn and co-workers [13] synthesized main-chain diarylethene homopolymer **p13** using a convenient polymerization method based on Friedel–Craft alkylation. 1,2-Bis(2-methyl-1-benzithiophen-3-yl)perfluorocyclopentene (**13**) in $CHCl_3$ was readily reacted with chloromethyl methyl ether and $TiCl_4$ at 0 °C. After 1 hour of stirring at room temperature under a nitrogen atmosphere, the solution was quenched with H_2O to afford the bis(benzothienyl)ethene polymer. The polymer was soluble in common organic solvents, such as THF and ethyl acetate. The molecular weight varied depending on the reaction conditions from 7970 to 40 000. The glass transition temperature T_g of the polymer was 110 °C. The polymer gave a characteristic photoswitching performance as the corresponding monomer unit in solution and in solid film. The fatigue resistance of the polymer to UV light was significantly improved compared with its monomer unit.

Another approach to synthesize polymers having high content of diarylethene units is oxidation polymerization. Uchida and co-workers [10–12] prepared cross-linked network diarylethene polymers, **p11** and **p12**, by oxidation polymerization of diarylethenes having phenol moieties, as shown in Scheme 8.2.

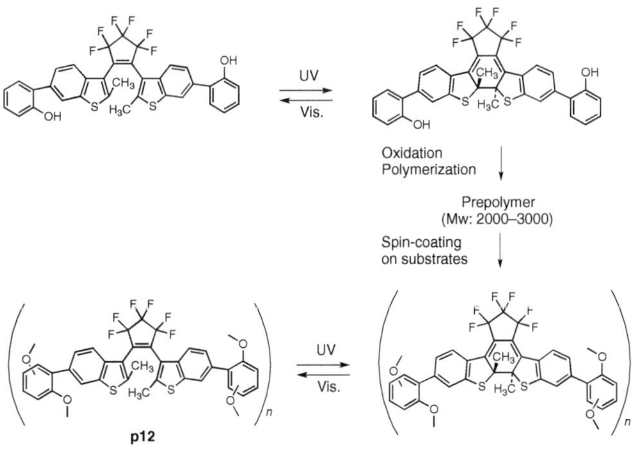

Scheme 8.2 Synthesis of polymer **p12** by oxidation polymerization of diarylethenes.

The prepolymers were formed by heating the closed-ring form diarylethene derivative with *N,N,N′,N′*-tetramethylethylenediamine and copper bromide in the presence of oxygen gas at 85 °C for 100 minutes. The prepolymers showed the molecular weight at *m/z* 3252, 2602, and 1953. The result indicates the formation of pentamer, tetramer, and trimer. The prepolymers were soluble in chloroform. By spin-coating the chloroform solution onto substrates, purple-colored films were formed. The films underwent cross-linking and turned insoluble in all solvents. In the cross-linked polymer film, diarylethene units are fixed in a photoactive antiparallel conformation. When the film was prepared from the open-ring form diarylethene, the spectral change of the film upon UV irradiation was very small. On the other hand, efficient coloration was observed when the film was prepared from the closed-ring form diarylethene. The quantum yield of the cyclization reaction was 70–80% of the monomer in ethyl acetate. The cycloreversion yield was almost the same as that of the monomer. The cross-linked polymer was successfully applied to optical memory media with nondestructive IR readout capability.

Polymers having high content of diarylethene units were prepared by ring-opening metathesis polymerization of diarylethene monomers with strained bicyclic 5-*exo*-norbornene [17, 30–32]. Copolymers **p41** are examples of such polymers, which undergo multicolor photoswitching. The multicolor polymers **p41** were prepared by copolymerization of different diarylethene monomers having different colors, yellow, red and blue, in the closed-ring isomer forms. The polymers had the molecular weight in the range of 40 600–53 700 and showed narrow molecular weight distribution. T_g of the copolymers was below 10 °C. The low T_g is beneficial to photoswitching, because there is adequate flexibility in the matrices. A three-component copolymer, for example, could be switched between 2^3 or 8 possible states. The polymer materials offer a new approach to diarylethene color-tuning for display technology and have potential applications in high-density optical data storage.

Such multicolor copolymers were prepared by conventional radical polymerization of vinyl monomers having diarylethene units, which show yellow, red, and blue colors in the closed-ring isomer forms. Kobatake and co-workers [27] synthesized copolymers **p33**, which contains three colored diarylethenes. Although the quantum yields of cyclization reactions of the homopolymers were as high as ~0.5 and similar to those in solution, the cyclization conversions of the polymer films were limited to 40–70% of those in solution. The decrease is ascribed to the presence of photoinactive parallel conformers, which cannot transform to photoactive antiparallel conformers in the solid state because of high T_g (~100 °C). The content of the parallel conformers was around 50%. Therefore, half of the diarylethene side groups are inactive. This limited the conversion yield. Polymers **p33** exhibited multicolor photoswitching by choosing appropriate wavelengths of illumination light. Upon irradiation with UV light, the polymer film turned dark black. The dark black color changed to yellow, red, and blue upon irradiation with $\lambda > 580$ nm, $\lambda > 434$ nm + 680 nm, and $\lambda > 434$ nm light, respectively. Such multicolor polymers have a potential for paper-like rewritable display media.

Bertarelli and co-workers [14] synthesized multicolor polyurethanes **p56** (Scheme 8.3) by the reaction between diarylethene end-capped with hydroxyl groups and aliphatic diisocyanate. Polycaprolactone diol and triol were also added, maintaining the overall stoichiometry with the diisocyanate. The reaction was catalyzed by dibutyltin dilaurate and 1,4-diazabicyclo[2,2,2]octane. The reaction mixture was spin-coated on substrates to form films and then heated at 130 °C. Due to the addition of triol, cross-linking took place and the films became totally insoluble in solvents. The amorphous films showed outstanding optical properties and photoswitching performance. Scheme 8.3 shows the structures of the multicolor polyurethane containing two kinds of diarylethenes, (1,2-bis(2-methyl-3-thienyl)perfluorocyclopentene (**3′o**) and 1,2-bis(2-methyl-5-phenyl-3-thienyl) perfluorocyclopentene (**20′o**)), in the main chain. Starting from **20′o–3′o,** the **20′c–3′c** state was obtained upon irradiation with 254 nm light. Upon irradiation with 670 nm light, **20′c–3′c** converted to **20′o–3′c**. The **20′c–3′o** state was reached upon irradiation of the initial **20′o–3′o** state with 366 nm light. Both closed-ring isomer units, **20′c** and **3′c,** returned to open-ring isomer units, **20′o** and **3′o**, upon irradiation with visible light in the range 500–600 nm.

Although the difference in IR spectra between the open- and closed-ring isomer states of **p56** was small, the Raman spectra showed marked signal difference. When the polymer contained the two monomer units in the ratio **3′o/20′o** of 1/5, the four states were clearly distinguished in the Raman spectra. The four states were readout nondestructively using a 1064 nm laser, which does not induce any photoswitching reaction, as the excitation light source. This is a promising nondestructive readout method in optical data storage.

8.2 Liquid Crystals

When molecular photoswitches are doped into liquid crystals, properties of the liquid crystals can be photochemically controlled. The first example is a cholesteric liquid crystal containing *trans–cis* photoisomerizable azobenzene [41]. Reversible color change in the cholesteric phase was observed by the photochemical *trans–cis* isomerization of the dissolved azobenzene. Nematic liquid crystals composed of *trans*-form azobenzene derivatives underwent a phase transition from the nematic phase to the isotropic phase upon irradiation with UV (366 nm) light [42]. The nematic phase was recovered when photoirradiation ceased because of thermal *cis* to *trans* isomerization. Photoswitching between chiral nematic (induced cholesteric) and apparent nematic phases was observed by doping chiral photorasemizable compounds into nematic liquid crystals [43].

As described in Section **3.5**, chiral properties of diarylethene derivatives are reversibly modulated upon irradiation with UV and visible light. When such chiral diarylethenes are doped into nematic liquid crystals, a phase transition in the liquid crystalline phase takes place [44]. Chiral cyclohexane **201** having

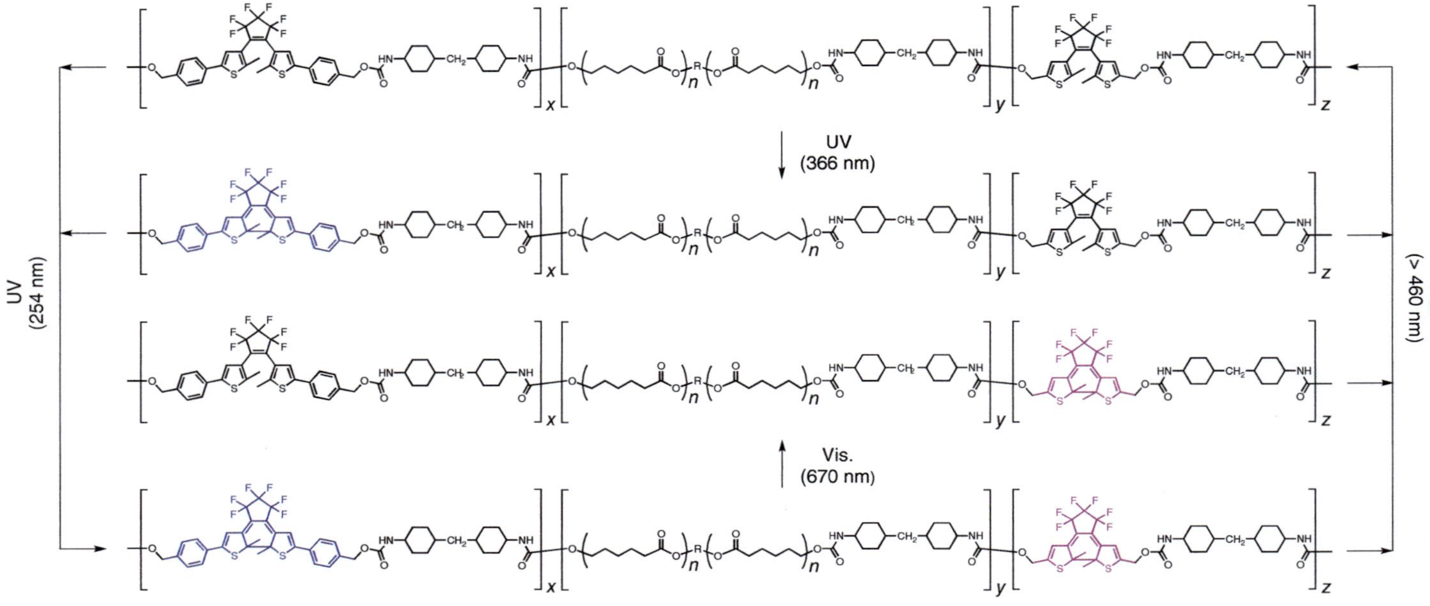

Scheme 8.3 Selective photoreactions of polyurethanes **p56** containing two kinds of diarylethene units, **3′** and **20′**.

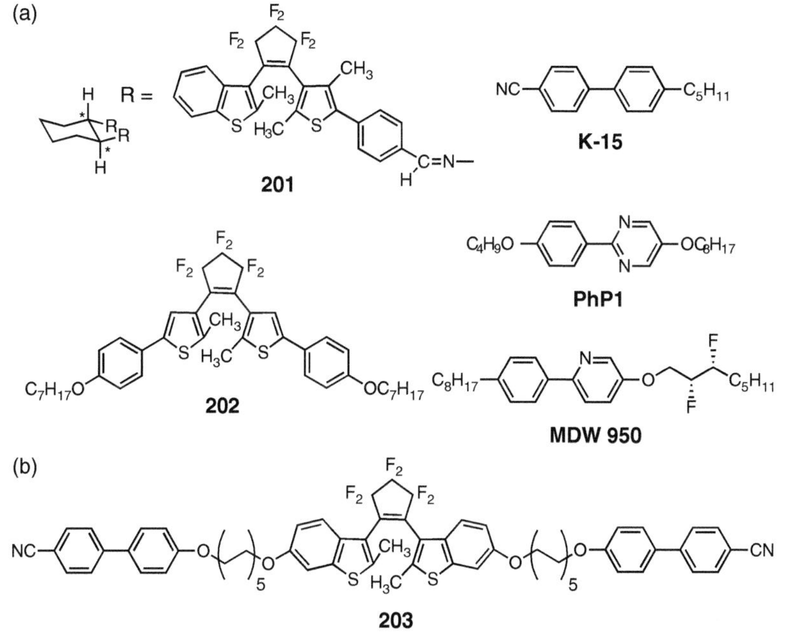

Scheme 8.4 (a) Diarylethene dopants for liquid crystals. (b) A diarylethene liquid crystal.

two diarylethene units was prepared and introduced into the liquid crystal of 4-cyano-4′-pentylbiphenyl (**K-15**) (Scheme 8.4). When the content of **201** was 1.0 wt%, the liquid crystal maintained a nematic texture. Upon irradiation with UV (313 nm) light, the two diarylethene units transformed into closed-ring isomers. At the photostationary state, the main product was both closed-ring **C–C** form (60%) and the rest was one closed-ring and one open-ring **C–O** form (40%). Upon irradiation with >480 nm light, the mixture returned to both open-ring **O–O** form. The phase change of the film was monitored using a polarization microscope. Upon irradiation with UV light, the nematic texture changed to a cholesteric fingerprint texture, which remained stable at room temperature for a long time. The fingerprint texture disappeared only when irradiated with visible light. Alternate irradiation with UV and visible light caused the reversible switching between the nematic and the chiral nematic (induced cholesteric) phases. The phase change is attributed to a large helical twisting power of the closed-ring isomer in comparison with that of the open-ring isomer. The pitch of the liquid crystalline phase was measured by a droplet method. The both open-ring **O–O** form showed the pitch larger than >40 μm, while the pitch decreased to 11.8 μm upon irradiation with UV light. Similar phase transition was observed when 3,3-binaphthyl derivatives having two diarylethene units were doped into nematic liquid crystals [45].

The chiral smectic C (SmC*) phase is ferroelectric and exhibits spontaneous electric polarization (P_S) that can be coupled to an electric field to produce an electro-optical switch. In most cases, photomodulation of P_S is achieved via *trans–cis* photoisomerization reactions of 4,4′-disubstituted azobenzene dopants, which cause a destabilization of the SmC* phase due to the change in shape of azobenzene units. However, the photoswitches are not stable due to thermal reversion of the *cis*-form back to the thermally stable *trans*-form. To provide stable photoswitching ability, it is indispensable to use thermally irreversible molecular photoswitches as the dopants. Lemieux and co-workers [46, 47] employed thermally irreversible diarylethene derivatives as the dopants and prepared photoresponsive ferroelectric SmC* liquid crystals. They doped diarylethene **202** in a ferroelectric liquid crystal host containing 10 mol% **MDW950** in achiral smectic liquid crystal **PhP1** (Scheme 8.4). The dopant caused a significant lowering of SmA*–SmC* phase transition temperature from 84 to 67 °C at a content of 3.0 mol%. The phase transition temperature further decreased from 67 to 57 °C upon UV irradiation. Alternate irradiation with UV and visible light resulted in the reversible modulation of *Ps* between 18.5 and 1.5 nC/cm^2 at 55 °C. The change in P_s is explained by the difference in conformational degrees of freedom of the open- and closed-ring isomers.

When mesogenic groups are attached, some of diarylethene derivatives form liquid crystals. Mehl and co-workers [48, 49] synthesized diarylethene **203**, which has cyanobiphenyl groups as mesogens via oligomethylene chain spacers. The highest stable phase was a nematic phase. The open-ring isomer showed the nematic phase between 78.5 and 74.5 °C. Below 74.5 °C, diarylethene **203** exhibited a SmC phase. Irradiation with UV (313 nm) light modulated the properties. In the photostationary state, the stability of the nematic phase was reduced and the nematic phase was observed between 75.9 and 54.6 °C. Below 54.6 °C, a smectic phase of unidentified structure appeared. The reduction of flexibility due to the cyclization reaction is considered to reduce the mesomorphic stability. The liquid crystalline properties were reversibly modulated by the photoisomerization of the diarylethene core.

Li and co-workers [50, 51] designed and synthesized chiral dithienylethene dopants with bridged binaphthyl groups at both sides, which exhibit unique characteristics with respect to their helical induction capability in the nematic liquid crystals. These compounds were used as chiral dopants to induce photoresponsive cholesteric liquid crystals. A dopant with a methylenedioxy bridge exhibited dramatic increase in the helical twisting power upon UV irradiation. When doped in an achiral liquid crystal with low concentration (0.4 wt%), it retained an achiral liquid crystal phase, whose helical superstructure was reversibly induced upon photoirradiation. When doped at relatively high concentration (7.7 wt%), three thermally stable reflection colors, red, green, and blue, could be attained in a single cholesteric film by varying the UV irradiation time. This is a convenient approach to construct multicolor display media.

References

1 Kawai, T., Kunitake, T., and Irie, M. (1999). Novel photochromic conducting polymer having diarylethene derivative in the main chain. *Chem. Lett.* 905-906.

2 Kawai, T., Nakashima, Y., Kunitake, T., and Irie, M. (2005). Photon-mode modulation of fluorescence and electric current with a photochromic conducting polymer. *Curr. Appl. Phys.* 5: 139–142.

3 Cho, H. and Kim, E. (2002). Highly fluorescent and photochromic diarylethene oligomer bridged by *p*-phenylenevinylene. *Macromolecules* 35 (23): 8684–8687.

4 Kim, E. and Lee, H.W. (2006). Photo-induced electrical switching through a mainchain polymer. *J. Mater. Chem.* 16: 1384–1389.

5 Bertarelli, C., Bianco, A., Boffa, V. et al. (2004). Poly(dithienylethene-*alt*-divinylenephenylene)s: increasing the molecular weights in diarylethene photochromic polymers. *Adv. Funct. Mater.* 14 (11): 1129–1133.

6 Wang, S., Li, X., Chen, B. et al. (2004). Photochromic copolymers containing bisthienylethene units. *Macromol. Chem. Phys.* 205: 1497–1507.

7 Lim, S.-J., An, B.-K., and Park, S.Y. (2005). Bistable photoswitching in the film of fluorescent photochromic polymer: enhanced fluorescence emission and its high contrast switching. *Macromolecules* 38 (15): 6236–6239.

8 Choi, H., Lee, H., Kang, Y. et al. (2005). Photochromism and electrical transport characteristics of a dyad and a polymer with diarylethene and quinoline units. *J. Org. Chem.* 70 (21): 8291–8297.

9 Lee, J., Kwon, T., and Kim, E. (2007). Electropolymerization of an EDOT-modified diarylethene. *Tetrahedron Lett.* 48: 249–254.

10 Uchida, K., Takata, S., Nakamura, S., and Irie, M. (2002). Photochromic films prepared by oxidation polymerization of a diarylethene having phenol moieties. *Chem. Lett.*: 476–477.

11 Uchida, K., Takata, A., Saito, M. et al. (2003). A novel photochromic film by oxidation polymerization of a bisbenzothienylethene with phenol groups. *Adv. Mater.* 15 (10): 121–125.

12 Uchida, K., Takata, A., Saito, M. et al. (2003). Synthesis of novel photochromic films by oxidation polymerization of diarylethenes containing phenol groups. *Adv. Funct. Mater.* 13 (10): 755–762.

13 Jeong, Y.-C., Park, D.G., Kim, E. et al. (2006). Polymerization of a photochromic diarylethenes by Friedel-Crafts alkylation. *Macromolecules* 39 (9): 3106–3109.

14 Pariani, G., Castagna, R., Dassa, G. et al. (2011). Diarylethene-based photochromic polyurethanes for multistate optical memories. *J. Mater. Chem.* 21: 13223–13231.

15 Sciascia, C., Castagnia, R., Dekermenjian, M. et al. (2012). Light-controlled resistance modulation in a photochromic diarylethene-carbon nanotube blend. *J. Phys. Chem. C* 116: 19483–19489.

16 Jeong, Y.-C., Yang, S.I., Kim, E., and Ahn, K.-H. (2006). A high-content diarylethene photochromic polymer for an efficient fluorescence modulation. *Macromol. Rapid. Commun.* 27: 1769–1773.

17 Wigglesworth, T.J. and Branda, N.R. (2004). Ultra-high-density photochromic main-chain 1,2-dithienylcyclopentene polymers prepared using ring-opening metathesis polymerization. *Adv. Mater.* 16 (2): 123–125.

18 Areephong, J., Kudemac, T., de Jong, J.J.D. et al. On/off photoswitching of the electropolymerizability of terthiophenes. *J. Am. Chem. Soc.* 130: 12850–12851.

19 Wesenhagen, P., Areephong, J., Landaluce, T.F. et al. (2008). Photochromism and electrochemistry of a dithienylcyclopentene elctroactive polymer. *Langmuir* 24 (12): 6334–6342.

20 Nakashima, H. and Irie, M. (1998). Synthesis of polystyrene and poly(alkyl methacrylate)s having photochromic dithienylethene pendant groups. *Polymer. J.* 30 (12): 985–989.

21 Kim, E., Choi, Y.-K., and Lee, M.-H. (1999). Photoinduced refractive index change of photochromic diarylethene polymer. *Macromolecules* 32 (15): 4855–4860.

22 Tian, H. and Tu, H.-Y. (2000). Synthesis and photochromic properties of new bisthienylethene derivatives and a copolymer. *Adv. Mater.* 12 (21): 1597–1600.

23 Kobatake, S. and Kuratani, H. (2006). Photochromism of diarylethene-functionalized polystyrene with high conversion in solid state. *Chem. Lett.* 35 (6): 628–629.

24 Lim, S.-J., Carling, C.-J., Warford, C.C. et al. (2011). Multifunctional photo- and thermo-responsive copolymer nanoparticles. *Dyes Pigm.* 89: 230–235.

25 Kobatake, S. and Yamashita, I. (2008). Synthesis of photochromic diarylethene polymers for a write-by-light/erase-by-heat recording system. *Tetrahedron* 64: 7611–7618.

26 Diaz, S.A., Monéndez, G.O., Etchehon, M.H. et al. (2011). Photoswitchable water-soluble quantum dots: pcFRET based on amphiphilic photochromic polymer coating. *ACS Nano* 5 (4): 2795–2805.

27 Nishi, H., Namari, T., and Kobatake, S. (2011). Photochromic polymers bearing various diarylethene chromophores as the pendant: synthesis, optical properties, and multicolor Photochromism. *J. Mater. Chem.* 21: 17249–17258.

28 Cho, S.Y., Yoo, M., Shin, H.-W. et al. (2002). Preparation of diarylethene copolymers and their photoinduced refractive index change. *Optical Materials* 21: 279–284.

29 Chen, B., Wang, M., Li, C. et al. (2003). New photochromic copolymers with bisthienylethene pendant groups. *Synthetic Metals* 135-136: 491–492.

30 Myles, A.J., Zhang, Z., Liu, G., and Branda, N.R. (2000). Novel Synthesis of photochromic polymers via ROMP. *Org. Lett.* 2 (18): 2749–2751.

31 Myles, A.J. and Branda, N.R. (2003). Novel photochromic homopolymers based on 1,2-bis(3-thienyl)cyclopentenes. *Macromolecules* 36 (2): 298–303.

32 Wigglesworth, T.J. and Branda, N.R. (2005). A family of multiaddressable, Multicolored photoresponsive copolymers prepared by ring-opening metathesis polymerization. *Chem. Mater.* 17 (22): 5473–5480.

33 Shen, L., Ma, C., Pu, S. et al. (2009). Synthesis and properties of novel photochromic poly(methyl methacrylate-*co*-diarylethene)s. *New J. Chem.* 33: 825–830.

34 Nishi, H. and Kobatake, S. (2008). Photochromism and optical property of gold nanoparticles covered with low-polydispersity diarylethene polymers. *Macromolecules* 41 (11): 3995–4002.

35 Nishi, H., Asahi, T., and Kobatake, S. (2011). Enhanced photocycloreversion reaction of diarylethene polymers attached to gold nanoparticles in the solid state. *J. Photochem. Photobiol. A: Chem.* 221: 256–260.

36 Nakashima, H. and Irie, M. (1999). Synthesis of silsesquioxanes having photochromic dithienylethene pendant groups. *Macromol. Chem. Phys.* 200: 683–692.

37 Li, X. and Tian, H. (2005). High-content pendant photochromic copolymer with dithienylethene/fluorene 2:1 mole ratio. *Macromol. Chem. Phys.* 206: 1769–1777.

38 Hayasaka, H., Tamura, K., and Akagi, K. (2008). Dynamic switching of linearly polarized emission in liquid-crystallinity-embedded photoresponsive conjugated polymers. *Macromolecules* 41 (7): 2341–2346.

39 Hayasaka, H., Miyashita, T., Tamura, K., and Akagi, K. (2010). Helically π-stacked conjugated polymers bearing photoresponsive and chiral moieties in side chains: reversible photoisomerization-enforced switching between emission and quenching of circularly polarized fluorescence. *Adv. Mater.* 20: 1243–1250.

40 Lim, S.-J., An, B.-K., Jung, S.D. et al. (2004). Photoswitchable organic nanoparticles and polymer film employing multifunctional molecules with enhanced fluorescence emission and bistable photochromism. *Angew. Chem. Int. Ed.* 43: 6346–6350.

41 Sackmann, E. (1971). Photochemically induced reversible color changes in cholesteric liquid crystals. *J. Am. Chem. Soc.* 93 (25): 7088–7090.

42 Ikeda, T. and Tsutsumi, O. (1995). Optical switching and image storage by means of azobenzene liquid-crystal films. *Science* 268: 1873–1875.

43 Udayakumar, B.S. and Schuster, G.B. (1993). 1-Bnezylidene-2,3:4,5:6,7-tribenzocyclohepta-2,4,6-trienes: synthesis and photochemistry of a series of optically active potential triggers for physical amplification of a photoresponse in liquid crystalline media. *J. Org. Chem.* 58 (15): 4165–4169.

44 Yamaguchi, T., Inagawa, T., Nakazumi, H. et al. (2000). Photoswitching of helical twisting power of a chiral diarylethene dopants: pitch change in a chiral nematic liquid crystal. *Chem. Mater.* 12 (4): 869–871.

45 Yamaguchi, T., Inagawa, T., Nakazumi, H. et al. (2001). Photoinduced pitch changes in chiral nematic liquid crystals formed by doping with chiral diarylethene. *J. Mater. Chem.* 11: 2453–2458.

46 Maly, K.E., Wand, M.D., and Lemieux, R.P. (2002). Bistable ferroelectric liquid crystal photoswitch triggered by a dithienylethene dopant. *J. Am. Chem. Soc.* 124 (27): 7898–7899.

47 Maly, K.E., Zhang, P., Wand, M.D. et al. (2004). Reversible photocyclization of achiral dithienylperfluorocyclopentene dopants in a ferroelectric liquid crystal: bistable SSFLC photoswitching. *J. Mater. Chem.* 14: 2806–2812.

48 Frigoli, M. and Mehl, G.H. (2003). A photochromic liquid crystal system. *ChemPhysChem.* 4: 101–103.

49 Frigoli, M. and Mehl, G.H. (2004). Modulation of the absorption, fluorescence, and liquid-crystal properties of functionalized diarylethene derivatives. *Chem. Eur. J.* 10: 5243–5250.

50 Li, Y., Urbas, A., and Li, Q. (2012). Reversible light-directed red, green, and blue reflection with thermal stability enabled by a self-organized helical superstructure. *J. Am. Chem. Soc.* 134: 9573–9576.

51 Li, Y., Wang, M., Urbas, A., and Li, Q. (2013). A photoswitchable and thermally stable axially chiral dithienylperfluorocyclopentene dopant with high helical twisting power. *J. Mater. Chem. C* 1: 3917–3923.

9 Applications

9.1 Organic Field-Effect Transistors (OFETs)

Molecular photoswitches modulate electrical properties, such as oxidation/reduction potentials, dipole moments, and carrier mobility, upon photoirradiation and are applicable in the fabrication of various types of electronic devices [1]. One such example is the photocontrollable Organic Field-Effect Transistor (OFET). The transistor can be gated in two independent ways by either using an electrical stimulus or an optical stimulus. The output current can be simultaneously and independently controlled through voltage and light.

Figure 9.1 shows two typical structures of OFETs. Photoswitchable molecules are incorporated in the semiconductor layer or the insulator layer. A very simple approach is the direct use of neat photoswitchable molecules as the channel semiconductor layer. Wakayama and co-workers [2] employed this approach and fabricated a device shown in Figure 9.1a. A top-contact and bottom-gate-type diarylethene transistor was fabricated on a highly doped p^+-Si (001) wafer with a 200 nm-thick insulator layer of SiO_2. The Si wafer and the oxide layer act as a gate electrode and a gate insulator, respectively. A thick amorphous 1,2-bis(2-methyl-6-biphenyl-3-thienyl)perfluorocyclopentene film (thickness: 45 nm) was grown on the substrate through a shadow mask with a vacuum deposition method. Subsequently, 150 nm-thick Au layers were deposited on top of the diarylethene film as source and drain electrodes. Initially, the sample was irradiated with UV light for 5 minutes and 75% of the open-ring isomers were converted to the closed-ring isomers. The drain current of the irradiated device was found to increase linearly at a low drain voltage. This indicates that the closed-ring diarylethene neat film works as a p-type channel transistor. The drain current was completely suppressed after visible light irradiation for 2 minutes. The reduced current again increased upon irradiation with UV light. The reversible change in the drain current was repeatedly observed by alternate irradiation with UV and visible light. The *on/off* ratio in the drain current at a fixed gate voltage of −100 V was about 10^2. Furthermore, the *on* and *off* states modulated by light irradiation were maintained even after one week under dark conditions.

Thin films made of neat diarylethene derivatives suffer from their poor charge transport characteristics. To overcome this problem, Samori and

Diarylethene Molecular Photoswitches: Concepts and Functionalities, First Edition. Masahiro Irie.
© 2021 WILEY-VCH GmbH. Published 2021 by WILEY-VCH GmbH.

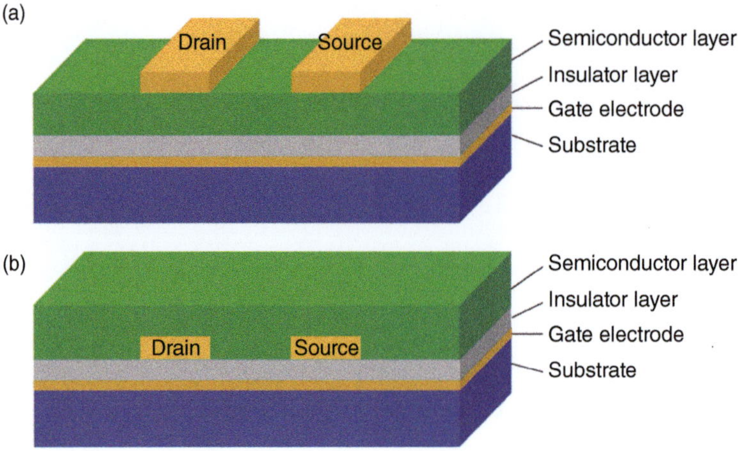

Figure 9.1 Two types (a and b) of structures of organic field-effect transistors (OFETs).

co-workers [3, 4] incorporated diarylethene molecular photoswitch **204** into an organic semiconductor layer, such as poly(3-hexylthiophene) (**P3HT**) or 2,7-dialkyl-benzothieno(3,2-b)benzothiophene (**BTBT**) (Scheme 9.1). Both **P3HT** and **BTBT** have relatively high field-effect hole mobility in thin films and easy processing in various organic solvents. They show comparably low ionization energies and ensure an ohmic hole injection and/or extraction with gold electrodes. To reveal the electrical properties of diarylethene-doped semiconductor systems, the blend of **P3HT** and diarylethene **204** was investigated in detail. An OFET device in bottom-contact bottom-gate geometry was fabricated as shown in Figure 9.1b. A solution of **P3HT** in chloroform containing **204** was spun to form a thin film on Si/SiO surface in-between the gold pre-patterned source and drain electrodes. The content of **204** in **P3HT** was prepared to be 20 wt%. The two-dimensional diffraction pattern of the film indicated that **204** molecules in both the open- and closed-ring isomers do not disrupt significantly the molecular packing of **P3HT**, but prefer to reside within the amorphous domains of the semiconductor polymer. Although the hole mobility decreased with increase in the content of **204**, the mobility remained relatively high. The modulation of drain current at drain voltage of $V_d = -10\,V$ and gate voltage of $V_g = -80\,V$ upon alternate irradiation with UV and white light was found to be as large as 80%. The current responses are governed by the different HOMO levels of the two isomers. The isomerization induced with UV light results in the formation of traps affecting the transport of holes through the device. Dynamic characterization of **P3HT/204**-based phototransistor was carried out by exposing the transistor to fast irradiation cycles. This experiment revealed that the current in the devise can be reversibly photomodulated on a time scale of seconds without showing fatigue. The devise showed a response time of a few microseconds on irradiation with pulsed UV light ($\lambda = 313\,nm$). Most importantly, the devises did not show any degradation of the photoswitchability, even after two months of storage under nitrogen. Based on the above photoswitchable charge trapping in the

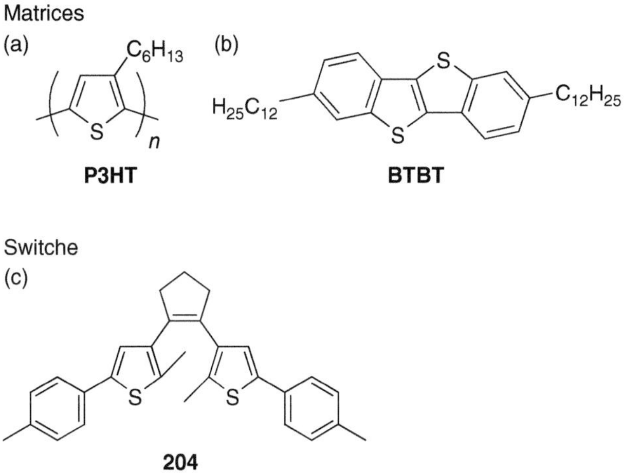

Scheme 9.1 Chemical structures of (a) polymeric and (b) small-molecule semiconducting matrices, and (c) a photoswitching component.

thin-film transistors, a multilevel nonvolatile flexible optical memory device was fabricated [5].

The above p-type semiconductors can be replaced with n-type conductors [6]. The electron transport in OFETs was photomodulated by blending n-type organic semiconductors, such as fullerene derivatives, with a suitable diarylethene. Photocontrollable ambipolar transistors were also developed by using a diarylethene derivative having dibenzo[*b*,*d*]thiophene substituents as the transistor channel [7]. In these transistors, both electron transport and hole transport take place. UV and visible light irradiation caused the modulation of both electron and hole currents. The light-induced *on*/*off* ratios were calculated to be in the 95–225 range for the hole transport and in the 6–22 range for the electron transport. Not only the semiconductor layers but also the insulator materials can be used as a phototunable layer [8].

9.2 Metal Organic Frameworks (MOFs)

Metal Organic Frameworks (MOFs) are attractive gas-adsorbent candidates. The large internal surface areas of MOF absorb gases, such as CO_2, CO, C_2H_2, C_2H_4, hydrogen, and methane. The release process of the absorbed gases mainly depends on vacuum or heat operation, which is very energy consuming. It is strongly desired to develop a new technology that does not require large energy cost. In this respect, light is the best choice. Incorporation of photoswitchable molecules into MOFs can achieve this goal.

As the first trial, azobenzene moieties were incorporated into MOFs as pendant groups of the linkers or as building components of the frameworks, as shown in Figure 9.2 [9, 10]. When the azobenzene unit was attached as the pendant group to the inner pore wall, the photoisomerization readily took place in the cavity and led to

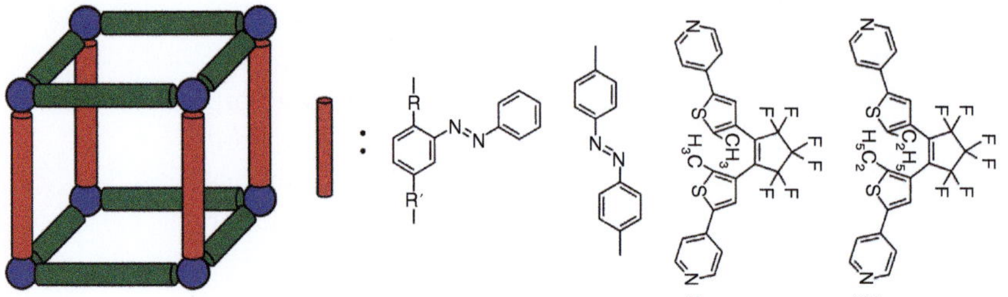

Figure 9.2 3D framework structures of photoresponsive metal organic frameworks (MOFs).

the changes in the cavity space and CO_2 uptake. Before UV irradiation, CO_2 uptake of the pristine sample at room temperature was 22.9 cm^3/g at 1 bar, while it decreased to 16.8 cm^3/g after UV irradiation for 1 hour, corresponding to 26.4% decrease in CO_2 uptake. The UV-irradiated state was thermally unstable. After heating at 60 °C for 20 hours in the dark the uptake was restored to the initial value. The result indicates that the geometrical structure change of the azobenzene moiety is responsible for the change in gas uptake.

Although the azobenzene moiety undergoes the *trans*-to-*cis* photoisomerization when attached as the pendant group, the isomerization is prohibited when incorporated into the framework component, because of its large geometrical structure change during the isomerization reaction. One of the drawbacks of azobenzene is the thermal instability of the photogenerated *cis*-form. Undesired thermal backward reactions or side reactions would be operative. To prepare efficient and thermally stable photocontrollable MOFs, diarylethene derivatives were employed as the framework components.

A flexible porous MOF was prepared by a solvothermal reaction of $Zn(NO_3)_2 \cdot 6H_2O$, 1,4-benzenedicaroboxylic acid (H_2bdc) and 1,2-bis(2-methyl-5-pyridyl-3-thienyl)-perfluorocyclopentene (**42**) in *N,N*-dimethylformamide (DMF) [11]. The paddlewheel-type 2D zinc complexes composed of bdc^{2-} and Zn^{2+} were connected by the coordination of **42** to construct 3D pillared layer structure with interlayer distance of 2.16 nm between the 2D sheets. The twofold interpenetrated frameworks afforded a channel structure that accommodates DMF and water molecules. The void volume was evaluated to be 30% of the total unit cell volume. The colorless crystal turned dark blue upon exposure to UV light and the conversion from **42o** to **42c** was almost quantitative. The void volume was significantly reduced from 30% to 21% with the decrease in the amount of guest molecules accommodated in the channels. The molecular structure change induced upon irradiation with UV light resulted in the global structural change of the frameworks and channels. Photochemical control of gas sorption was investigated for the heat-treated guest-free sample. The CO_2 sorption experiment proved that the void volume change caused the change in the absorbed CO_2 volume. The amount of adsorbed CO_2 at $P/P_0 = 0.98$ for nonirradiated crystal reached 136 ml (stp) g^{-1}. After UV irradiation, the adsorption at $P/P_0 = 0.98$

decreased to 108 ml (stp) g^{-1}. The adsorption recovered to 129 ml (stp) g^{-1} upon irradiation with visible light. Reversible modulation of CO_2 sorption was observed by using this flexible MOF. Similar photomodulation of CO_2 adsorption was observed for MOF composed of 1,2-bis(2-ethyl-5-pyridyl-3-thienyl)perfluorocyclopentene (**205**), $Zn(NO_3)_2$ and 4,4-oxybisbenzoic acid (oba) [12]. Incorporation of sufficiently flexible ligand oba further facilitated the change in the framework geometry by the photoisomerization of **205**.

The micro-size particles of MOF composed of **42**, $Zn(NO_3)_2{\cdot}6H_2O$ and biphenyl-4,4′-dicarboxylic acid were used to separate C_2H_2 and C_2H_4 [13]. The particles were prepared by ball milling and the particle size was around 1 μm. At 100 kPa and 293 K the C_2H_2 absorption capacity of the particles was 36.1 $cm^3\ g^{-1}$, which is 1.3 times larger than C_2H_4 uptake of 27.4 $cm^3\ g^{-1}$. On the other hand, at 100 kPa and 195 K the C_2H_2 uptake largely increased to 120.3 $cm^3\ g^{-1}$, which is 1.85 times larger than the corresponding value of 64.9 $cm^3\ g^{-1}$ for C_2H_4. The result indicates that at low temperature, C_2H_2 is preferentially adsorbed. Upon irradiation with UV light, the ratio of C_2H_2/C_2H_4 was dramatically reduced to 1.08. The selectivity of C_2H_2/C_2H_4 (50/50) mixture at 195 K and 100 kPa was found to be 47.1. The high selectivity of the pristine sample was reduced to 3.0 upon UV irradiation. The UV irradiation induced the geometrical structure change of the MOF and the change caused the decrease of the selectivity.

The above organic ligand-to-metal coordinations are not only useful for constructing infinite 3D molecular frameworks, but also applicable in the preparation of a single coordination cage. Figure 9.3 shows structures of diarylethene ligands **206** and metal organic cages composed of two square-planar-coordinated Pd^{2+} and four bis-monodentate pyridyl ligands **206** based on a dithienylethenes photoswitch [14]. Upon irradiation with UV light, the photocyclization reaction transformed the complete set of isomers into their closed-ring forms. The closed-ring isomers in the cage reversibly returned to the open-ring isomers upon irradiation with visible light. The hydrodynamic radius (r_h) of the cages was found to change depending on the isomer states of the **206** ligands. When four **206** ligands were in the open-ring isomer states, r_h was 0.704 nm. On the other hand, when the four ligands converted to the closed-ring isomer states, r_h increased to 0.867 nm. The size of the open-ring isomer cage was smaller than that of the closed-ring isomer cage. This is ascribed to the conformational flexibility of the open-ring isomer cage in contrast to the rigid and stretched structure of the closed-ring isomer cage. The open-ring isomer cage showed a much stronger affinity for the guest anion, dodecafluorododecaborate ($[B_{12}F_{12}]^{2-}$). The open-ring isomer cage showed a large binding constant ($K_{293} = 3.2\times10^4\ M^{-1}$), while the value of the closed-ring isomer cage decreased ($K_{293} = 6.7\times10^2\ M^{-1}$). The significant difference in guest binding strength is explained by the structural flexibility of the open-ring isomer cage, which is able to accept the guest anion in the fashion of an induced fit, whereas the rigid closed-ring isomer cage is not able to structurally adapt the guest. Photoswitching by irradiation with UV and visible light allows the control of the mole fraction of the bound guest. The modulation of the affinity may find application in the field of drug delivery.

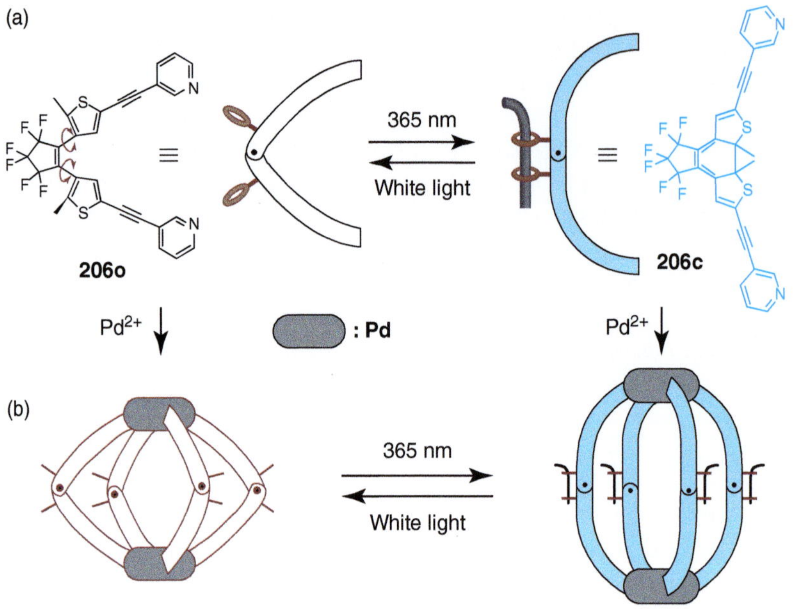

Figure 9.3 The metal organic cage composed of two Pd^{2+} ions and four bis-monodentate pyridyl ligand **206.** (a) Chemical structures of diarylethene ligands. (b) Schematic illustration of the shape change of the photoresponsive cage.

9.3 Super-Resolution Fluorescence Microscopy

Remarkable progress of fluorescence microscopy has revolutionized optical imaging, especially spatial resolution. The diffraction limit, as described in Section **5.2**, no longer restricts the spatial resolution in optical microscopy by using photoswitching of fluorescent probes between dark and bright states [15, 16]. The super-resolution techniques are grouped into two categories: coordinate-targeted and coordinate-stochastic methods. In the coordinate-targeted method, such as STED (stimulated emission depression) and RESOLFT (reversible saturable [switchable] optical linear fluorescence transition) microscopies, the position of the *on* and *off* states is determined by a doughnut-shaped pattern of light, which induces a turn-*on* or a turn-*off* transition in the sample except at the central position with vanishing intensity. In the coordinate-stochastic method, such as PALM (photoactivated localization microscopy) and STORM (stochastic optical reconstruction microscopy), fluorescence *off* and *on* events of molecules located at random positions take place individually at different intervals of time. PALM and STORM are based on single-molecule fluorescence detection. Each localized probe at the single-molecule level is visualized in the form of a fluorescent spot and the center of the spot can be approximated to the position of the molecule. The coordinate of each localized probe is stored and accumulated. The image is reconstructed by accumulating the localized fluorescent spot of each molecule.

The turn-on mode fluorescent diarylethene derivatives are widely applied to both RESOLFT microscopy and PALM/STORM. The application of turn-off mode photoswitchable molecules to PALM/STORM is limited, because the fluorescence quenching, in many cases, is imperfect and the PALM and STORM techniques require an absolutely dark background to detect single fluorescent molecules. For the application to super-resolution fluorescence microscopy, it is indispensable to develop fluorescent diarylethene derivatives, which fulfill the following photoswitching performance.

1. Appropriate cycloreversion quantum yield.
2. Resistance to photofatigue.
3. Water solubility for their application to biological systems.

9.3.1 Control of Cycloreversion Quantum Yield

Although the turn-on mode fluorescent diarylethene derivatives, described in Section 3.4.2, undergo efficient photocyclization reactions, the photocycloreversion quantum yields are relatively low. The low cycloreversion quantum yields are favorable for their application to coordinate-stochastic super-resolution microscopy, such as PALM/STORM, while coordinate-targeted microscopy, such as RESOLFT, claims easy photoswitching of probe molecules from *on* state to *off* state upon irradiation with a doughnut-shaped low-power laser beam. According to numerical simulations, the optimal switching-*off* quantum yield is concluded to be in the order of 10^{-2}–10^{-3} [17]. It is desired to reveal a molecular design principle to control the cycloreversion quantum yield without affecting the cyclization and fluorescence quantum yields, and fatigue resistance.

To obtain appropriate cycloreversion quantum yields, various substituents were introduced at 6- and 6′-poistions and also 2- and 2′-positions of sulfone derivatives of 1,2-bis(2-alkyl-6-aryl-1-benzothiophen-3-yl)perfluorocyclopentene, as shown in Scheme 9.2 [18–21]. Table 9.1 shows their photophysical and photochemical properties. The derivatives having electron-donating substituents at 6- and 6′-positions, such **142**, **207**, **208** and **209**, exhibited very low cycloreversion quantum yields. The quantum yields of the derivatives having thiophene substituents were less than 1×10^{-5}. Introduction of methoxy groups at *para*-positions of the side phenyl groups, such as **208**, also decreased the yield in comparison with unsubstituted derivative **8**. On the other hand, introduction of methyl substituents at *ortho*-positions of the phenyl groups, such as **210** and **170**, increased the yield. The derivative **170c** having *ortho*-dimethylphenyl groups showed the yield of 1.8×10^{-2}. The increase is ascribed to decrease in the coplanarity of the closed-ring isomers, which causes the decrease in the extension of π-conjugation length and results in a hypsochromic shift of the absorption spectrum. *Ortho*-substitution is useful to increase the cycloreversion quantum yield to 10^{-2}–10^{-3}. Modification of the aryl substituents did not affect the fluorescence quantum yield. All derivatives showed a yield higher than 0.74.

Another strategy to increase the cycloreversion quantum yield is to introduce isobutyl substituents at 2- and 2′-positions of the sulfone derivatives. Table 9.1

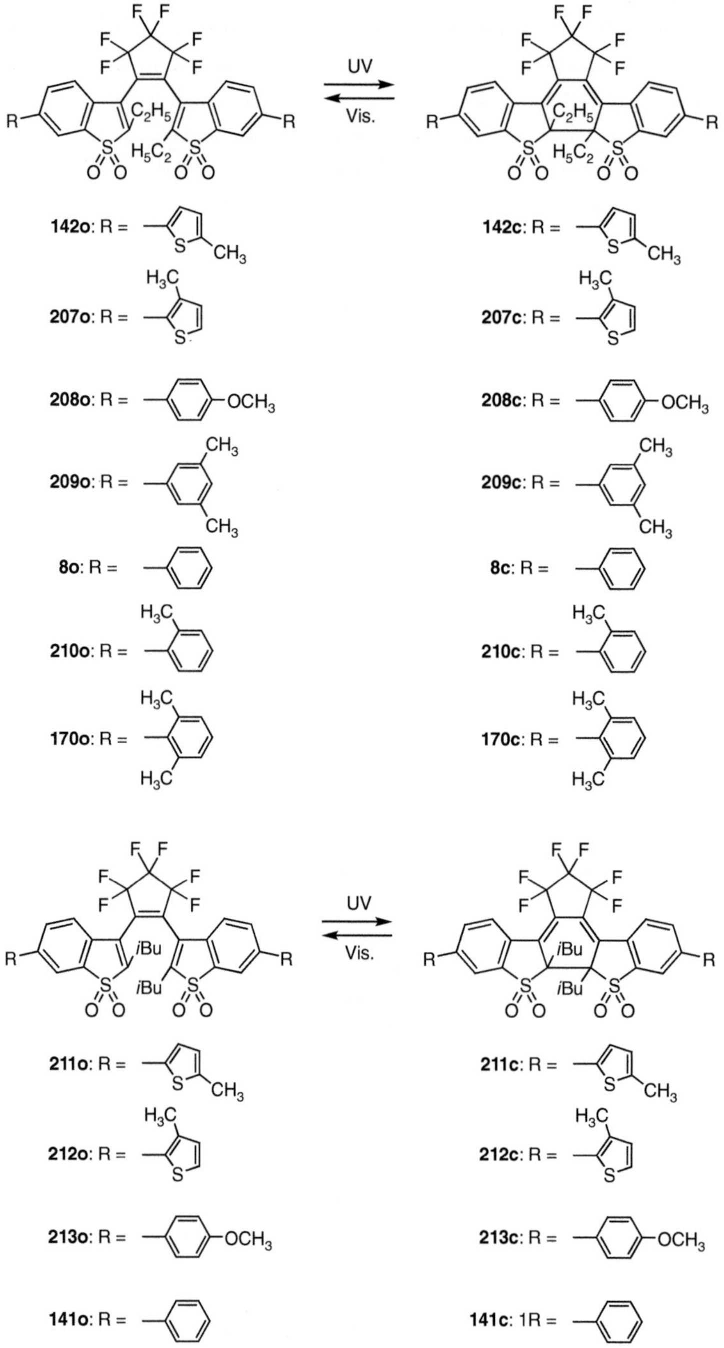

Scheme 9.2 Turn-on mode fluorescent diarylethene derivatives.

Table 9.1 Photophysical and photochemical properties of turn-on mode fluorescent diarylethene derivatives in 1,4-dioxane.

	Open-ring isomer		Closed-ring isomer		
	λ_{max} (nm) (ε/10^4 M^{-1} cm^{-1})	Φ_{oc}	λ_{max} (nm) (ε/10^4 M^{-1} cm^{-1})	Φ_{co}	Φ_f
142	330 (2.4), 374 (2.5)	0.23	506 (5.8)	$<1.0\times10^{-5}$	0.78
207	316 (1.6), 356 (1.6)	0.32	484 (4.7)	$<1.0\times10^{-5}$	0.77
208	313 (2.1), 355 (2.0)	0.28	486 (5.1)	$<1.0\times10^{-5}$	0.83
209	300 (1.9), 340 (1.6)	0.65	461 (4.6)	2.2×10^{-4}	0.74
8	298 (1.9), 336 (1.5)	0.62	456 (4.6)	5.9×10^{-4}	0.87
210	289 (1.2), 326 (1.0)	0.63	441 (3.7)	3.2×10^{-3}	0.84
170	285 (0.97), 320 (0.82)	0.59	430 (3.5)	1.8×10^{-2}	0.79
211	330 (2.3), 375 (2.3)	0.29	510 (5.5)	3.4×10^{-5}	0.75
212	318 (1.7), 358 (1.6)	0.39	488 (4.5)	2.4×10^{-4}	0.83
213	315 (2.2), 356 (2.0)	0.38	487 (5.1)	2.5×10^{-4}	0.87
141	300 (1.9), 336 (1.5)	0.53	460 (4.2)	1.0×10^{-2}	0.80

λ_{max}, absorption maximum; ε, absorption coefficient; Φ_{oc}, cyclization quantum yield; Φ_{co}, cycloreversion quantum yield; Φ_f, fluorescence quantum yield.

shows the cycloreversion quantum yields of the isobutyl-substituted diarylethene derivatives. The cycloreversion quantum yield of **141c** was as large as 1.0×10^{-2}, which is 17 times larger than the value of ethyl-substituted derivative **8c**. The cycloreversion quantum yields of **212c** and **213c** also increased by 20–30 times in comparison with **207c** and **208c**. These closed-ring isomers were thermally stable even at 80 °C for more than 20 hours and practically never returned to the open-ring isomers at room temperature. The introduction of isobutyl substituents at the central reactive carbon atoms is very effective to increase the cycloreversion quantum yield without affecting fluorescence and cyclization reaction quantum yields. The derivatives having isobutyl substituents are favorable for the application to RESOLFT microscopy.

9.3.2 Fatigue Resistance

Fatigue resistance upon alternate irradiation with UV and visible light is an indispensable property for the application to RESOLFT. Figure 9.4 shows fluorescence intensity changes of **8** and **141** in acetonitrile upon alternate irradiation with UV (365 nm) and visible (470 nm) light. Although the fluorescence intensity of **8** decreased to 80% of the initial value after 10^3 cycles, **141** kept the same intensity even after 10^3 cycles. It is safe to say that the cyclization/cycloreversion cycle of the sulfone derivatives can be repeated as many as 10^3 times maintaining adequate fluorescence photoswitching ability. These sulfone derivatives fulfill the requirements for the application to super-resolution microscopy.

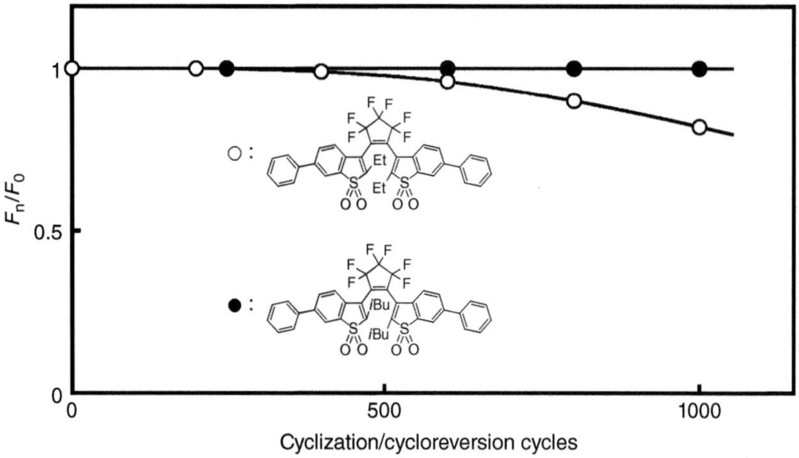

Figure 9.4 Fatigue-resistant property of sulfone derivatives of 1,2-bis(2-alkyl-6-phenyl-1-benzothiophen-3-yl)perfluorocyclopentene having ethyl or isobutyl substituents at 2- and 2′-positions **8** and **141**. Acetonitrile solutions of the derivatives were alternatively irradiated with UV (365 nm) and visible (450 nm) light. The solutions were irradiated with UV light until the conversion from the open-ring isomers to the closed-ring isomers reached 90% of the photostationary state. Then the colored solutions were almost completely bleached (>95%).

It was found that fatigue resistance of the sulfone derivatives of bis(benzothienyl)-ethenes depends on the substituents at 6- and 6′-positions of the benzothiophene-1,1-dioxide groups [21]. Introduction of electron-donating substituents improved the fatigue resistance, while the fatigue resistance was lost when electron-withdrawing substituents were introduced. The substituent effect is opposite to that observed in unoxidized dithienylethenes (see Section 3.3).

9.3.3 Photoswitching with Single-Wavelength Visible Light

Visible light responsiveness is desirable for application to biological systems. Although the absorption tails of the open-ring isomers of diarylethenes having thiophene substituents, such as **142o**, **207o**, **211o**, and **212o**, extend to visible wavelength regions and can be activated with 405 nm light, the absorption bands of the open-ring isomers of other derivatives are shorter than 400 nm. During the course of study of photoswitchable fluorescent diarylethenes, an unprecedented photoreaction was found [22]. Upon irradiation with visible light longer than the 0–0 transition, a sparse population of **8o** was reversibly activated to the *on* state and both cyclization and cycloreversion reactions simultaneously took place.

As usual, diarylethene **8** undergoes the cyclization reaction upon irradiation with UV (365 nm) light and the cycloreversion reaction takes place upon irradiation with visible (450 nm) light. Although both absorption and fluorescence of the closed-ring isomer considerably decreased upon irradiation with 450 nm light, noticeable intensity of fluorescence ascribed to **8c** remained even after prolonged irradiation with 450 nm light. Diarylethene **8o** has no apparent absorption at 450 nm. This means all closed-ring isomers are expected to completely revert back to the open-ring isomers. But, a small amount of **8c** survived. This result indicates that 450 nm light induces the cyclization reaction of **8o** in addition to the cycloreversion reaction of **8c**.

It was assumed that the reversible photoisomerizations are induced by irradiation with single-wavelength (λ) light.

$$\text{Open-ring isomer } \mathbf{8o}\ (\mathbf{A}) \underset{\lambda}{\overset{\lambda}{\rightleftharpoons}} \text{Closed-ring isomer } \mathbf{8c}\ (\mathbf{B}) \tag{9.1}$$

$$\frac{dC_B}{dt} = (\Phi_{AB}\varepsilon_A C_A - \Phi_{BA}\varepsilon_B C_B)F$$

$$F = \frac{1 - 10^{-(\varepsilon_A C_A + \varepsilon_B C_B)d}}{\varepsilon_A C_A + \varepsilon_B C_B} \mathrm{x} \frac{10^3 I_0}{d}, \tag{9.2}$$

where Φ_{AB}, Φ_{BA}, ε_A, ε_B, C_A, C_B, d, and I_0 are the cyclization quantum yield, the cycloreversion quantum yield, the absorption coefficient of **8o** at λ, the absorption coefficient of **8c** at λ, the concentration of **8o** (M), the concentration of **8c** (M), the cell length (cm), and the light intensity (einstein/cm^2 s), respectively. When $\varepsilon_A C_A + \varepsilon_B C_B << 1$, the formation of the closed-ring isomers is expressed as follows, where C_O is the total concentration of **8o** and **8c**:

$$C_B = \frac{\Phi_{AB}\varepsilon_A}{\Phi_{AB}\varepsilon_A + \Phi_{BA}\varepsilon_B} C_O (1 - e^{-(\Phi_{AB}\varepsilon_A + \Phi_{BA}\varepsilon_B) \times 10^3 I_0 t}) \tag{9.3}$$

In the photostationary state,

$$\frac{C_B}{C_O} = \frac{\Phi_{AB}\varepsilon_A}{\Phi_{AB}\varepsilon_A + \Phi_{BA}\varepsilon_B} \tag{9.4}$$

Based on the analysis of the formation curves of closed-ring isomers upon irradiation with 450 nm light according to Eq. (9.3), the conversion ratios at the photostationary state in n-hexane, CCl_4, and 1,4-dioxane under irradiation were determined to be 0.060%, 0.34%, and 0.91%, respectively. Fluorescence intensity change was used to detect the very small amount of **8c** less than 10^{-7} M. The absorption coefficients of **8c** were estimated from Eq. (9.4) at 450 nm in n-hexane, CCl_4, and 1,4-dioxane to be 0.084 M^{-1} cm^{-1}, 0.19 M^{-1} cm^{-1}, and 0.39 M^{-1} cm^{-1}, respectively. Table 9.2 summarizes the physical and chemical properties of diarylethene **8**.

According to Wondrazek and co-workers [23] and Kinoshita and co-workers [24], the absorption spectra of fluorescent dyes show an exponential frequency dependence in the off-resonance region of the absorption tails. The tails obey the Urbach rule as shown by the following formula [25]:

$$\varepsilon(E) \propto \varepsilon(E_{00}) \exp\left[-\frac{\sigma(E_{00} - E)}{k_B T}\right] \tag{9.5}$$

where ε, k_B, T, and σ are the absorption coefficient, the Boltzmann constant, the absolute temperature of the sample, and the steepness parameter, respectively.

Table 9.2 Photochemical and photophysical properties of **8**.

	Solvent		
	n-Hexane	CCl_4	1,4-Dioxane
Φ_f	0.92	0.89	0.87
Φ_{AB}	0.42	0.50	0.62
Φ_{BA}	1.3×10^{-3}	6.6×10^{-4}	5.9×10^{-4}
$\varepsilon_{A\,(450\,nm)}/M^{-1}\,cm^{-1}$	0.084	0.19	0.39
$\varepsilon_{B\,(450\,nm)}/M^{-1}\,cm^{-1}$	4.5×10^4	4.3×10^4	4.5×10^4
Conversion (%)	0.06	0.34	0.91

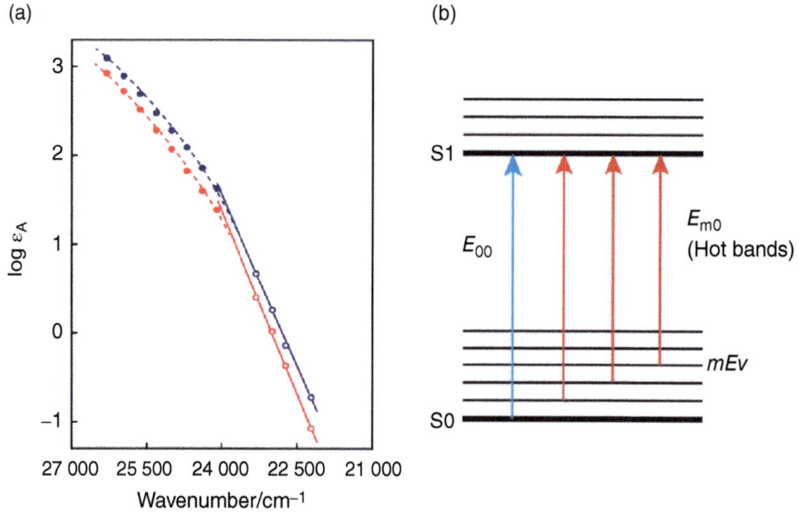

Figure 9.5 (a) Logarithmic plots of molar absorption coefficients of **8o** in *n*-hexane (red) and CCl_4 (blue). (b) Schematic illustration of 0–0 transition (E_{00}, blue arrow) and hot bands (E_{m0}, red arrows).

The absorption coefficients estimated from Eq. (9.4) are plotted in Figure 9.5 along with the absorption tails of the open-ring isomer **8o**. The coefficients obey the exponential frequency dependence at wavelengths longer than ~415 nm (24 000 cm^{-1}). The open-ring isomer in the vibrationally excited states undergoes optical transition into electronically excited states by absorbing the very weak hot bands, as shown in Figure 9.5b and the cyclization reaction from **8o** to **8c** takes place. The reversible fluorescence photoswitching with single-wavelength visible light is advantageously applicable to PALM/STORM as shown in the next section.

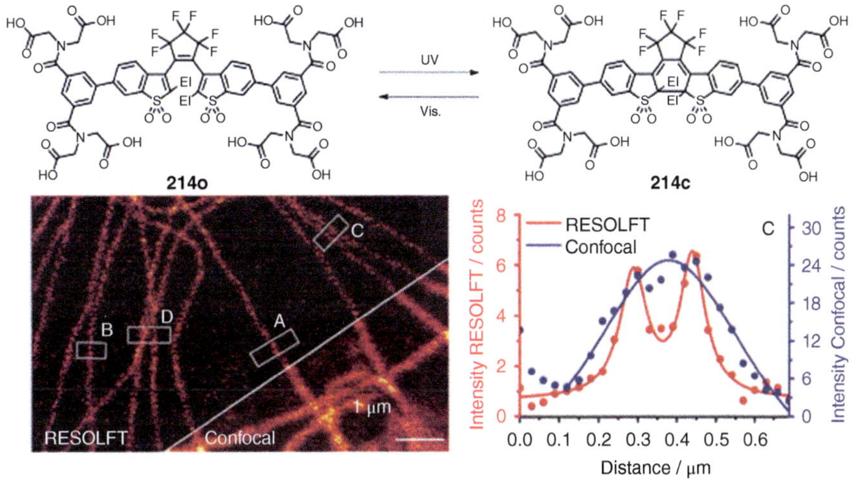

Figure 9.6 RESOFT and confocal images of Vero cells. The line profiles of region C in the images are shown in the right-hand side (blue line: confocal image, red line: RESOLFT image). Scale bar: 1 μm.

9.3.4 Super-Resolution Bioimaging

As described in Section 9.3.3, properly designed sulfone derivatives of bis(ben zothienyl)ethene fulfill the requirements for the application to super-resolution microscopy. For the application to biological systems, another indispensable property is water solubility. The sulfone derivatives shown in Scheme 9.2 are insoluble in water. To provide both water solubility and bioconjugation ability to the fluorescent diarylethenes without affecting the photoswitching performance, carboxylic acid groups were introduced at the side groups. The diarylethenes having 8 or 12 carboxylic acid groups, such as **214–216**, are readily soluble in aqueous buffer solutions at $pH > 5$ and retain adequate photoswitching and fluorescent properties [26–28].

Figure 9.6 demonstrates the applicability of fluorescent diarylethene **214** for RESOLFT microscope. Fixed immunolabeled Vero cells were imaged using a modified 1C RESOLFT QUAD Scanning microscope (Abberior Instruments, Göttingen). The RESOLFT imaging was performed by applying the following pulse sequences: switching **214o** to **214c** with 355 nm light (130 W cm^{-2}) for 50 μs, 200 μs delay, application of doughnut-shaped 488 nm beam (36 kW cm^{-2}) for 1.2 ms to bleach **214c** to open-ring isomer **214o** in the periphery of focal spot. Finally, the remaining fluorescent **214c** was probed for 80 μs with a Gaussian-shaped 488 nm laser beam (9.7 kW cm^{-2}). Using this imaging scheme, it was possible to acquire RESOLFT images with a resolution of 74 nm FWHM (full width at half minimum), as shown in the line profiles of Figure 9.6. The success of the use of diarylethene **214** is attributed to the relatively large cycloreversion quantum yield, $2.0 \times 10^{-3} \pm 3 \times 10^{-4}$.

On the contrary, PALM and STORM require a low cycloreversion quantum yield, because the resolution of these techniques depends on the localization process.

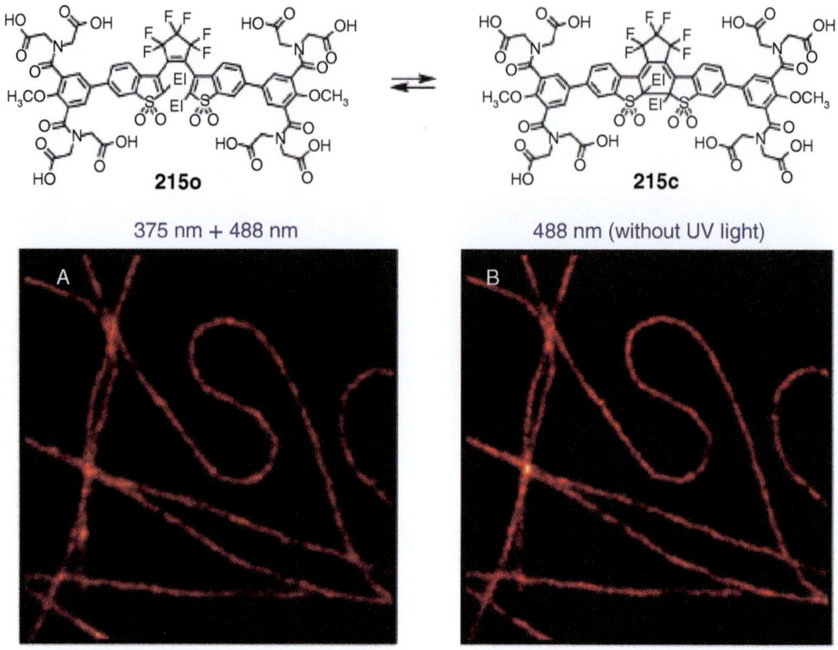

Figure 9.7 STORM images of Vero cells acquired (A) with and (B) without UV activation light. *Source*: Adapted with permission from ref. [29]. Copyright 2017 American Chemical Society.

The preferable cycloreversion quantum yield is in the order of 10^{-4} or less. To decrease the yield, methoxy substituents were introduced at *para*-positions of the side phenyl groups in diarylethene **215**. The cycloreversion and fluorescence quantum yields of **215c** were measured to be 1.8×10^{-4} and 0.45 in water, respectively. Figure 9.7 shows the STORM images of Vero cells labeled with **215**. The STORM image (A) was acquired using UV (375 nm) activation, while the image (B) was obtained using only a 488 nm laser beam without UV activation. This imaging technique, which uses a single-wavelength 488 nm excitation beam without harmful UV light, is favorable for application to biological systems. As described in the previous section, the activation in the absence of UV light is ascribed to the optical transition induced by the absorption of very weak hot bands or Urbach tails. Images acquired without and with UV photoactivation were identical within experimental errors, as shown in images (A) and (B).

Although **214** and **215** undergo reversible photoswitching between the open- and closed-ring isomers upon irradiation with UV and visible light, the repeatable cycle number is limited. The cycle number of **214** maintaining adequate photoswitching ability is less than 50 times. For the application to RESOLFT, it is desirable to increase the fatigue resistance. Figure 9.8 shows one of the fatigue-resistant and water-soluble fluorescent diarylethenes having 12 carboxylic acid groups. When short alkyl chains were introduced between the side phenyl substituents and carboxylic acid groups, as

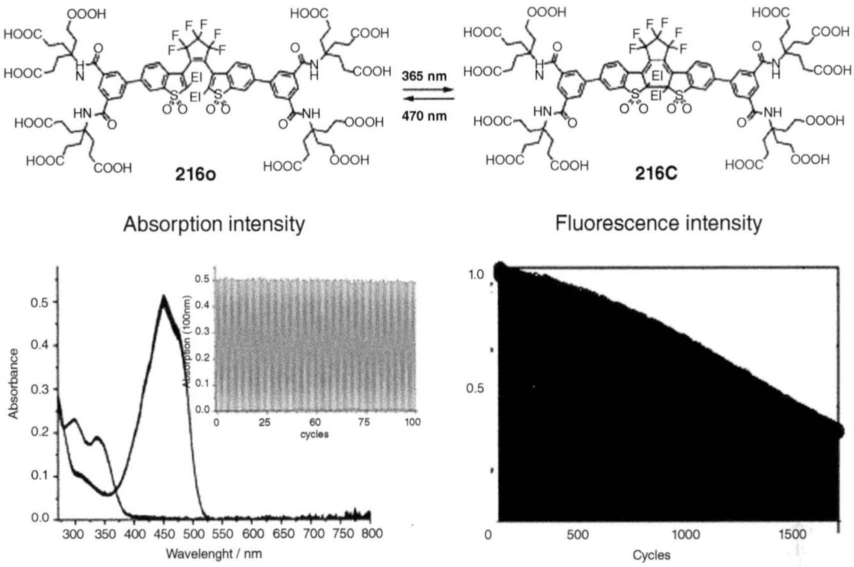

Figure 9.8 Fatigue resistance of diarylethene **216** in an aqueous phosphate buffer solution upon alternate irradiation with UV (365 nm) and visible (470 nm) light. Absorption intensity changes in the initial 100 cycles (left side) and fluorescence intensity changes during 1700 cycles (right side).

shown in **216**, the fatigue resistance was dramatically improved. An aqueous phosphate buffer solution containing diarylethene **216** was irradiated alternatively with UV (365 nm) and visible (470 nm) light. UV light irradiation was continued until the photostationary state was achieved. As can be seen, no noticeable decrease was observed in the absorption intensity of **216o** during the first 100 cycles. Diarylethene **216** maintained adequate fluorescence photoswitching ability even after 10^3 cycles in the aqueous phosphate buffer solution.

Super-resolution microscopy using photoswitchable fluorescent diarylethenes can be applied not only to biological systems, but also to materials science [30–32]. It is capable of resolving nanostructures in *block*-copolymers and others by using super-resolution PALM/STORM.

Progress to improve the resolution in the super-resolution microscopy is continuing. Hell and co-workers [33] reported a new nanometer-resolution imaging method named MINFLUX. The method can image individual molecules separated by just 1 nanometer. The present photoswitchable fluorescent diarylethene derivatives have a potential to be used in MINFLUX microscopy.

9.4 Chemical Reactivity Control

Pericyclic reactions, such as cycloaddition and electrocyclization, induce rearrangements of positions of single and double bonds, as shown in Scheme 9.3. In the Diels–Alder reaction, the central single bond of the furan ring converts to a double

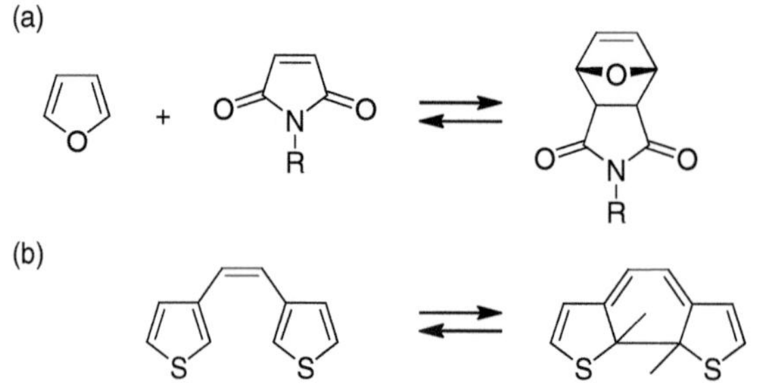

Scheme 9.3 Rearrangement of positions of single and double bonds in (a) Diels–Alder cycloaddition and (b) diarylethene electrocyclic reactions.

bond during the thermal cycloaddition reaction, while in the diarylethene electrocyclization reaction the central double bond changes to a single bond upon irradiation with UV light. When the furan unit is combined to a diarylethene structure, the Diels–Alder reactions are able to be controlled by photoirradiation [34].

Branda and co-workers [35] prepared photoresponsive polymers having Diels–Alder units. The molecular design principle is illustrated in Scheme 9.4. Monomer **M1** contains two dithienylfuran units in both sides of a short alkyl chain. Upon heating at 85 °C in the presence of **M2**, which contains two maleimides in both sides of a short alkyl chain, the two diene units of **M1** undergo the Diels–Alder cycloaddition reaction with the maleimides of **M2** and a linear polymer **P1o** is formed. The central single bond of the furan unit converts to a double bond, as indicated in Scheme 9.3. In **M1,** the two dithienylfuran units with the central single bond are photoinactive, while in polymer **p57o** the newly created double bond converts the photoinactive dithienylfuran unit into a photoreactive hexatriene structure. Upon irradiation with UV light, **p57o** undergoes cyclization reactions to produce **p57c**. The cyclization reactions convert the hexatriene units to the cyclohexadiene ones and remove the hexatriene structures in **p57o**. The absence of the hexatriene structure prevents a reverse Diels–Alder reaction and the system is locked. On exposing polymer **p57c** to visible light, polymer **p57o** is regenerated and the system is unlocked. Then, polymer **p57o** can undergo the reverse Diels–Alder reactions to produce **M1** and **M2**.

The above polymer acts as an adhesive. The tensile shear strength of two pieces of plates glued together by the polymer was measured at two different temperatures. The shear force for the strips glued together with **p57o** at 90 °C was 66% of that at room temperature. This indicates that at 90 °C, Diels–Alder equilibrium is established and the adhesive property is poor. When the strip was exposed to UV light to convert polymer **p57o** to polymer **p57c**, the adhesive property of the polymer was enforced. The average shear force was 2.15 times higher than that for **p57o** at room temperature. This difference increased almost 3.26 times at 90 °C. The adhesive property was reversibly controlled by irradiation with UV and visible light.

The furan/maleimide pair undergoes the forward cycloaddition reaction between 25 and 50 °C, while above 70 °C the reverse Diels–Alder reaction takes place. Hecht and co-workers [36] prepared a diarylethene having a furan group at one side of

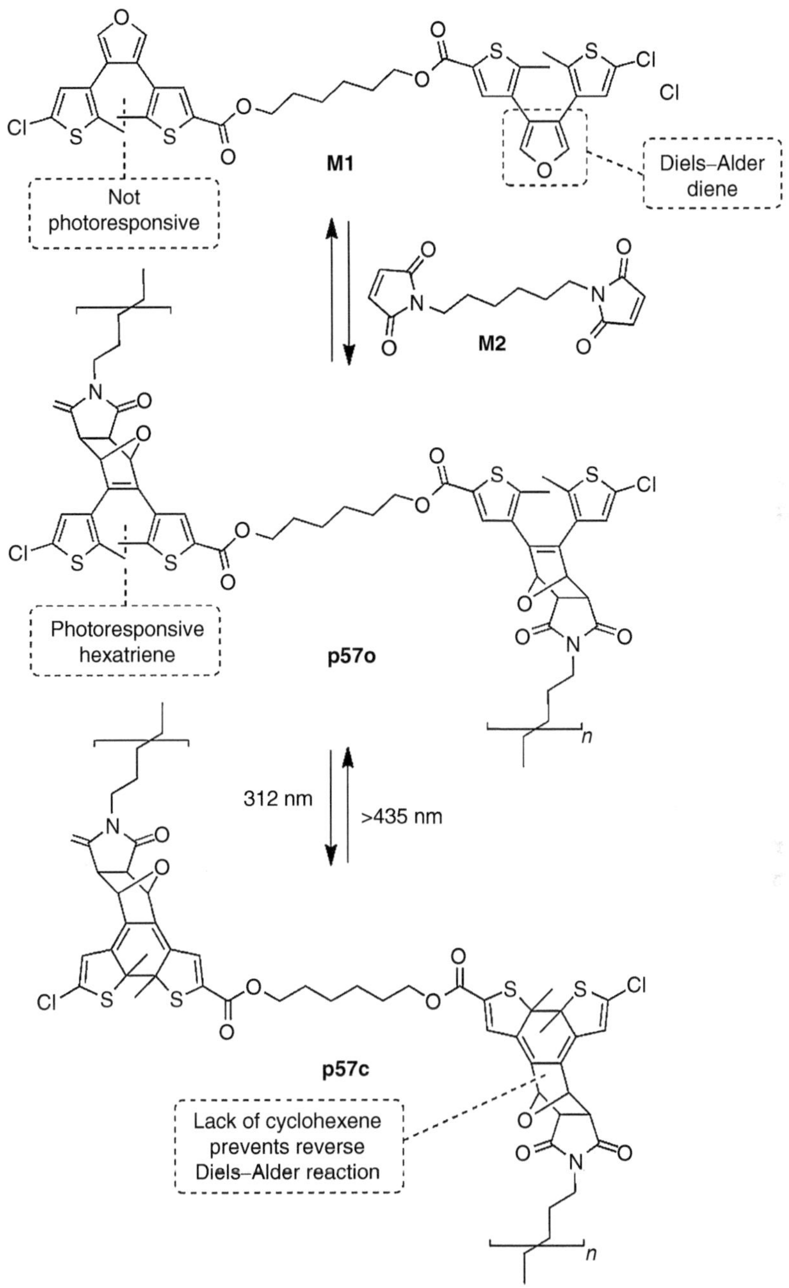

Scheme 9.4 The molecular design principle of photoresponsive polymers having Diels–Alder units. Absorption of visible light by polymer **p57c** triggers photochemical bond-breaking reactions that produce polymer **p57o**, which can undergo the reverse Diels–Alder reactions and produce **M1** and **M2**.

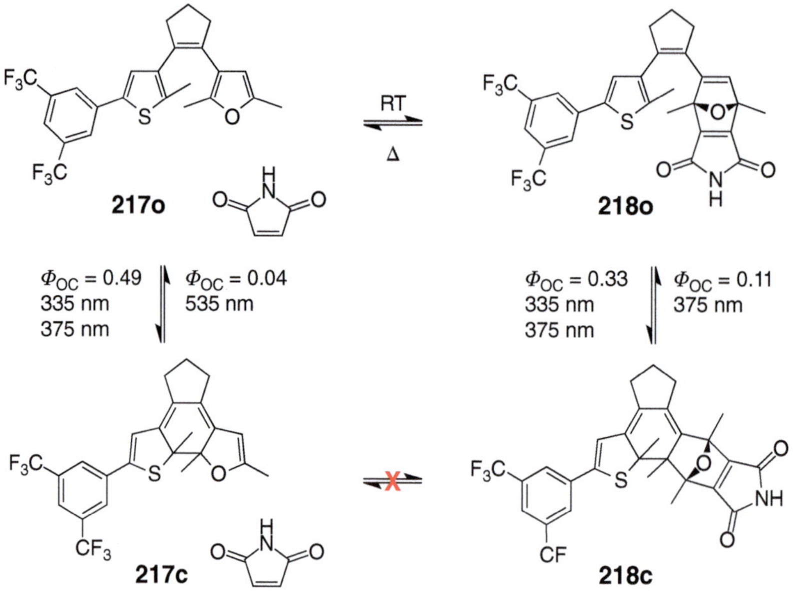

Scheme 9.5 Photocontrol of formation and scission of bonds using diarylethene photoswitches as locks. **217o**, **217c**, **218o**, and **218c** show the chemical structures in the photocontrolled Diels–Alder reactions.

the aryl groups and examined the Diels–Alder reaction with a maleimide derivative under irradiation with light of appropriate wavelengths, as shown in Scheme 9.5. UV-induced photocyclization reaction of **217o** to **217c** removed the diene functionality of furan ring and prevented the molecule from participating in the Diels–Alder reaction. Diarylethene **217o** underwent the Diels–Alder reaction with maleimide and formed **218o**. The chemical bond produced between the diarylethene **217o** and the maleimide derivative by the Diels–Alder reaction was locked upon irradiation with UV light. The closed-ring adduct **218c** was unreactive and stable even at higher temperatures. On exposing diarylethene **218c** to visible light, **218c** returned to **218o** and the chemical bond created by the Diels–Alder reaction was broken at higher temperatures. The closed-ring isomers **217c** and **218c** exhibited well-separated visible absorption spectra. Therefore, it is possible to selectively irradiate one of them. Upon irradiation with 335 and 535 nm light, the cyclization reactions of both **217o** and **218o** took place, while only **217c** underwent the ring-opening reaction. The selective reactions effectively led to an enrichment of **218c**. Under such irradiation conditions, the formation of Diels–Alder adduct **218c** was amplified in comparison with the dark Diels–Alder reaction.

The enrichment of Diels–Alder adduct formation can be applied to prodrugs or protags. The adduct of a water-soluble furyl-substituted diarylethene and *N*-substituted maleimide (a prodrug or a protag) was prepared and transformed to the ring-closed locked form upon UV irradiation [37]. The locked prodrug or protag adduct was stable and efficiently impeded the reverse Diels–Alder reaction back to the starting materials. On the other hand, upon irradiation with visible

(400 nm < λ < 500 nm) light, the adduct readily regenerated the maleimide under physiological conditions. Maleimide-based drugs are released only in the area where visible light is irradiated. The present system potentially provides photoswitchable prodrugs for maleimide-based reactive inhibitors.

Hecht and co-workers [38] further applied the above Diels–Alder reaction of furyl-substituted diarylethenes with maleimides for self-healing polymers. The ability to repair inflicted damage was achieved by using the Diels–Alder dynamic covalent chemistry.

9.5 Biological Activity

One of the exciting advances in pharmacology is the emergence of a branch that uses compounds whose biological activity can be controlled by external stimuli. Among various stimuli, light is the most convenient one because of its high spatiotemporal resolution [39]. Several general approaches to the photocontrol of biological activity have been developed and they can be classified as "irreversible" and "reversible." An example of the irreversible approach is the use of caged compounds. The biological activity of caged compounds can be controlled by irreversible photochemical cleavage of photolabile groups. The drawback of the caged compounds is the possible formation of unwanted side-products. The reversible photocontrol of biological activity is attained by incorporating molecular photoswitches into the compounds. The compounds can avoid the formation of side-products. The ultimate goal is to develop compounds that can be applied to living systems in their inactive photoform, and then to convert these compounds into their active form using light with high spatiotemporal resolution.

At first, it is worth noting that the toxicity test study (Ames-test) of diarylethenes. The study confirmed that both diarylethenes **3** and **5** do not show any positive response in the mutagenic tests, which means that they are non-mutagenic [40]. The basic diarylperfluorocyclopentene frameworks are not mutagenic, possibly due to their hydrophobic nature. However, some derivatives shown in Scheme 9.6 were found to exhibit biological activity.

Branda and co-workers [41] reported photocontrol of paralysis of *Caenorhabditis elegans* (*C. elegans*) by diarylethene **43** having *N*-methylpyridinium cation groups (Scheme 9.6a). *C. elegans* has been extensively used as a model to examine neuroactivity and metabolism. The in vivo effect of **43** on *C. elegans* was demonstrated by feeding them diarylethene **43** and toggling the photoswitch back-and-forth between two isomers using UV and visible light. The presence of **43** showed a remarkable effect on the mobility of the nematodes depending on which isomer was present. The nematodes were paralyzed and became immobile by cyclizing the diarylethene with UV light, but the mobility was restored by the cycloreversion reaction with visible light. The paralysis is ascribed to the result of interruption of the metabolic electron pathways due to the unique electron-accepting ability of the closed-ring isomer **43c**.

There is an ongoing interest in the regulation of enzyme activity in the field of chemical biology. Among various regulations, photocontrolling the activity of

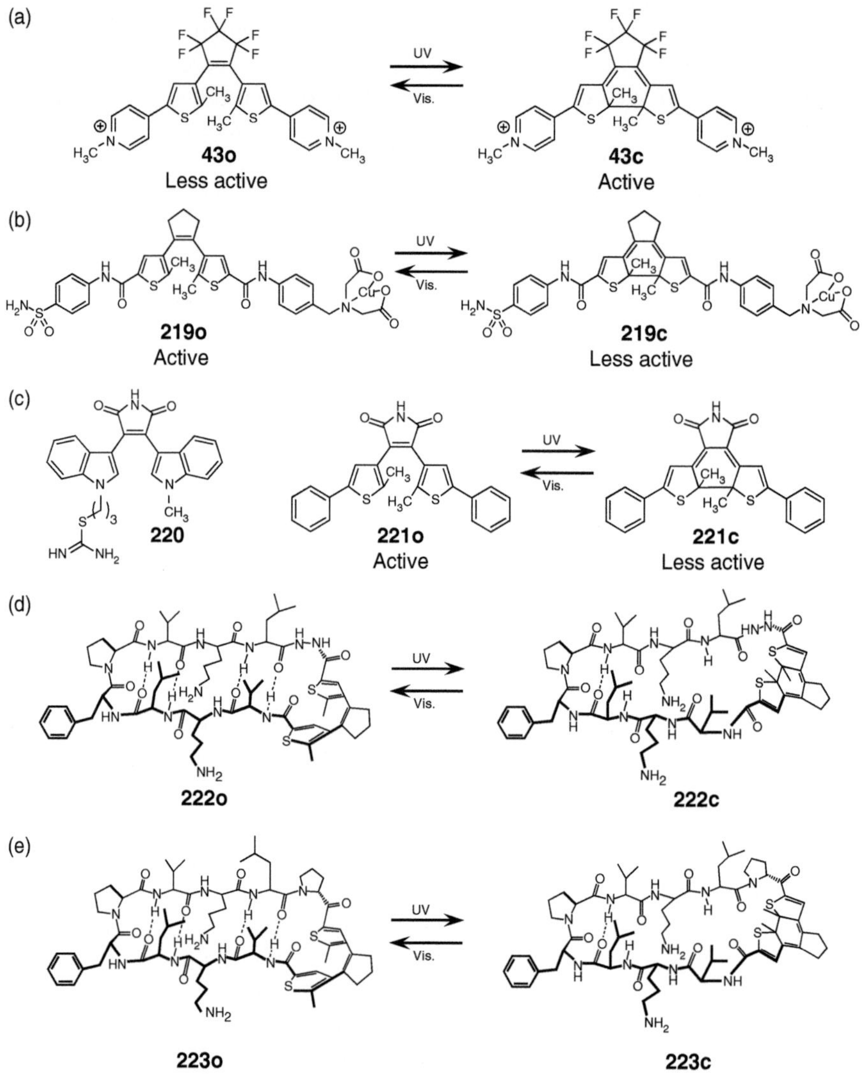

Scheme 9.6 Examples of biologically active diarylethene derivatives.

carbonic anhydrase is of special interest as it is an enzyme central to both cellular transport and metabolic processes. Diarylethene structure was used to reversibly photocontrol the activity of carbonic anhydrase by decorating the diarylethene architecture with sulfonamide and copper (II) iminodiacetate Cu(ida) moieties, as shown in Scheme 9.6b [42]. While the sulfonamide groups of **219** act as the inhibitor, the role of Cu(ida) component is to reversibly coordinate to imidazole side chains of the histidine residues close to the Zn (II) active site. Irradiating an aqueous solution of **219o** (5% DMSO, Tris sulfate buffer, 20 mM, pH 8.3 at 25 °C) with 320 nm light generated the closed-ring isomer **219c** almost quantitatively. The inhibition effect of the closed-ring isomer **219c** (IC_{50} [half maximal inhibitory concentration] = 0.4 μM) was comparable to that of sulfanilamide. This is because the rigid backbone in **219c** prevents the simultaneous binding of Cu(ida) and sulfonamide components to the enzyme active site. On the other hand, the open-ring isomer **219o** inhibited the enzyme much more significantly, and its inhibition activity was 50 times higher (IC_{50} = 8 nM) than that of the closed-ring isomer **219c**. This increase is attributed to the structural flexibility of **219o**, which allows both recognition components to bind to the active site and leads to higher overall binding ability. The reversible cyclization and cycloreversion cycles were also possible in the presence of enzyme. The enzyme activity was reversibly and significantly changed by toggling diarylethene **219** between the open- and closed-ring isomers. The thermal stability and visible light activation make the system suitable for the reversible regulation of enzyme activity by light.

Another example of a photoresponsive enzyme inhibitor is shown in Scheme 9.6c [43]. Histone deacetylases termed sirtuins are enzymes, which deacetylate hoistones and certain non-histone proteins. Bisindolylmaleimide derivative **220** is known to act as an efficient sirtuin inhibitor (IC_{50} = 0.8 μM) to human Sirt2 (hSirt2). Although 2,3-bis(2-methyl-1-benzothiophen-3-yl)maleimide is a structurally close photoswitch to bioactive **220**, both open- and closed-ring isomers showed relatively low inhibition. When both benzothiophene rings were replaced with phenyl-substituted thiophene rings, such as **221**, the open-ring diarylethene **221o** exhibited relatively high enzyme inhibition corresponding to an IC_{50} value of 4.2 μM. The closed-ring isomer **221c** showed a 20-fold lower enzyme inhibition with 92.3 μM. The enzyme activity of hSirt2 was controlled by light under assay conditions. The less active **221c** was incubated with the enzyme under assay conditions and irradiated with green LED light (2.5 W, 530 nm). The irradiated samples switched from less potent closed-ring isomer **221c** to significantly more potent open-ring isomer **221o** and the inhibition increased with increase in the content of the open-ring isomer. The remote control of the enzyme activity was achieved upon irradiation with the green light.

The conformation of peptides can be photocontrolled by incorporating molecular photoswitches into the backbone or side chains. A dithienylcyclopentene photoswitch was introduced into a cyclic peptide, antibiotic gramicidin S (GS), that is stabilized by intramolecular hydrogen-bonds in their β-sheet conformation, as shown in **222** [39, 44, 45]. The sequence of GS is structured as a C_2-symmetric antiparallel β-pleated conformation, connected by two type-II' β-turns. One of the β-turns was replaced with a dithienylcyclopentene moiety. The open-ring form resembles GS more closely than the closed-ring form in terms of several parameters that reflect the overall distribution of hydrophobic and hydrophilic patches on the molecule. The hydrogen-bonded network that stabilizes the β-sheet was more disturbed in the closed-ring form than in the open-ring form. The conformational change of the GS analogue upon photoswitching is crucial for their membranolytic action. The minimal inhibitory concentration (MIC) of the diarylethene-modified GS **222**, both against Gram-positive and Gram-negative bacteria, was efficiently modulated by light. The bactericidal activity for the open-ring form **222o** generated by irradiation with 530 nm light was 16 times higher than that of corresponding UV-generated closed-ring form **222c**. The open-ring form was almost as active as the parent antibiotic itself.

An approach to tumor-selective and minimally invasive cancer treatment relies on photodynamic therapy (PDT). Currently used photodynamic therapy agents are not active cytotoxic compounds, but they act as photosensitizers to generate singlet oxygens. It is awaited to develop photoactivatable cytotoxic agents that can be used for the therapy. One of the GS analogues **223** demonstrated excellent photocontrolled performance in vivo in mammals [45]. To provide near-infrared (NIR) sensitivity, the dithienylcyclopentene group was connected to the peptide backbone via a keto group. Then, the absorption shifted by around 46 nm to NIR region. Irradiation at 664 nm caused around 50% transformation of **223** from the closed-ring form to the open-ring one up to 12 mm deep into the model tissue. A BALB/c mouse aortic endothelial cell line (MAEC) was used to estimate the possible damage that may be caused by **223**. IC_{50} values of **223o** were very close to those of wild-type GS ($IC_{50} \sim 6\,\mu M$), while the closed-ring form **223c** showed about 5.5- to 8.0-fold lower cytotoxicity. An in vivo experiment was set up using Lewis lung carcinoma (LLC) allograft model in C57B1/6 mice. Seven groups, each consisting of eight animals with palpable tumors, received a daily intraperitoneal injection of either vehicle, GS, or the closed-ring form **223c**. The experimental results showed that animal survival in the groups that received visible light irradiation was improved by 60% compared to the control. Significant tumor shrinkage was observed on the LLC model in C57B1/6 mice, when **223** was administered in the closed-ring form and locally photoactivated by irradiation of the tumor with visible light. The open-ring form **223o** generated by visible light showed much stronger toxicity against tumor cells than the UV-generated close-ring form **223c**. These compounds are promising candidates as future PDT agents for cancer.

9.6 Color Dosimeters

Radiation dosimetry for industrial and medical purposes has evolved over the last few decades with the introduction of new detectors. In addition to physical dosimeters, such as calorimeters or ion chambers, various chemical dosimeters, which use radiation-induced chemical reactions to detect absorbed dose, have been developed and become widely used in medical fields [46]. One such chemical dosimeter is radiochromic film dosimeters, which estimate the absorbed dose from a color change or an intensity change of the color. The dosimeters have a high spatial resolution and do not require any special development procedure. They can be handled and prepared under normal room light. The image of the absorbed dose is produced as a dye forming or a polymerization process. A typical example is a vinyl chloride–vinyl acetate copolymer film containing 4-(dimethylamino)azobenzene. Upon γ-irradiation the yellow film turns red. The red color is due to the reaction of 4-(dimethylamino)azobenzene with acid produced upon exposure to high-energy radiation. A new type of film dosimeters is based on radiation-induced polymerization of diacetylenes, which produces blue-colored polyconjugated polymer chains. The film dosimeters have found extensive use in medical applications where low doses are required to be measured. The drawbacks of the dosimeters are storage instability and post-coloration after stopping the irradiation. Diarylethene derivatives would be used as stable chemical dosimeters if the derivatives could undergo the cyclization reactions to produce the colored isomers upon high-energy irradiation.

As the first step, radiation-induced coloration of single crystals of diarylethene derivatives was examined [47]. A single crystal of diarylethene **3** was prepared by recrystallization from a *n*-hexane solution and irradiated with ^{60}Co γ-rays. Upon γ-irradiation, the colorless crystal turned red and the absorption maximum was observed at 535 nm. The color change indicates that γ-irradiation induces the cyclization reaction to produce red-colored closed-ring isomer **3c** (Scheme 9.7). In order to obtain information concerning the dose dependence, the absorption intensity change of the crystal upon γ-irradiation was measured. The optical density linearly increased with increase in the absorbed dose below 12.5 kGy. The absorption intensity correlates with the crystal thickness. Absorbed dose as small as 160 Gy was detected (Abs. = 0.01) using a single crystal with a thickness of 5 mm. The colored crystal could be reused after photobleaching.

Although the crystalline materials are useful as sensitive dosimeters, the crystals cannot be used for large-area detection. For the large-area detection, bulk amorphous or polymer film dosimeters were developed. Diarylethene **224** forms an amorphous film by spin coating the toluene solution. Upon γ-irradiation (absorbed dose: 500 Gy), the colorless amorphous film (film thickness: 130 μm) turned blue with the absorbance of 0.033 at 570 nm [48]. The color change was detected even when the absorbed dose was decreased to as small as 150 Gy.

In the above crystals and amorphous films, high-energy radiation directly induces the excitation of diarylethene molecules. On the other hand, in polymer films doped with diarylethene derivatives only limited numbers of diarylethene molecules are directly excited with the high-energy radiation, because the radiation homogeneously irradiates the films. For example, when the content of the derivative is 2.0 mol%, only 2.0% of absorbed dose is used to excite diarylethene molecules and the rest is used to excite the polymer matrices. Matrices play an important role in the polymer film dosimeters. To understand the role of matrices, radiation-induced coloration of diarylethene **3** was examined in various solvents at 20 °C, and −196 °C [47]. Table 9.3 summarizes the result. No color change was discerned in *n*-hexane, 2-methyltetrahydrofuran (MTHF), and *n*-butyl chloride

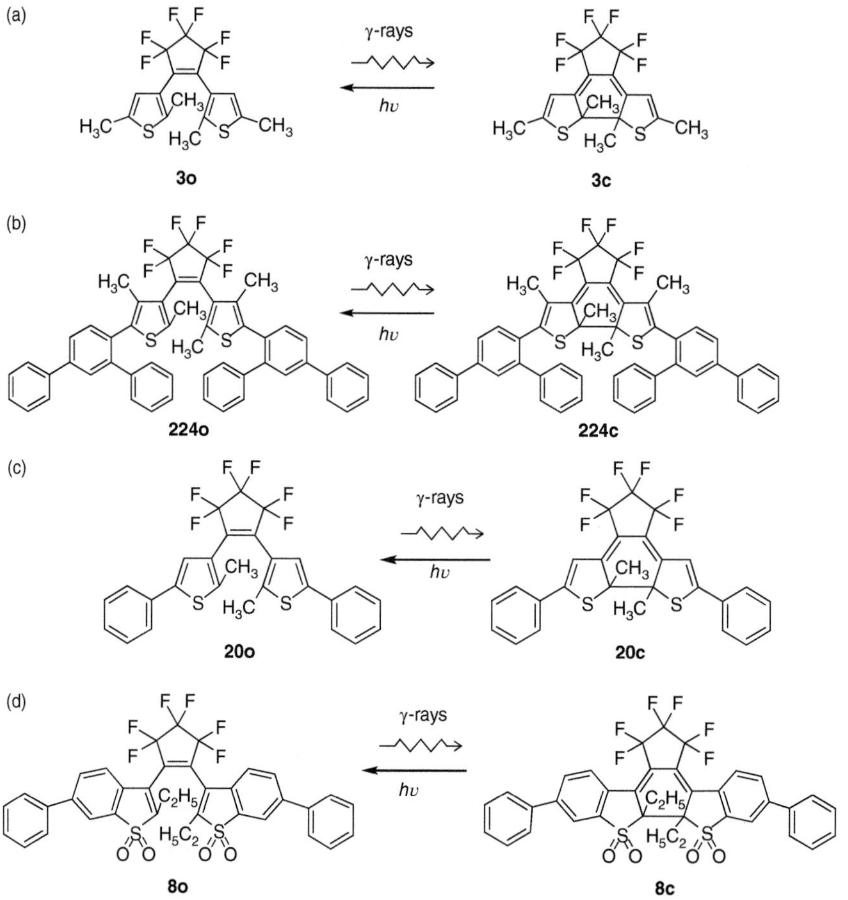

Scheme 9.7 (a) Diarylethene crystal **3,** which turned red upon γ-rays irradiation. (b) An amorphous diarylethene **224,** which turned blue upon γ-rays irradiation. (c) Diarylethene **20,** which underwent the radiation-induced photoswitching reaction in polystyrene films. (d) Diarylethene **8,** which turned fluorescent upon γ-rays irradiation.

(*n*-BuCl) solutions. In radiation chemistry, it is well known that radical anions and cations of solute molecules are produced in MTHF and *n*-BuCl solutions, respectively. In *n*-hexane, both radical anion and cations are produced. The absence of radiation-induced coloration in *n*-hexane, MTHF, and *n*-BuCl indicates that ionic species of diarylethene **3o** do not contribute to the production of colored isomer **3c**. On the other hand, benzene solutions and solids containing **3o** turned red upon γ-irradiation at 20 °C, and −196 °C. The energy transfer from γ-irradiated benzene matrices to **3o** is considered to cause the coloration. At first, benzene is excited to electronically excited states by γ-irradiation, and the excited energy migrates among benzene molecules and is finally transferred to solute **3o**. The excited **3o** (in singlet or triplet state) converts to the closed-ring isomer **3c**, which exhibits a red color. It is worth noting that the red color appeared in polycrystalline *n*-hexane upon γ-irradiation at −196°C. The color change indicates that excited energy produced

Table 9.3 Radiation-induced coloration of diarylethene **3** in solutions and solids (total dose: 10 kGy, concentration: 5×10^{-5} mol/l).

Solvent	State	Coloration
Hexane (20 °C)	Solution	No
Hexane (−196 °C)	Solid	Yes
MTHF (20 °C)	Solution	No
MTHF (−196 °C)	Solid	No
n-BuCl (20 °C)	Solution	No
n-BuCl (−196 °C)	Solid	No
Benzene (20 °C)	Solution	Yes
Benzene (−196 °C)	Solid	Yes

by γ-irradiation can be transferred to **3o** in the polycrystalline phase, though such energy migration and transfer are negligible in solution. The result provides new insight into the energy transfer mechanism in alkane matrices.

The above results suggest that proper selection of polymer matrices is indispensable to prepare efficient film dosimeters. In fact, efficient coloration was observed at 605 nm when a polystyrene film containing 2.0 wt% diarylethene **20o** was irradiated with ^{60}Co γ-rays, while coloration was scarcely observed when the polymer matrix was changed to poly(methyl methacrylate) (PMMA) [49]. The absorbance of **20c** generated upon γ-irradiation in a PMMA film (**20**: 2.0 wt%, Dose: 1 kGy) was only 22% of the absorbance observed in the polystyrene film containing the same amount of **20**. The dose dependence of the polystyrene film containing 2.0 wt% diarylethene **20o** (film thickness: 0.75 mm) was examined. The absorbance change as large as 0.03 was observed even when the absorbed dose was only 100 Gy and the absorption intensity linearly increased with increasing the absorbed dose below 3 kGy. The polystyrene dosimeter doped with 2.0 wt% diarylethene **20** showed similar sensitivity to that of crystal and amorphous film dosimeters.

The sensitivity was found to increase by the addition of fluorescent metal complexes, such as BaFCl:Eu and $CeMgAl_{11}O_{19}$, which exhibit fluorescence in the 300–400 nm wavelength region, into the polystyrene film [50, 51]. It is known that the radiation sensitivity increases by the addition of metal complexes of a large atomic number. The fluorescent metal complexes have large atomic numbers and act at the same time as sensitizers. A film dosimeter was prepared by casting a toluene solution containing polystyrene, diarylethene **20** (10 wt%), and BaFCl:Eu (30 wt%) on a Teflon plastic plate. The film was dried under vacuum and then the film (thickness: 0.5 mm) was irradiated with ^{60}Co γ-rays at room temperature. Upon γ-irradiation, the film turned blue. The blue color was clearly observed even when the irradiation dose was as low as 10 Gy. By the addition of fluorescent metal complexes, the sensitivity of the polystyrene dosimeter increased by ~10 times. The absorbed high energy is deactivated by such processes as the photoelectric effect, the Compton effect, and pair production, and finally excites the phenyl side groups of polystyrene and the metal complexes. The excited phenyl rings and metal complexes transfer the energy to diarylethene molecules non-radiatively as well as radiatively. The excited diarylethene molecules undergo the cyclization reactions to produce the colored closed-ring isomers and the polystyrene film changes the color.

As described in Section 3.4.2, sulfone derivatives of bis(benzothienyl)ethene undergo cyclization reactions to produce fluorescent closed-ring isomers upon UV irradiation. The derivatives were also applied to radiochromic dosimeters. When the absorbed dose is estimated from the absorbance change of color, the intensity change as large as 0.01 is required. This change corresponds to radiation-induced production of the closed-ring isomer (ε: 10^4 M^{-1} cm^{-1}) as high as 10^{-5} M (film thickness is 1 mm). On the other hand, when the dose is estimated from the fluorescence intensity change, very low concentration of 10^{-8} M can be easily detected. Therefore, the sensitivity is considered to increase as much as 10^3 times.

Typical fluorescent diarylethene **8** was solubilized in toluene (7×10^{-3} mol/l) and irradiated with ^{60}Co γ-rays [52]. Figure 9.9a shows the relation between the absorbed dose and the fluorescence intensity of the solution excited with 488 nm light. Fluorescence intensity increases with an increase in the absorbed dose below 1 Gy. The relative fluorescence intensity is 0.76 at 530 nm when the dose is 0.05 Gy, while it increases to 1.65 when the dose is 0.1 Gy. The fluorescence dosimeter can detect the absorbed dose as low as 0.05 Gy. For practical applications, it is required to make film dosimeters. A polystyrene film containing diarylethene **8** (5.0 wt%) was prepared and irradiated with ^{60}Co γ-rays. Figure 9.9b shows the fluorescence spectral changes upon irradiation with 0.1, 0.5, and 1.0 Gy. Even when the absorbed dose is as small as 0.1 Gy, a clear fluorescence spectrum was observed. In this experiment, the film was sandwiched between two BaFCl:Eu films to increase the sensitivity. The sensitivity of the fluorescence dosimeter doped with fluorescent sulfone derivatives is 10^2–10^3 times higher than that of color dosimeters. The diarylethene dosimeters,

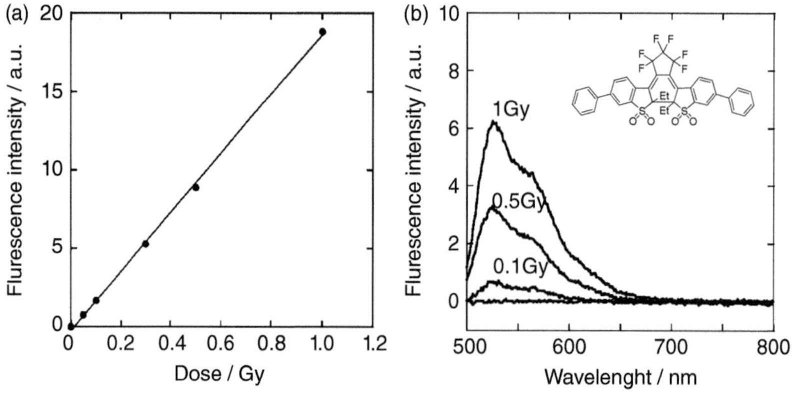

Figure 9.9 Fluorescence intensity increase upon γ-irradiation. (a) Dependence of fluorescence intensity of diarylethene **8** in toluene on the absorbed dose. Concentration of **8** in toluene was 7×10^{-3} mol/l (b) Fluorescence intensity changes of polystyrene films containing **8** upon γ-irradiation. The excitation wavelength was 488 nm.

both color and fluorescence dosimeters, are stable even at 80 °C and do not show any post-coloration after stopping the irradiation. The spatial resolution is very high (~ μm) and the image of absorbed dose is obtained without any special development procedure. The color dosimeters can be used in radiation processing, such as radiation curing and sterilization, where relatively high-dose irradiation is required. On the other hand, the fluorescence dosimeters have the potential to be applied in medical fields where low doses are required to be measured.

References

1 Tsujioka, T. and Irie, M. (2010). Electrical functions of photochromic molecules. *J. Photochem. Photobiol. C: Photochem. Rev.* 11: 1–14.

2 Hayakawa, R., Higashiguchi, K., Matsuda, K. et al. (2013). Optically and electrically driven organic thin film transistors with diarylethene photochromic channel layers. *Appl. Mater. Interfaces* 5: 3625–3630.

3 Orgiu, E., Crivillers, N., Herder, M. et al. (2012). Optically switchable transistor via energy-level phototuning in a bicomponent organic semiconductor. *Nat. Chem.* 4: 675–679.

4 Gemayel, M.E., Börjesson, K., Herder, M. et al. (2015). Optically switchable transistors by simple incorporation of photochromic systems into small-molecule semiconducting matrices. *Nat. Commun.* 6: 6330.

5 Leydecker, T., Herder, M., Pavlica, E. et al. (2016). Flexible non-volatile optical memory thin-film transistor device with over 256 distinct levels based on an organic bicomponent blend. *Nat. Nanotechnol.* 11 (9): 769–775.

6 Börjesson, K., Herder, M., Grubert, L. et al. (2015). Optically switchable transistors comprising a hybrid photochromic molecule/n-type organic active layer. *J. Mater. Chem. C* 3: 4156–4161.

7 Kurokawa, Y., Hayakawa, R., Shimada, S. et al. (2019). Photocontrollable ambipolar transistors with π-conjuagted diarylethene photochromic channels. *Jpn. J. Appl. Phys.* 58: SDDH03.

8 Yoshida, M., Suemori, K., Uemura, S. et al. (2010). Development of field-effect transistor-type photorewritable memory using photochromic interface layer. *Jpn. J. Appl. Phys.* 49: 04DK09.

9 Modrow, A., Zargarani, D., Herges, R., and Stock, N. (2011). The first porous MOF with photoswitchable linker molecules. *Dalton Trans.* 40: 4217–4222.

10 Park, J., Yuan, D., Pham, K.T. et al. (2012). Reversible alteration of CO_2 adsorption upon photochemical or thermal treatment in a metal-organic framework. *J. Am. Chem. Soc.* 134: 99–102.

11 Zheng, Y., Sato, H., Wu, P. et al. (2017). Flexible interlocked porous frameworks allow quantitative photoisomerization in a crystalline solid. *Nat. Commun.* 8: 100.

12 Nikolayenko, V., Herbert, S.A., and Barbour, L.J. (2017). Reversible structural switching of a metal-organic framework by photoirradiation. *Chem. Commun.* 53: 11142–11145.

13 Fan, C.B., Gong, L.L., Huang, L. et al. (2017). Significant enhancement of C_2H_2/C_2H_4 separation by a photochromic diarylethene unit: a temperature- and light-responsive separation switch. *Angew. Chem. Int. Ed.* 56: 7900–7906.
14 Han, M., Michel, R., He, B. et al. (2013). Light-triggered guest uptake and release by a photochromic coordination cage. *Angew. Chem. Int. Ed.* 52: 1319–1323.
15 Hell, S.L. (2007). Far-field optical nanoscopy. *Science* 316: 1153–1158.
16 Hell, S.L. (2009). Microscopy and its focal switch. *Nat. Method* 6 (1): 24–32.
17 Dedecker, P., Hotta, J., Flors, C. et al. (2007). Subdiffraction imaging through the selective donut-mode depletion of thermally stable photoswitchable fluorophores: Numerical analysis and application to fluorescent protein dronpa. *J. Am. Chem. Soc.* 129 (51): 16132–16141.
18 Irie, M. and Morimoto, M. (2018). Photoswitchable turn-on mode fluorescent diarylethenes: strategies for controlling the switching response. *Bull. Chem. Soc. Jpn.* 91 (2): 237–250.
19 Takagi, Y., Kunishi, T., Katayama, T. et al. (2012). Photoswitchable fluorescent diarylethene derivatives with short alkyl chain substituents. *Photochem. Photobiol. Sci.* 11: 1661–1665.
20 Takagi, Y., Morimoto, M., Kashihara, R. et al. (2017). Turn-on mode fluorescent diarylethenes: control of the cycloreversion quantum yield. *Tetrahedron* 73: 4918–4924.
21 Iwai, R., Morimoto, M., and Irie, M. (2020). Turn-on mode fluorescent diarylethenes: effect of electron-donating and electron-withdrawing substituents on photoswitching performance. *Photochem. Photobiol. Sci.* 19: 783–789.
22 Kashihara, R., Morimoto, M., Ito, S. et al. (2017). Fluorescence photoswitching of a diarylethene by irradiation with single-wavelength visible light. *J. Am. Chem. Soc.* 139: 16498–16501.
23 Wondrazek, F., Seilmeier, A., and Kaiser, W. (1984). Ultrafast intramolecular redistribution and intermolecular relaxation of vibrational energy in large molecules. *Chem. Phys. Lett.* 104 (2,3): 121–128.
24 Kinoshita, S., Nishi, N., Saitoh, A., and Kushida, T. (1987). Urbach tail of organic dyes in solution. *J. Phys. Soc. Jpn.* 56 (11): 4162–4175.
25 Urbach, F. (1953). Long-wavelength edge of photographic sensitivity and of the electronic absorption of solids. *Phys. Rev.* 92: 1324.
26 Roubinet, B., Bossi, M.L., Alt, P. et al. (2016). Caroboxylated photoswitchable diarylethenes for biolabeling and super-resolution RESOLFT microscopy. *Angew. Chem. Int. Ed.* 55: 15429–15433.
27 Roubinet, B., Weber, M., Shojaei, H. et al. (2017). Fluorescent photoswitchable diarylethenes for biolabeling and single-molecule localization microscopies with optical superresolution. *J. Am. Chem. Soc.* 139: 6611–6620.
28 Uno, K., Bossi, M., Irie, M. et al. (2019). Reversibly photoswitchable fluorescent diarylethenes resistant against photobleaching in aqueous solutions. *J. Am. Chem. Soc.* 141: 16471–16478.
29 Lim, S.-J., Carling, C.-J., Warford, C.C. et al. (2011). Multifunctional photo- and thermo-responsive copolymer nanoparticles. *Dyes Pigm.* 89: 230–235.

30 Nevskyi, O., Sysoiev, D., Oppermann, A. et al. (2016). Nanoscopic visualization of soft matter using fluorescent diarylethene photoswitches. *Angew. Chem. Int. Ed.* 55: 12698–12702.
31 Nevskyi, O., Sysoiev, D., Dreier, J. et al. (2018). Fluorescent diarylethene photoswitches – a universal tool for super-resolution microscopy in nanostructured materials. *Small* 14: 1703333.
32 Qiang, Z., Shebek, K.M., Irie, M., and Wang, M. (2018). A polymerizable photoswitchable fluorophore for super-resolution imaging of polymer self-assembly and dynamics. *ACS Macro Lett.* 7: 1432–1437.
33 Balzarotti, F., Eilers, Y., Gwosch, K.C. et al. (2017). Nanometer resolution imaging and tracking of fluorescent molecules with minimal photon fluxes. *Science* 355: 606–612.
34 Erno, Z., Asadirad, A.M., Lemieux, V., and Branda, N.R. (2012). Using light and a molecular switch to "lock" and "unlock" the Diels-Alder reaction. *Org. Biomol. Chem.* 10: 2787–2792.
35 Asadirad, A.M., Boutault, S., Erno, Z., and Branda, N.R. (2014). Controlling a polymer adhesive using light and a molecular switch. *J. Am. Chem. Soc.* 136: 3024–3027.
36 Göstl, R. and Hecht, S. (2014). Controlling covalent connection and disconnection with light. *Angew. Chem. Int. Ed.* 53: 8784–8787.
37 Göstl, R. and Hecht, S. (2015). Photoreversible prodrugs and protags: switching the release of maleimides by using light under physiological conditions. *Chem. Eur. J.* 21: 4422–4427.
38 Fuhrmann, A., Göstl, R., Wendt, R. et al. (2014). Conditional repair by locally switching the thermal healing capability of dynamic covalent polymers with light. *Nat. Commun.* 7: 13623.
39 Komarov, I.V., Afonin, S., Babii, O. et al. (2028). Efficiently photocontrollable or not? Biological activity of photoisomerizable diarylethenes. *Chem. Eur. J.* 24: 11245–11254.
40 The study was performed by Envigo CRS GmbH, Rossdorf, Germany.
41 Al-Atar, U., Femandes, R., Johnsen, B. et al. (2009). A photocontrolled molecular switch regulates paralysis in a living organism. *J. Am. Chem. Soc.* 131 (44): 15966–15967.
42 Vomasta, D., Högner, C., Branda, N.R., and König, B. (2008). Regulation of human carbonic anhydrase I (hCAI) activity by using a photochromic inhibitor. *Angew. Chem. Int. Ed.* 47: 7644–7647.
43 Falenczyk, C., Schiedel, M., Karaman, B. et al. (2014). Chromo-pharmacophores: photochromic diarylmaleimide inhibitors for sirtuins. *Chem. Sci.* 5: 4794–4799.
44 Babii, O., Afonin, S., Berditsch, M. et al. (2014). Controlling biological activity with light: diarylethene-containing cyclic peptidomimetics. *Angew. Chem. Int. Ed.* 53: 3392–3395.
45 Babii, O., Afonin, S., Garmanchuk, L.V. et al. (2016). Direct photocontrol of peptidomimetics: an alternative to oxygen-dependent photodynamic cancer therapy. *Angew. Chem. Int. Ed.* 55: 5493–5496.

46 Butson, M.J., Yu, P.K.N., Cheung, T., and Metcalfe, P. (2003). Radiochromic film for medical dosimetry. *Mater. Sci. Eng. R* 41: 61–120.

47 Irie, S., Yamaguchi, T., Nakazumi, H. et al. (1999). Radiation- induce coloration of photochromic dithienylethene derivatives. *Bull. Chem. Soc. Jpn.* 72 (5): 1139–1142.

48 Irie, S., Kim, M.-S., Kawai, T., and Irie, M. (2004). The radiation-induced coloration of amorphous photochromic dithienylethene films. *Bull. Chem. Soc. Jpn.* 77 (5): 1037–1040.

49 Irie, S. and Irie, M. (2000). Radiation-induced coloration of photochromic dithienylethene derivatives in polymer matrices. *Bull. Chem. Soc. Jpn.* 73 (10): 2385–2388.

50 Irie, S. and Irie, M. (2002). The effect of additives on the radiation-induced coloration of photochromic dithienylethene derivatives. *Bull. Chem. Soc. Jpn.* 75 (9): 2071–2074.

51 Irie, S. and Irie, M. (2006). Ultrahigh sensitive color dosimeters composed of photochromic diarylethenes and fluorescent metal complexes. *Chem. Lett.* 35 (12): 1434–1435.

52 Yamaguchi, T. and Irie, M. (2020). Ultrahigh-sensitive fluorescence dosimeters that use turn-on mode fluorescent diarylethenes. *Tetrahedron Lett.* 61: 152232.

Appendix A

Synthesis Procedures of Typical Diarylethenes

Typical diarylethene derivatives, 1,2-bis(2,4-dimethyl-5-phenyl-3-thienyl)perfluorocyclopentene (**7**) and 1,2-bis(2-ethyl-6-phenyl-1-benzothiophene-1,1-dioxide-3-yl) perfluorocyclopenetene (**11**), are synthesized as follows [1–4].

A.1 1,2-Bis(2,4-dimethyl-5-phenyl-3-thienyl)perfluorocyclopentene (7) [5–7]

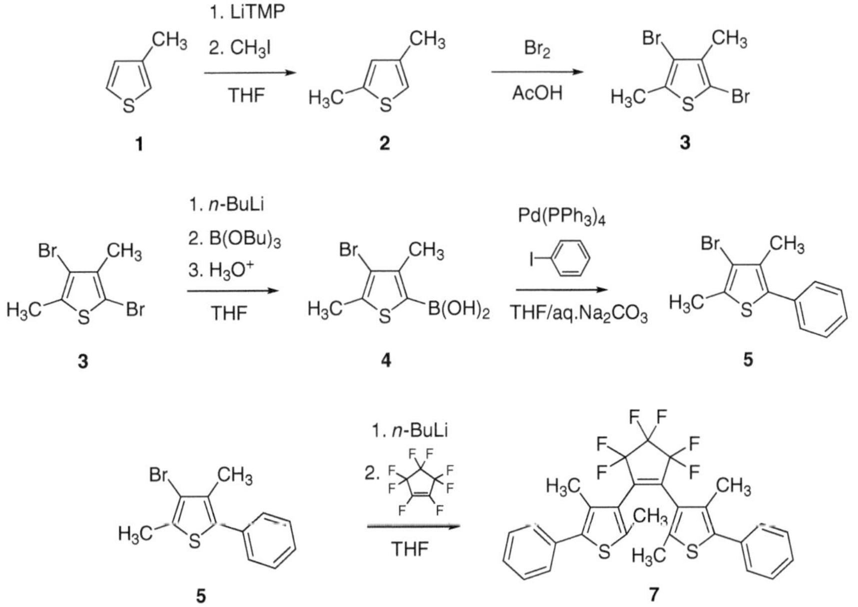

2,4-Dimethylthiophene (2) was prepared from 3-methylthiophene (**1**) according to the procedure described in Ref. [6].

Diarylethene Molecular Photoswitches: Concepts and Functionalities, First Edition. Masahiro Irie.
© 2021 WILEY-VCH GmbH. Published 2021 by WILEY-VCH GmbH.

2,4-Dimethyl-3,5-dibromothiophene (3)

To an acetic acid solution (50 ml) containing 2,4-dimethylthiophene (**2**) (5.0 g, 44.6 mmol), bromine (5.0 ml, 97 mmol) was added at 0 °C and the mixture was stirred overnight. The resulting mixture was poured into water (saturated with NaCl) and extracted with chloroform. The organic layer was washed with a Na_2CO_3 aqueous solution and water. Bromine was eliminated by washing the chloroform layer with a $Na_2S_2O_3$ aqueous solution. The organic layer was washed again with water 3 times, dried over $MgSO_4$, filtered, and evaporated. The residue was purified by silica gel column chromatography (hexane) to give 7.8 g of 2,4-dimethyl-3,5-dibromothiophene (**3**) in 65% yield as a colorless oil (7.8 g, 65%). ^1H NMR (200 MHz, $CDCl_3$, TMS) δ 2.17 (s, 3H), 2.34 (s, 3H); MS (EI) *m*/*z*: 268 $[M]^+$.

2,4-Dimethyl-3-bromo-5-phenylthiophene (5)

To a stirred solution of 2,4-dimethyl-3,5-dibromothiophene (**3**) (15.0 g, 55.5 mmol) in 300 ml of diethyl ether at −78 °C, 1.6 M *n*-BuLi hexane solution (40 ml, 64 mmol) was added dropwise, and the reaction mixture was stirred for 30 minutes at −78 °C. To the reaction mixture, tributyl borate (10 ml, 81 mmol) was added, and the mixture was stirred for 2 hours and was allowed to stand to room temperature. To the mixture, 100 ml of 4% HCl was added, and the organic phase was separated. The organic phase was extracted with 100 ml of 4% NaOH aqueous solution 2 times. The aqueous phase was separated and acidified with 10% HCl. The white precipitate was collected by filtration and washed with water. Drying of the white powder in vacuo afforded 10.3 g of 3,5-dimethyl-4-bromothiophene-2-boronic acid (**4**) in 80% yield. The boronic acid was used without further purification. ^1H NMR (200 MHz, DMSO-d_6, TMS) δ 2.32 (s, 3H), 2.37 (s, 3H).

3,5-Dimethyl-4-bromothiophene-2-boronic acid (**4**) (3.3 g, 14 mmol) was added into a flask containing THF (100 ml), 20 wt% Na_2CO_3 aqueous solution (35 ml), iodobenzene (2.4 g, 12 mmol), and a small amount of $Pd(PPh_3)_4$ (0.83 g, 0.72 mmol). The mixture was refluxed for 5 hours at 70 °C. The product was extracted with chloroform. The organic layer was dried over anhydrous $MgSO_4$, filtered, and concentrated. The residue was purified by silica gel column chromatography and by recrystallization from *n*-hexane to give 2.6 g of 2,4-dimthyl-3-bromo-5-phenylthiophene (**5**) in 83% yield as colorless crystals. ^1H NMR (200 MHz, $CDCl_3$, TMS) δ 2.25 (s, 3H), 2.42 (s, 3H), 7.2–7.5 (m, 5H); MS (EI) *m*/*z*: 266 $[M]^+$.

1,2-Bis(2,4-dimethyl-5-phenyl-3-thienyl)perfluoro-cyclopentene (7)

To a stirred THF solution (15 ml) containing 2,4-dimthyl-3-bromo-5-phenylthiophene (**5**) (1.76 g, 6.60 mmol), 6.21 ml of 1.6 M *n*-BuLi hexane solution (9.94 mmol) was slowly added at −78 °C under a nitrogen atmosphere, and the solution was stirred for 1 hour at that temperature. Octafluorocyclopentene (0.219 ml, 1.65 mmol) was slowly added at −78 °C twice (for a total of 0.438 ml), and the mixture was stirred overnight at the low temperature. The reaction mixture was poured into water and extracted with diethyl ether. The organic layer was washed with 1 N HCl aqueous solution and water. The organic layer was dried over anhydrous $MgSO_4$, filtered, and evaporated. The residue was purified by column chromatography on silica gel (*n*-hexane) to give 948 mg of 1,2-bis(2,4-dimethyl-5-phenyl-3-thienyl)perfluorocyclopentene (**7**) in 52% yield as a white solid.

White solid: mp 136–137 °C; ^{1}H NMR (200 MHz, $CDCl_3$) δ 2.08 (s, 3H), 2.10 (s, 3H), 2.35 (s, 3H), 2.36 (s, 3H), 7.35 (m, 10H); MS *m/z*: 548 $[M]^+$. Anal. Calcd for $C_{29}H_{22}F_6S_2$: C, 63.49; H. 4.04. Found: C, 63.58; H, 4.28.

A.2 1,2-Bis(2-ethyl-6-phenyl-1-benzothiophene-1,1-dioxide-3-yl)-perfluorocyclopenetene (11) [8–10]

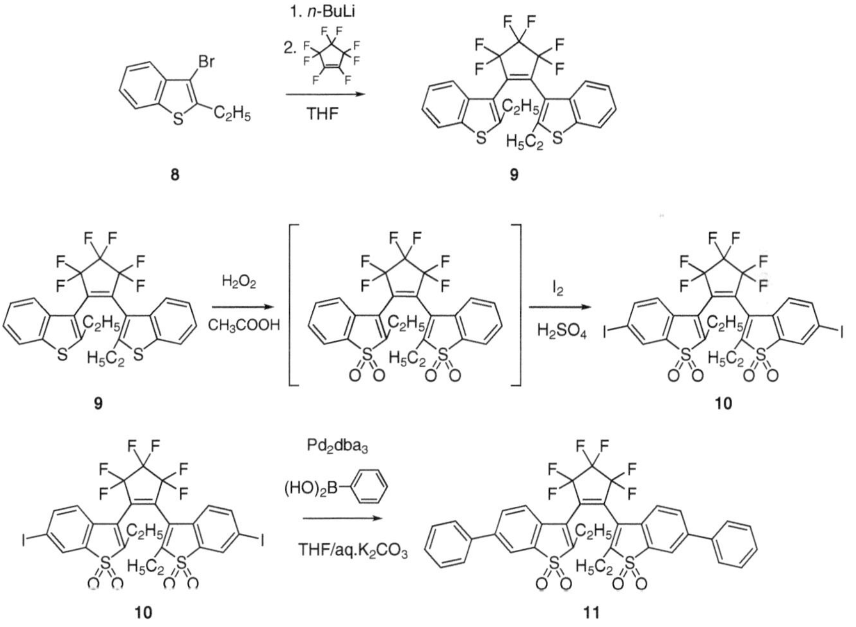

1,2-Bis(2-ethyl-1-benzothiophen-3-yl)perfluoro-cyclopentene (9)

To a stirred THF solution (40 ml) containing 3-bromo-2-ethyl-1-benzothiophene (**8**) (2.00 g, 8.29 mmol), 5.70 ml of 1.6 M *n*-BuLi hexane solution (9.12 mmol) was slowly added at −78 °C, and the solution was stirred for 30 minutes at −78 °C. Then, octafluorocyclopentene (0.549 ml, 4.15 mmol) was added slowly to the reaction mixture at −78 °C and left to stand overnight with stirring at the low temperature. The solution was slowly warmed-up to room temperature with stirring. Then, the reaction mixture was poured into water (saturated with NaCl) and extracted with diethyl ether. The organic layer was dried over anhydrous $MgSO_4$ and evaporated in vacuo. The crude product was purified by column chromatography on silica gel (*n*-hexane) to give 0.943 g of 1,2-bis(2-ethyl-1-benzothiophen-3-yl)perfluorocyclopentene (**9**) in 46% yield.

Colorless crystal; mp 133–134 °C; ^{1}H NMR (200 MHz, $CDCl_3$) δ 0.78 (t, J = 7.8 Hz, 4.2H), 1.29 (t, J = 7.8 Hz, 1.8H), 2.38–2.91 (m, 4H), 7.16–7.42 (m, 4H), 7.60–7.74 (m, 4H); MS (EI) *m/z*: 496 $[M]^+$. Anal. Calcd for $C_{25}H_{18}F_6S_2$:c 60.47, H 3.65. Found: 60.54, H 3.67.

1,2-Bis(2-ethyl-6-iodo-1-benzothiophene-1,1-dioxide-3-yl)perfluorocyclopentene (10)

A solution of 1,2-bis(2-ethyl-1-benzothiopehen-3-yl)perfluorocyclopentene (**9**) (4.0 g, 8.1 mmol) in acetic acid was heated to the boiling temperature (118 °C). To the solution 20 ml of 50% hydrogen peroxide solution was slowly added, and the mixture was refluxed for 30 minutes. The resulting mixture was poured into cold water (200 ml) to give a precipitate. The precipitate was filtered, washed with distilled water several times, and dried in vacuo. The white powder product was added into cold sulfuric acid (80 ml), and the mixture was stirred rigorously at −5 °C for 20 minutes to dissolve the powder perfectly. Orthoperiodic acid (2.3 g, 10 mmol) and iodine (4.5 g, 18 mmol) were added to the solution, and the mixture was stirred at −5 °C for 2 hours. After the pale green mixture turned dark purple, the mixture was warmed to 0 °C and further stirred for 2 hours, and then carefully poured into cold ice water (160 ml). The mixture was extracted with chloroform (300 ml) 3 times. Additionally, the aqueous layer was extracted with ethyl acetate (200 ml) twice. The organic layer (chloroform and ethyl acetate) was washed with aqueous $NaHCO_3$, $Na_2S_2O_3$, and brine, dried over anhydrous $MgSO_4$, and evaporated in vacuo. The crude product was purified by silica gel column chromatography (*n*-hexane/ethyl acetate = 10/1 − 8/1) to give 4.7 g of 1,2-bis(2-ethyl-6-iodo-1-benzothiophene-1,1-dioxide-3-yl)perfluoro-cyclopentene (10) in 72% yield as a white solid.

White solid; mp 123–124 °C; ^{1}H NMR (400 MHz, $CDCl_3$) δ 1.06 (t, J = 7.6 Hz, 3.4H (ap)), 1.39 (t, J = 7.6 Hz, 2.6H (p)), 2.29–2.37 (m, 1.7H (p)), 2.42–2.59 (m, 2.3H(ap)), 6.80 (d, J = 8.4 Hz, 0.9H (p)), 6.91 (d, J = 8.4 Hz, 1.1 H(ap)), 7.79 (dd, J = 8.4 and 1.6 Hz, 0.9H (p)), 7.95 (dd, J = 8.4 and 1.6 Hz, 1.1H(ap)), 8.00 (d, J = 1.6 Hz, 0.9 H(p)), 8.06 (d, J = 1.6 Hz, 1.1H(ap)); MS (EI) *m/z*: 812 $[M]^+$. Anal Calcd for $C_{25}H_{16}F_6I_2O_4S_2$: C, 36.96; H, 1.99; S, 7.89. Found: C, 36.97; H, 1.91; S, 7.85.

1,2-Bis(2-ethyl-6-phenyl-1-benzothiophene-1,1-dioxide-3-yl) perfluorocyclopenetene (11)

To a THF solution (10 ml) containing 1,2-bis(2-ethyl-6-iodo-1-benzothiophene-1,1-dioxide-3-yl)perfluorocyclopentene (**10**) (150 mg, 0.185 mmol) and phenylboronic acid (57 mg, 0.47 mmol), saturated aqueous K_2CO_3 (10 ml), tris(dibenzylidene-acetone)-dipalladium(0) (30 mg, 0.033 mmol) and 18% tricyclohexylphosphine toluene solution (0.1 ml) were added, and the mixture was stirred at room temperature for 20 minutes. The resulting mixture was neutralized with diluted hydrochloric acid and extracted with chloroform. The organic layer was evaporated in vacuo, and the residue was purified by silica gel column chromatography (*n*-hexane/ethyl acetate = 9/1 – 7/1) to give 126 mg of 1,2-bis(2-ethyl-6-phenyl-1-benzothiophene-1,1-dioxide-3-yl)perfluorocyclopenetene (**11**) in 96% yield as a white solid.

White solid; mp 199–200 °C; ^{1}HNMR (400 MHz, 1,4-dioxane-d_8) δ 1.04 (t, J = 7.6 Hz, 3.7H (ap)), 1.37 (t, J = 7.6 Hz, 2.3H (p)), 2.45–2.51 (m, 1.6H (p)), 2.59–2.70 (m, 2.4H (ap)), 7.34–7.50 (m, 8H), 7.62 (d, J =7.2 Hz, 1.6H (p)), 7.67 (d, J =7.2 Hz, 0.8H (p)), 7.71 (d, J = 7.2 Hz, 2.4H (ap)), 7.87 (d, J = 7.2 Hz, 1.2H (ap)), 8.16 (s, 0.8H (p)), 8.22 (s, 1.2H (ap)); MS (EI) *m/z*: 712 $[M]^+$. Anal Calcd for $C_{37}H_{26}F_6O_4S_2$: C, 62.35; H, 3.68; S, 9.00. Found: C, 62.55; H, 4.02; S, 8.92.

References

1 Szalóki, G. and Pozzo, J.-L. (2013). Synhesis of symmetrical and nonsymmetrical bisthienylcyclopentenes. *Chem. Eur. J.* 19: 11124–11132.

2 Dixon, S. (1956). Elimination reaction of fluoroölefins with organolithium compounds. *J. Org. Chem.* 21: 400–403.

3 Ushiogi, Y., Hase, T., Iinuma, Y. et al. (2007). Synthesis of photochromic diarylethenes using a microflow system. *Chem. Commun.* 2947-2949.

4 Hiroto, S., Suzuki, K., Kamiya, H., and Shinokubo, H. (2011). Synthesis protocol for diarylethenes through Suzuki-Miyaura coupling. *Chem. Commun.* 47: 7149–7151.

5 de Meijere, A., Zhao, L., Belov, V.N. et al. (2007). 1,3-Bicyclo[1,1,1]pentanediyl: the shortest rigid linear connector of phenylated photochromic units and 1,5-dimethoxy-9,10-di(phenylethynyl)anthracene fluorophore. *Chem. Eur. J.* 13: 2503–2516.

6 Smith, K. and Barratt, M.L. (2007). Highly selective 5-substitution of 3-methylthiophene via directed lithiation. *J. Org. Chem.* 72: 1031–1034.

7 Irie, M., Sakemura, K., Okinaka, M., and Uchida, K. (1995). Photochromism of dithienylethenes with electron-donating substituents. *J. Org. Chem.* 60 (25): 8305–8309.

8 Hanazawa, M., Sumiya, R., Horikawa, Y., and Irie, M. (1992). Thermally irreversible photochromic systems. Reversible photocyclization of 1,2-bis(2-methylbenzo[*b*]thiophene-3-yl)perfluorocycloalkane derivatives. *J. Chem. Soc., Chem. Commun.* 3: 206–207.

9 Uno, K., Niikura, H., Morimoto, M. et al. (2011). In situ preparation of highly fluorescent dyes upon photoirradiation. *J. Am. Chem. Soc.* 133: 13558–13564.

10 Yamaguchi, T. and Irie, M. (2006). Photochromism of bis(2-alkyl-1-benzothiophen-3-yl)perfluoroethene derivatives. *J. Photochem. Photobiol. A: Chem.* 178: 162–169.

Index

Diarylethene Molecular Photoswitches: Concepts and Functionalities, First Edition. Masahiro Irie.
© 2021 WILEY-VCH GmbH. Published 2021 by WILEY-VCH GmbH.